Role of Microorganisms and Their Metabolites in Agriculture, Food and the Environment: 2nd Edition

Role of Microorganisms and Their Metabolites in Agriculture, Food and the Environment: 2nd Edition

Guest Editor

Barbara Sokolowska

MDPI

Basel • Beijing • Wuhan • Barcelona • Belgrade • Novi Sad • Cluj • Manchester

Guest Editor
Barbara Sokolowska
Department of Microbiology
Prof. Wacław Dąbrowski
Institute of Agricultural and
Food Biotechnology
Warsaw
Poland

Editorial Office
MDPI AG
Grosspeteranlage 5
4052 Basel, Switzerland

This is a reprint of the Special Issue, published open access by the journal *International Journal of Molecular Sciences* (ISSN 1422-0067), freely accessible at: https://www.mdpi.com/journal/ijms/special_issues/UOG8H09EVE.

For citation purposes, cite each article independently as indicated on the article page online and as indicated below:

Lastname, A.A.; Lastname, B.B. Article Title. *Journal Name* **Year**, *Volume Number*, Page Range.

ISBN 978-3-7258-6616-8 (Hbk)
ISBN 978-3-7258-6617-5 (PDF)
https://doi.org/10.3390/books978-3-7258-6617-5

Contents

About the Editor

Barbara Sokolowska

Barbara Sokolowska, D. Sc. is a Professor at the Institute of Agricultural and Food Biotechnology in Warsaw, Poland. She is a microbiologist specializing in food quality and safety. From 2018 to 2024, she served as Head of the Microbiology Department.

Her research focuses on microbial hazards in the food chain, particularly acidophilic spore-forming bacteria of the genus *Alicyclobacillus*, which are responsible for spoilage in juices and beverages. She also works on innovative, non-thermal food processing technologies, including high hydrostatic pressure. In recent years, her scientific interests have expanded to the isolation and characterization of bacteriophages and their potential use as biocontrol agents in food production. It also deals with the possibilities of practical use of microbial resources collected in the Collection of Industrial Microorganisms at the Department of Microbiology of the Institute.

From 2015 to 2024, she was additionally affiliated with the Institute of High-Pressure Physics of the Polish Academy of Sciences.

Prof. Sokołowska is an active member of the National Standards Body in Poland (Technical Committee No. 3 on Microbiology of the Food Chain). She also has extensive experience in quality assurance, having served as a technical auditor for the Polish Centre for Accreditation from 2001 to 2023.

Preface

Microorganisms and their metabolites are central to agricultural productivity, food quality, and environmental sustainability. Their diverse metabolic capabilities influence soil health, plant resilience, pollutant degradation, and the safety and sensory properties of food. Advances in molecular and omics technologies now allow us to better understand and harness these microbial functions.

This Reprint, "Role of Microorganisms and Their Metabolites in Agriculture, Food and the Environment: 2nd Edition", brings together studies that highlight the versatility and applied potential of microbial systems. The articles explore efficient microbial consortia for pesticide and plastic degradation, the role of soil microbiomes in agroforestry and biochar-amended systems, and microbial inoculants that enhance plant growth under stress conditions. Contributions on bacteriophages as food biocontrol agents, microbial enzymes involved in flavor generation, and fungal pathogenicity further illustrate the breadth of microbial influence across biological and industrial processes.

Collectively, these works underscore the importance of microbial research in developing sustainable strategies for agriculture, environmental protection, and food innovation. We thank all authors and reviewers for their valuable contributions to this Reprint.

Barbara Sokolowska
Guest Editor

International Journal of
Molecular Sciences

MDPI

Article

Biological Characterization and Genomic Analysis of Three Novel *Serratia*- and *Enterobacter*-Specific Virulent Phages

Dziyana Shymialevich [1], Stanisław Błażejak [2], Paulina Średnicka [3], Hanna Cieślak [1], Agnieszka Ostrowska [4], Barbara Sokołowska [3,*] and Michał Wójcicki [3,*]

1 Culture Collection of Industrial Microorganisms—Microbiological Resources Center, Department of Microbiology, Prof. Wacław Dąbrowski Institute of Agricultural and Food Biotechnology—State Research Institute, Rakowiecka 36 Str., 02-532 Warsaw, Poland; diana.szymielewicz@ibprs.pl (D.S.); hanna.cieslak@ibprs.pl (H.C.)

2 Department of Biotechnology and Food Microbiology, Institute of Food Sciences, Warsaw University of Life Sciences (WULS–SGGW), Nowoursynowska 166 Str., 02-776 Warsaw, Poland; stanislaw_blazejak@sggw.edu.pl

3 Department of Microbiology, Prof. Wacław Dąbrowski Institute of Agricultural and Food Biotechnology—State Research Institute, Rakowiecka 36 Str., 02-532 Warsaw, Poland; paulina.srednicka@ibprs.pl

4 Department of Nanobiotechnology, Institute of Biology, Warsaw University of Life Sciences (WULS–SGGW), Ciszewskiego 8 Str., 02-786 Warsaw, Poland; agnieszka_ostrowska@sggw.edu.pl

* Correspondence: barbara.sokolowska@ibprs.pl (B.S.); michal.wojcicki@ibprs.pl (M.W.)

Abstract: Due to the high microbiological contamination of raw food materials and the increase in the incidence of multidrug-resistant bacteria, new methods of ensuring microbiological food safety are being sought. One solution may be to use bacteriophages (so-called phages) as natural bacterial enemies. Therefore, the aim of this study was the biological and genomic characterization of three newly isolated *Serratia*- and *Enterobacter*-specific virulent bacteriophages as potential candidates for food biocontrol. Serratia phage KKP_3708 (vB_Sli-IAFB_3708), Serratia phage KKP_3709 (vB_Sma-IAFB_3709), and Enterobacter phage KKP_3711 (vB_Ecl-IAFB_3711) were isolated from municipal sewage against *Serratia liquefaciens* strain KKP 3654, *Serratia marcescens* strain KKP 3687, and *Enterobacter cloacae* strain KKP 3684, respectively. The effect of phage addition at different multiplicity of infection (MOI) rates on the growth kinetics of the bacterial hosts was determined using a Bioscreen C Pro growth analyzer. The phages retained high activity in a wide temperature range (from −20 °C to 60 °C) and active acidity values (pH from 3 to 12). Based on transmission electron microscopy (TEM) imaging and whole-genome sequencing (WGS), the isolated bacteriophages belong to the tailed bacteriophages from the *Caudoviricetes* class. Genomic analysis revealed that the phages have linear double-stranded DNA of size 40,461 bp (Serratia phage KKP_3708), 67,890 bp (Serratia phage KKP_3709), and 113,711 bp (Enterobacter phage KKP_3711). No virulence, toxins, or antibiotic resistance genes were detected in the phage genomes. The lack of lysogenic markers indicates that all three bacteriophages may be potential candidates for food biocontrol.

Keywords: virulent bacteriophages; biological characterization; genomic analysis; functional annotation; unconventional food preservation

1. Introduction

Consuming a diet that includes a variety of fruits and vegetables is crucial for maintaining good health, reducing the risk of diseases, and promoting overall well-being. For this reason, there has been an increase in demand for minimally processed food products [1–3]. The production of minimally processed food is limited to basic procedures (washing, peeling, packaging), which allows the obtaining of a ready-to-eat product with a short shelf life (about 4–7 days) [4,5]. The environment of raw plant materials promotes faster growth of microorganisms, which leads to the production of undesirable chemical compounds, reduces the nutritional value, and may become harmful to the consumer's health [6,7]. Recent reports from the EFSA and ECDC have shown an increasing trend in the number of cases and epidemiological outbreaks caused by the consumption of contaminated raw plant materials [8]. Recently, fresh vegetables have been indicated as an important reservoir of *Enterobacteriaceae*-producing extended-spectrum β-lactamase (ESBL) and AmpC, which raises concerns [9,10]. The spread of β-lactamase-producing bacteria in the food chain may lead to the acquisition of resistance to this type of antibiotic, which is associated with limited therapeutic options and the need to implement/search for innovative treatment methods [11].

Bacteria from the *Serratia* genus belong to the *Yersiniaceae* family and *Enterobacterales* order. *Serratia* are Gram-negative, facultatively anaerobic, non-spore-forming, straight rod-shaped bacteria [12,13]. *Serratia liquefaciens* and *Serratia marcescens* are widespread and occur in soil, water, and the digestive tract of insects, fish, rodents, and humans [14]. They can grow at low temperatures and are capable of secreting thermostable proteases and lipases, which may consequently lead to the production of biogenic amines (e.g., cadaverine and putrescine) [15–17]. *S. liquefaciens* and *S. marcescens* often cause spoilage of meat products and frozen vegetables [18,19]. These species can also cause food poisoning or cause illness (gastroenteritis) if they occur in food products [20,21]. A relatively low number of *S. marcescens* (even 10–100 CFU) can be enough to cause an infection; however, the infectious dose may vary depending on the infected individual [20]. *S. marcescens* easily adapts to changing physicochemical conditions, thus demonstrating the ability to survive in various environments, including disinfection solutions, which makes it one of the important nosocomial pathogens [22,23].

Bacteria from the *Enterobacteriaceae* family (*Enterobacterales* order) are considered indicator bacteria of the microbiological quality of food and the hygienic production process [24,25]. This family includes, among others, rods of the *Enterobacter* genus. *Enterobacter* are Gram-negative, facultatively anaerobic, non-spore-forming, rod-shaped bacteria. They constitute part of the intestinal microbiota in 40–80% of the human population [26,27]. These species were often isolated from food processing plants, meat and vegetable products, and rice [28,29]. *Enterobacter cloacae* can cause opportunistic infections, including infections of the eyes and blood, wounds, soft tissues, skin, pneumonia, and urinary tract infections [30,31]. This species has been reported to be resistant to several antibiotics, such as ampicillin, rifampicin, erythromycin, and sulfamethoxazole [32,33]. Moreover, *S. marcescens* and *E. cloacae* exhibit a natural resistance to some antibacterial drugs, including macrolides and penicillins (ampicillin, amoxicillin, amoxicillin with clavulanic acid) [34].

The difficulty in eliminating *S. liquefaciens*, *S. marcescens* and *E. cloacae* lies in the ability of these bacteria to secrete extracellular polymers, which allows the creation of dense biofilms that protect bacterial cells against environmental factors and hinders their elimination from raw materials and surfaces of food production lines [21,35,36]. A promising technique for extending the shelf life of food is the use of strictly lytic bacteriophages or their enzymes, which do not affect the nutritional value of the product and do not change the sensory characteristics [37–39]. Before practical application in biocontrol or phage therapy, a comprehensive characterization of bacteriophages is needed. Therefore, in this study, three bacteriophages specific for saprophytic bacteria from the *Enterobacterales* order were isolated from plant-based food products.

Bacteriophage particles were visualized using transmission electron microscopy (TEM), biological characteristics were determined, and genomic analyses were performed. This study aimed to enrich the Culture Collection of Industrial Microorganisms—Microbiological Resources Center of the Department of Microbiology at the Prof. Wacław Dąbrowski Institute of Agricultural and Food Biotechnology—State Research Institute (IAFB) resources with new phages targeting saprophytic bacteria from the *S. liquefaciens*, *S. marcescens*, and *E. cloacae* species to obtain potential candidates for food biocontrol.

2. Results and Discussion

2.1. Phage Morphology

Three bacteriophages were isolated from municipal sewage: Serratia phage KKP_3708 (phage titer 2.23×10^8 PFU mL^{-1}), Serratia phage KKP_3709 (phage titer 9.41×10^9 PFU mL^{-1}), and Enterobacter phage KKP_3711 (phage titer 1.23×10^9 PFU mL^{-1}) targeting *Serratia liquefaciens* strain KKP 3654, *Serratia marcescens* strain KKP 3687, and *Enterobacter cloacae* strain KKP 3684, respectively. Bacteriophages formed plaques with a transparent halo zone that indicates the production of enzymes from the group of polysaccharide depolymerases, degrading capsular and structural polysaccharides, including exopolysaccharides (EPS), which are the dominant component of the bacterial biofilm. EPS depolymerases are believed to be effective in dispersing biofilms [40]. Photos of phage plaques on bacterial lawn, the procedure for isolating bacteriophage enzymes, and assessing their impact on limiting bacterial growth were described in an earlier article by Shymialevich et al. [41]. The morphology of the isolated bacteriophages was visualized by TEM, and the electron micrographs are shown in Figure 1.

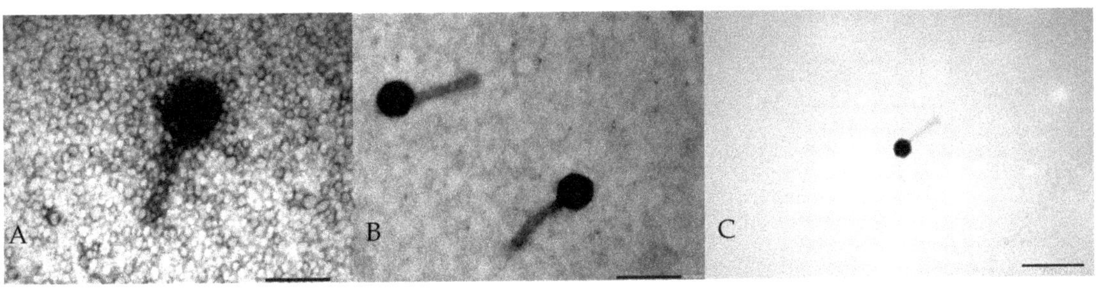

Figure 1. Electron micrographs from the TEM showing the morphology of bacteriophages: (**A**)—Serratia phage KKP_3708; (**B**)—Serratia phage KKP_3709; (**C**)—Enterobacter phage KKP_3711. Scale bar of 200 nm (**A**,**C**) or 100 nm (**B**).

All three isolated bacteriophages belong to the *Caudoviricetes* class (tailed bacteriophages) and contain an icosahedral head and long non-contractile tails (typical siphovirus morphology). Tailed bacteriophages represent the most widespread and numerous group of bacterial viruses [42].

2.2. Bacterial Host Range of Bacteriophages

Twenty-five strains from the *Enterobacterales* order were used to determine the host range of newly isolated phages. This bacterial host range was also checked in the study by Wójcicki and co-workers [39] to determine the effectiveness of a phage cocktail for biocontrol against saprophytic bacteria in ready-to-eat plant-based food [39]. The results obtained for Serratia phage KKP_3708, Serratia phage KKP_3709, Enterobacter phage KKP_3711 and the calculated efficiency of plating (EOP) are presented in Table 1.

Table 1. Range of bacterial hosts for three newly isolated phages.

Bacterial Host Strain (n = 25)	GenBank Accession Number	Year of Isolation	Source of Isolation	Phage Strain (EOP)		
				KKP_3708	KKP_3709	KKP_3711
Enterobacter cloacae KKP 3082	MZ827006	2018	Salad mix with carrot	−	−	+ (0.215)
Enterobacter ludwigii KKP 3083	MZ827002	2018	Salad mix with carrot	−	+ (0.033)	−

Table 1. *Cont.*

Bacterial Host Strain (*n* = 25)	GenBank Accession Number	Year of Isolation	Source of Isolation	Phage Strain (EOP)		
				KKP_3708	KKP_3709	KKP_3711
Serratia fonticola KKP 3084	MZ827668	2018	Salad mix with beetroot	–	–	–
Escherichia coli KKP 3650	OM287487	2018	Salad mix with beetroot	–	–	–
Pantoea agglomerans KKP 3651	OP978292	2018	Washed spinach	–	–	+ (0.091)
Serratia fonticola KKP 3652	OM287486	2018	Washed spinach	–	–	–
Serratia liquefaciens KKP 3654	OP978313	2018	Salad mix with beetroot	++ (1)	–	–
Citrobacter freundii KKP 3655	MZ827001	2018	Rucola	–	–	–
Enterobacter cloacae KKP 3656	OM304355	2018	Unwashed spinach	+ (0.274)	–	–
Enterobacter cloacae KKP 3684	OM281790	2018	Salad mix with beetroot	–	–	++ (1)
Serratia fonticola KKP 3685	OM281802	2018	Unwashed spinach	–	–	–
Enterobacter cloacae KKP 3686	OM281778	2018	Unwashed spinach	–	–	+ (0.305)
Serratia marcescens KKP 3687	OK103977	2018	Salad mix with beetroot	–	++ (1)	–
Escherichia coli KKP 3688	OM281784	2018	Salad mix with beetroot	–	–	–
Raoultella terrigena KKP 3689	OK085529	2018	Salad mix with beetroot	–	–	–
Escherichia coli KKP 3691	OM281773	2018	Washed spinach	–	–	–
Enterobacter cloacae KKP 3692	OM281803	2018	Unwashed spinach	++ (0.553)	–	+ (0.00003)
Escherichia coli KKP 3705	OM212647	2018	Salad mix with beetroot	–	–	+ (0.614)
Enterobacter cloacae KKP 3706	OM278533	2018	Salad mix with carrot	–	–	+ (0.024)
Escherichia coli KKP 3707	OM281777	2018	Washed spinach	–	–	–
Escherichia coli KKP 3800	OM250392	2022	Rucola	–	–	–
Escherichia coli KKP 3801	OM250391	2022	Rucola	–	–	–
Escherichia coli KKP 3802	OM250393	2022	Washed spinach	–	–	–
Escherichia coli KKP 3824	ON303636	2020	Salad mix with carrot	–	–	–
Escherichia coli KKP 3825	ON303626	2020	Rucola	–	–	–

Notes: When strain infected with phage: "++"—transparent plaques; "+"—cloudy plaques; "–"—no plaques (insensitive bacterial strain). Each spot test was performed in triplicate (*n* = 3).

Among the tested bacteriophages, Enterobacter phage KKP_3711 targeting *Enterobacter cloacae* strain KKP 3684 (as primary host) demonstrated the widest host range and infected 28% of the tested bacteria (7/25). Based on the calculated EOP, it was found that the *Escherichia coli* strain KKP 3705 enables high production of Enterobacter phage KKP_3711 virions (EOP > 0.5). Serratia phage KKP_3708 targeting *Serratia liquefaciens* strain KKP

3654 (as primary host) infected 13% (3/25) of the tested bacterial hosts. According to the obtained EOP, *Enterobacter cloacae* strain KKP 3692 allows for effective infection and phage production (EOP > 0.5). Other bacterial strains will not enable high efficiency of plating. In the case of Serratia phage KKP_3709 targeting *Serratia marcescens* strain KKP 3687 (as primary host), activity was observed only against one strain (EOP < 0.1) The EOP evaluates the seeding efficiency, i.e., the ability of a bacteriophage to form plaque in a given strain compared to a primary bacterial host [43,44]. According to another division, EOP values can be divided into several categories based on their values. An EOP value below 0.001 indicates that the phage is not effective against the target bacteria or is avirulent. EOP values ranging from 0.001 to 0.099 are considered moderately virulent, EOP values ranging from 0.1 to 1.0 indicate that the bacteriophages are highly virulent, indicating the strong effectiveness of the phage against the target bacteria [45]. In the case of Serratia phage KKP_3708, this phage was classified as highly virulent. Depending on the bacterial host, Serratia phage KKP_3709 was classified as moderately or highly virulent, while Enterobacter phage KKP_3711 as avirulent, moderate, or virulent. The variation in efficiency against target bacteria could be caused by inadequate bacteriophage adsorption into target bacteria. Moreover, the resistance mechanisms of the target bacteria could hinder the bacteriophage replication, leading to a lower EOP value [46].

Assessment of the host range is an important step in the characterization of bacteriophages, which enables the selection of bacterial viruses for the development of a broad-spectrum biopreparation [47]. The phage cocktail also prevents bacteria from developing resistance to phage infection, which results in the effective elimination of undesirable bacterial microbiota [48]. Although bacteriophages are very specific, it has been shown that some bacteriophages can infect bacteria within a complete genus or even a family. As an example, salmophages isolated by Wójcicki and co-workers [38], in addition to *Salmonella enterica* species, also infected saprophytic bacterial strains from *Escherichia coli* and *Enterobacter cloacae* species [38]. Understanding the complex dynamics of bacteria–phage interaction will ensure the development of effective biocontrol and phage therapy [49].

2.3. Bacterial Growth Inhibition after Phage Infection

Determining the equation of the dependence of optical density on the concentration of bacterial cells (data unpublish) made it possible to adjust the appropriate values of the infection coefficient (MOI = 1000; MOI = 100; MOI = 10; MOI = 1.0; MOI = 0.1; MOI = 0.01; and MOI = 0.001). The results of the dependence of the change in optical density of the bacterial cultures on the MOI used are graphically presented in Figure 2.

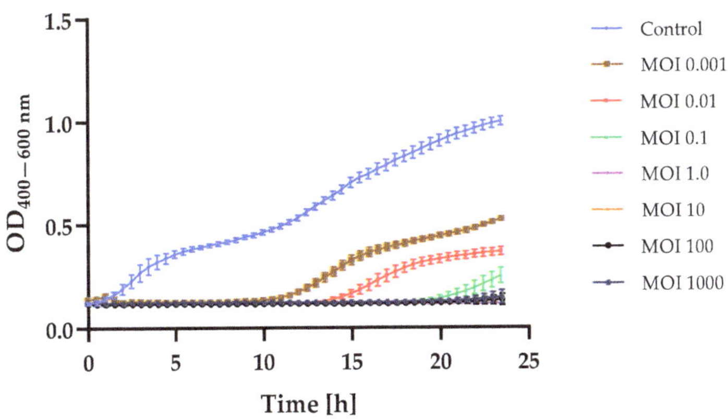

(A)

Figure 2. *Cont.*

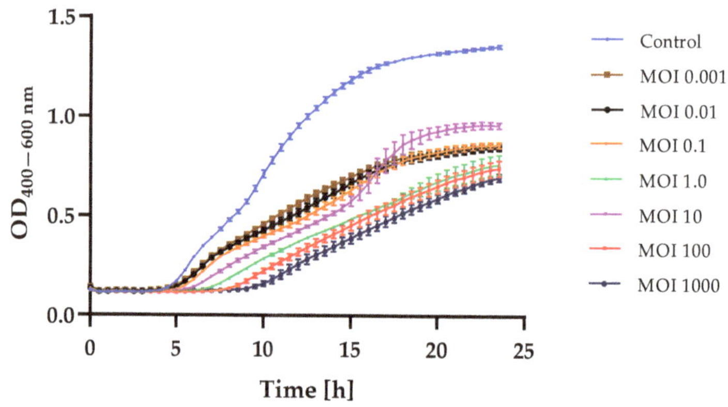

(**B**)

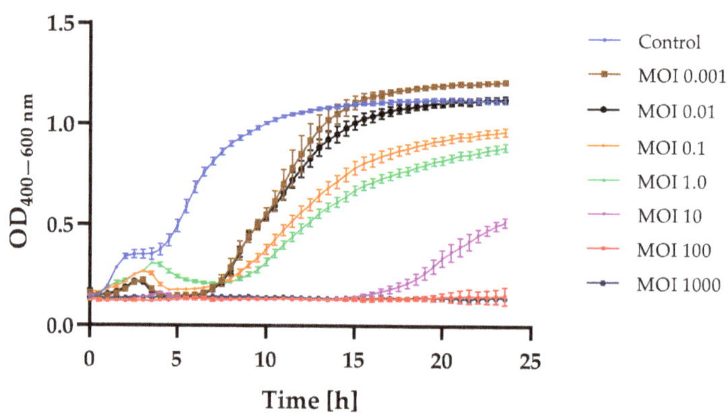

(**C**)

Figure 2. Growth curves of bacterial host strain (n = 10) treated with phages at different MOIs compared to the control culture (blue line): (**A**)—*Serratia liquefaciens* strain KKP 3654 infected with Serratia phage KKP_3708; (**B**)—*Serratia marcescens* strain KKP 3687 infected with Serratia phage KKP_3709; (**C**)—*Enterobacter cloacae* strain KKP 3684 infected with Enterobacter phage KKP_3711. Error bars represent the standard deviation (±SD) in optical density values of bacterial cultures.

In each case, infection with the tested bacteriophages resulted in a delay in the logarithmic growth phase of the bacterial culture compared to the control culture (bacterial strain not treated with phage). Serratia phage KKP_3708 caused complete inhibition of *Serratia liquefaciens* strain KKP 3654 growth at an MOI of 1.0 and above (Figure 2A). In the case of Serratia phage KKP_3709, the complete elimination of *Serratia liquefaciens* strain KKP 3687 was not achieved at the MOIs used (Figure 2B). However, a reduction in optical density was observed depending on the MOI rates. The use of the lowest MOI values (i.e., 0.001 and 0.01) for the Enterobacter phage KKP_3711 resulted in an initial reduction in optical density, which reached control culture values during incubation. However, MOIs of 1000 and 100 completely limited the growth of *Enterobacter cloacae* strain KKP 3684 (Figure 2C).

By analyzing the minimum inhibitory multiplicity of infection (miMOI) of a given bacteriophage, the most effective MOI value to use during application can be determined. The miMOI is defined as the minimum bacteriophage concentration required to inhibit bacterial growth (host), while MOI itself is the ratio of bacteriophages to its reference host bacteria. Based on bacterial growth curve analyses, the miMOI of Serratia phage

KKP_3708 was determined to be 1, while the miMOI of Enterobacter phage KKP_3711 was 100. The miMOI value could not be determined for Serratia phage KKP_3709. Hence, Serratia phage KKP_3708 was considered to be more effective since it required a lower phage concentration to inhibit bacterial growth. The obtained results also indicate that the decrease in bacterial growth happened due to an increase in bacteriophage concentration. According to literature data, the different miMOI values in bacteriophages result from their adsorption rate [50].

Determining bacterial growth curves after phage infection is used to predict its effectiveness during application. The study conducted by Cieślik and co-workers [51] showed that Enterobacter phage Entb_43 and Enterobacter phage Entb_45 significantly inhibited the growth of the *E. cloacae* strain 30345 and *E. cloacae* strain 29796, respectively, at a low infection rate (MOI of 0.001). During incubation, the optical density did not change significantly for bacteriophage-treated samples and did not exceed 0.150 [51]. In a study conducted by Wójcicki and co-workers [52], the effect of the addition of bacteriophages was determined at MOI = 1.0 and MOI = 0.1. For all tested bacterial host strains (i.e., *Citrobacter freundii* strain KKP 3655, *Enterobacter cloacae* strain KKP 3082, *Enterobacter ludwigii* strain KKP 3083, and *Serratia fonticola* strain KKP 3084), the use of bacteriophages (i.e., Citrobacter phage KKP 3664, Enterobacter phage KKP 3262, Enterobacter phage KKP 3263, and Serratia phage KKP 3264, respectively) at a higher MOI more strongly limited host cell division (lower optical density values). At the same time, the use of these two MOI values did not completely limit the growth of bacterial hosts during 24-hour culture [52]. According to a study conducted by Shang and co-workers [53] of Salmonella phage vB_SalP_TR2, the inhibitory effect at MOI = 0.01 and MOI = 0.001 was better than that at MOI = 0.1. Application of this phage to milk resulted in a significant reduction in the number of bacterial cells compared to the phage-free control group (reduction of 1.8 log CFU mL^{-1}). In the test of salmophage application on the surface of chicken meat, the maximum reduction (0.9 log CFU mL^{-1}) was achieved after 6 h of incubation. However, after 12 and 24 h of incubation, no significant statistical differences in pathogen reduction were observed [53].

Regrowth of bacteria after phage treatment may be related to the development of many defense mechanisms in bacteria that protect against phage infection [54]. The basic defense mechanism is masking the receptor through conformation or creating a shell that limits phage interaction [54]. Often, after a phage infection, a bacterial cell can destroy foreign genetic material. In this case, the phage DNA undergoes endonucleolytic digestion in the cell cytoplasm [55,56]. This process involves the recognition of foreign DNA by restrictionases and its cutting and modification through the access of methyl groups [54,57].

2.4. *Influence of Temperature and pH on the Phage Stability*

The influence of temperature in the range from −20 °C to 80 °C on the phage activity was determined (Figure 3). Serratia phage KKP_3709 and Enterobacter phage KKP_3711 retained high titers after 1 h incubation at temperatures up to 60 °C, while the temperature of 80 °C significantly reduced their activity (Figure 3B,C). In the case of Serratia phage KKP_3708, their high activity was retained up to a temperature of 40 °C, but at temperatures of 70 °C and 80 °C, no bacteriophage plaques were observed (Figure 3A).

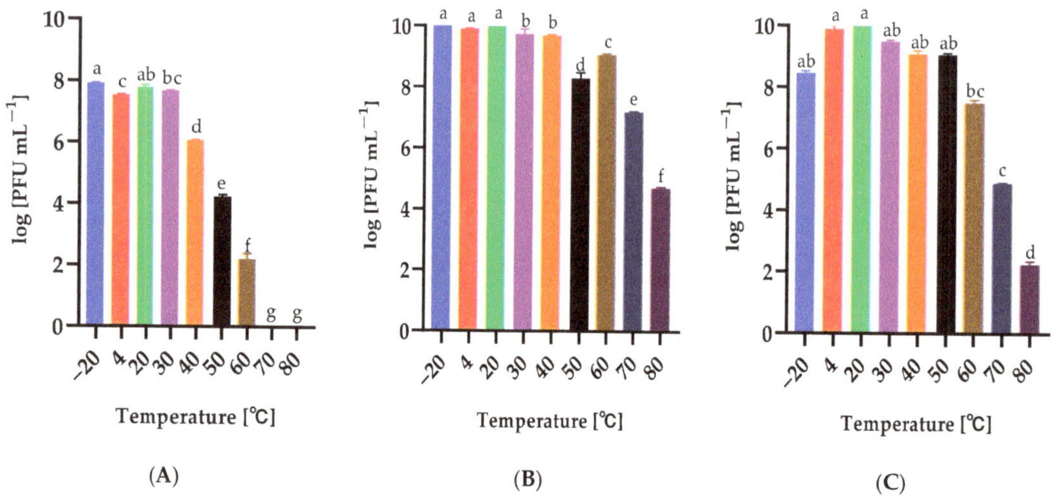

Figure 3. The activity of phages against bacterial host strains after exposure to a wide range of temperatures: (**A**)—Serratia phage KKP_3708, (**B**)—Serratia phage KKP_3709, (**C**)—Enterobacter phage KKP_3711. Letters a, b, c, d, e, f, and g indicate homogeneous groups at a significance level of $p \leq 0.05$, $n = 3$.

The effect of a 1 h incubation of phages in solutions with a pH ranging from 3 to 12 is shown in Figure 4. It was observed that incubation at extreme active acidity values (pH 3 and 12) slightly reduced the titer of the tested *Serratia-* (Figure 4A,B) and *Enterobacter-* specific phages (Figure 4C). Bacteriophages retained high activity over the entire pH range tested.

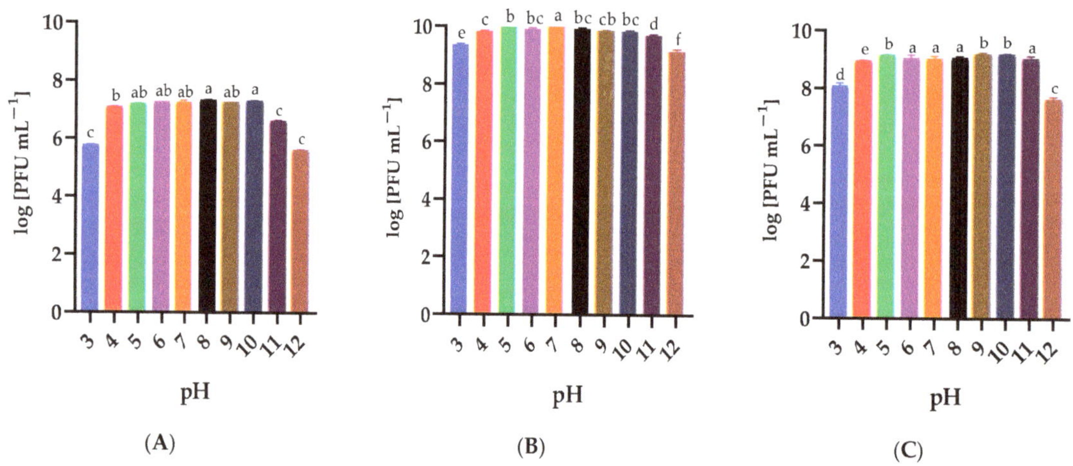

Figure 4. The activity of phages against bacterial host strains after exposure to a wide range of pH values: (**A**) Serratia phage KKP_3708, (**B**) Serratia phage KKP_3709, (**C**) Enterobacter phage KKP_3711. Letters a, b, c, d, e, and f indicate homogeneous groups at a significance level of $p \leq 0.05$, $n = 3$.

Biological characteristics are a key element allowing the prediction of phage activity in the product environment and its proper selection. The bacteriophage's stability in environmental conditions is an individual feature [52]. For example, bacteriophages specific to thermophilic bacteria are usually resistant to high temperatures. Research confirmed that bacteriophage AP45 targeting *Aeribacillus* sp. showed activity after 24 h of incubation at 85 °C [58]. Different results were observed in a study conducted by Shymialevich and co-workers [37]. The lysogenic Alicyclobacillus phage KKP 3916 retained high activity in the temperature range from 4 °C to 30 °C and pH from 3 to 11 [37]. While *A. acidoterrestris* grows in a strongly acidic environment and its optimal growth temperature is 42–53 °C, it may also range 35–55 °C [59,60]. In another study, stability testing for the *Salmonella*-specific bacteriophage vB_SalP_TR2 showed limited activity after incubation for an hour at 70 °C and 80 °C and at extreme pH values (i.e., pH 2–3 and pH 11–12) [53]. Phages targeting saprophytic bacteria characterized in a study conducted by Wójcicki and co-workers [52] retained their activity in a wide range of temperatures (from −20 °C to 50 °C) and active acidity values (pH from 4 to 11). All tested phages retained at least 70% of activity after incubation at 60 °C. After 1-hour incubation at 80 °C, no phage activity was observed against the tested bacterial host strains [52]. Phage stability/activity changes during long-term storage. Studies conducted by Cieślik and co-workers [51] showed that phage lysates retained initial titers even during six months of storage at both −70 °C and 4 °C, while storage at 37 °C resulted in a complete loss of their activity.

2.5. Phage Genome Sequencing and Analysis Using Bioinformatics Approaches

The complete genomes of the Serratia phage KKP_3708, Serratia phage KKP_3709, and Enterobacter phage KKP_3711 have been sequenced, annotated, and deposited in the GenBank database under the accession numbers OR067836, OR067833, and PP579741, respectively. Moreover, the newly isolated phages were deposited in the Culture Collection of Industrial Microorganisms—Microbiological Resources Center of the Department of Microbiology at the Prof. Wacław Dąbrowski Institute of Agricultural and Food Biotechnology—State Research Institute (IAFB; Warsaw, Poland).

For the genomic characterization of the bacteriophages, we used the latest guidelines on the taxonomy of bacterial viruses published in January 2023 [61]. Similar to TEM, genome analysis confirmed that all three phages belonged to the tailed complex phages from the *Caudoviricetes* class.

The proteomic tree of Serratia phage KKP_3708 generated by the BIONJ program-based TBLASTX genomic sequence comparisons of other phage genomes deposited in the Virus-Host DB [62] is presented in Figure 5A.

The phylogenetic relationship of Serratia phage KKP_3708 to other described prokaryotic viruses on a genome-wide basis was analyzed using ViPTree v4.0 server [63]. Based on the obtained phylogenetic tree, Serratia phage KKP_3708 was classified as a member from the *Autographiviridae* family, *Studiervirinae* subfamily, and *Przondovirus* genus (Figure 5A). Whole-genome sequencing (WGS) revealed that the Serratia phage KKP_3708 genome consists of 40,461 bp linear double-stranded DNA (dsDNA) with a total G+C content of 52.9% (Figure 5B). The described functional proteins have been divided into several groups depending on their functions: related to lysis, structure, DNA metabolisms and replication, and DNA genome packaging (Figure 5B). Out of the 55 predicted open reading frames (ORFs), 32 ORFs are associated with genes encoding proteins with known functions and 23 ORFs encode hypothetical proteins with unknown functions (Figure 5B and Table S1A).

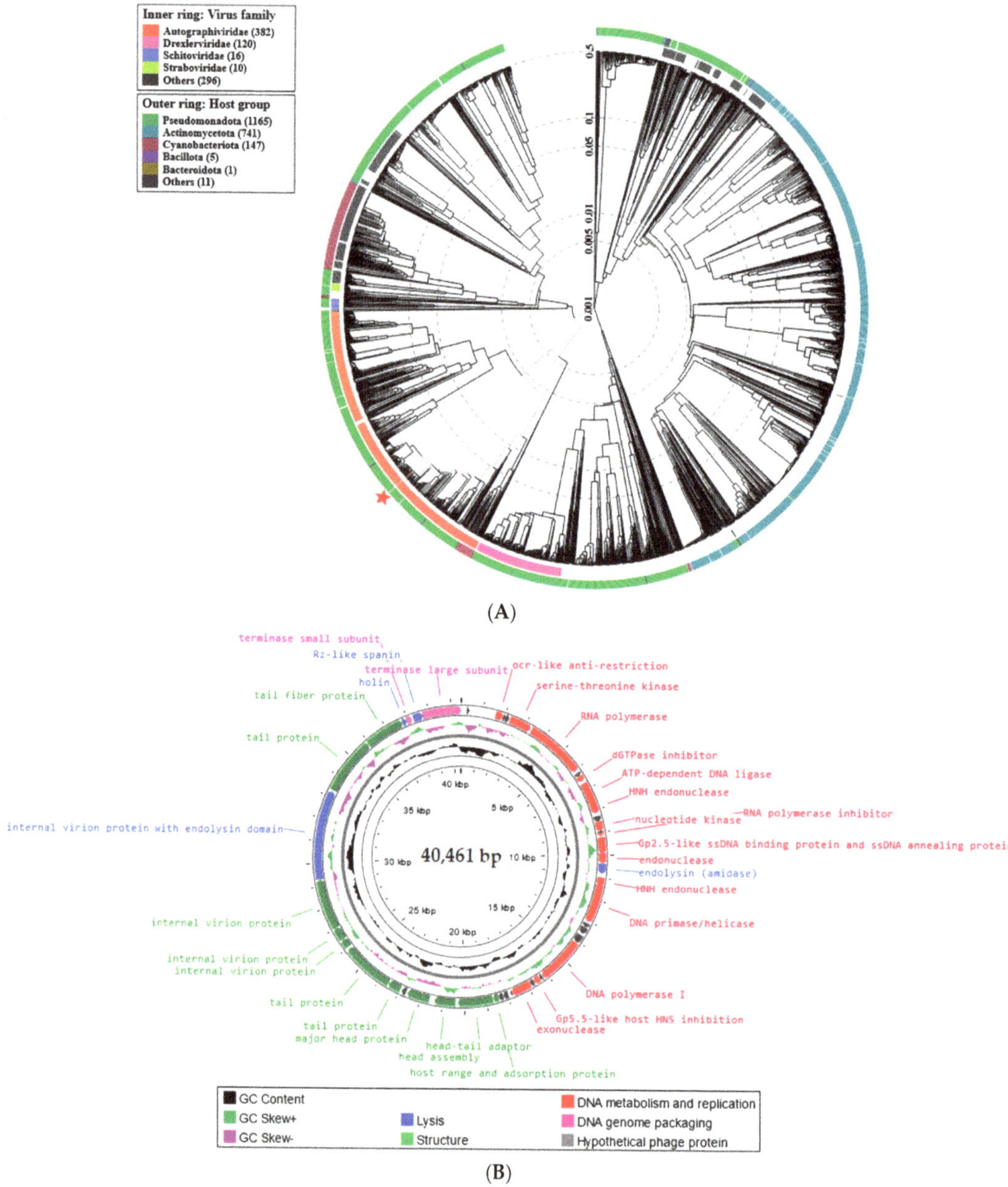

Figure 5. Genome analysis of Serratia phage KKP_3708: (**A**) The viral proteomic tree of Serratia phage KKP_3708 and other phage genomes is shown in a circular view. The branch represented studied phage is marked by an asterisk. Colored rings indicate virus families (inner ring) and host groups (at phylum level; outer ring). The tree was calculated by BIONJ based on genomic distance matrixes and mid-point rooted. Branch lengths are log-scaled. The sequence and taxonomic data were based on Virus-Host DB [62]. The tree shown was generated using the ViPTree v4.0 server [63]. (**B**) Map of the genomic organization of Serratia phage KKP_3708 generated using the Proksee program [64].

Structural proteins annotated in the Serratia phage KKP_3708 genome included: tail fiber protein, internal virion protein, tail protein, major head protein, head-tail adaptor, head assembly, host range, and adsorption protein. Among the genes associated with lysis, holine, endolysin (amidase), Rz-like spanin, and internal virion proteins with endolysin domain were found. Holins are small hydrophobic proteins that accumulate in the inner cell membrane and, after oligomerization, form pores [65]. This leads to the activation of peptidoglycan hydrolase (endolysin), which accumulates in the periplasm or cytoplasm and causes the degradation of the cell wall at the end of the lytic cycle [65,66]. Lytic enzymes consist of a catalytic domain, which hydrolyzes specific sites in the peptidoglycan, and a cell wall domain. Based on the catalytic domain, five classes of lytic enzymes are distinguished, i.e., glucosaminidase (cut bonds between murein amino sugars), muramidase, amidase (cut amide bonds between the glycan and the peptide bridge), endopeptidase (cut peptide bridges) and lytic transglycosylase [67,68]. Spanins are phage lysis proteins required to disrupt the outer membrane in the final step of Gram-negative host lysis [69]. Among the lytic proteins, an internal virion protein with endolysin domain was annotated, which forms the inner core of the virion and a channel in the host cell envelope, enabling the introduction of phage DNA [70]. Small and large terminase subunits were also detected in the Serratia phage KKP_3708 genome, which plays an important role in packing the viral genome into a pre-formed empty capsid using an ATP-powered molecular motor [71]. It is hypothesized that the small terminase subunit recognizes and binds to the viral genome substrate, while the large subunit is the main component of the terminase holoenzyme and is required for the cleavage of DNA into single genome lengths [71,72]. Serine/threonine kinase detected in the Serratia phage KKP_3708 genome directs the metabolism of the bacterial host for viral replication [73]. In turn, detected Ocr protein prevents the degradation and modification of phage DNA by inhibiting restriction endonuclease [74,75]. dGTPase inhibitor interacts and inhibits host dGTPase/dgt [76]. Two HNH endonuclease coding regions were detected in the Serratia phage KKP_3708 genome, which plays a key role in phage DNA packaging [77]. Among the genes related to replication and metabolism, the following proteins were annotated in the Serratia phage KKP_3708 genome: RNA polymerase, DNA primase/helicase, DNA polymerase I, and exonuclease. Moreover, no toxin or antibiotic resistance genes were detected in the Serratia phage KKP_3708 genome.

BLASTn similarity searches performed for Serratia phage KKP_3708 and related phages deposited in the GenBank database revealed 89.85% nucleotide similarity with 90% query coverage to Klebsiella phage Amrap (GenBank Acc. No. OQ579031.1), 92.36% nucleotide similarity with 88% query coverage to Klebsiella phage vB_Ko_K26PH128C1 (GenBank Acc. No. OY978813.1), 92.79% nucleotide similarity with 85% query coverage to Klebsiella phage VLCpiA3b (GenBank Acc. No. ON602742.1), 92.47% nucleotide similarity with 85% query coverage to Klebsiella phage P79_1 (GenBank Acc. No. OR256027.1), and 92.45% nucleotide similarity with 85% query coverage to Klebsiella phage KP32_isolate 194 (GenBank Acc. No. NC_047969.1) (Figure 6 and Table S2).

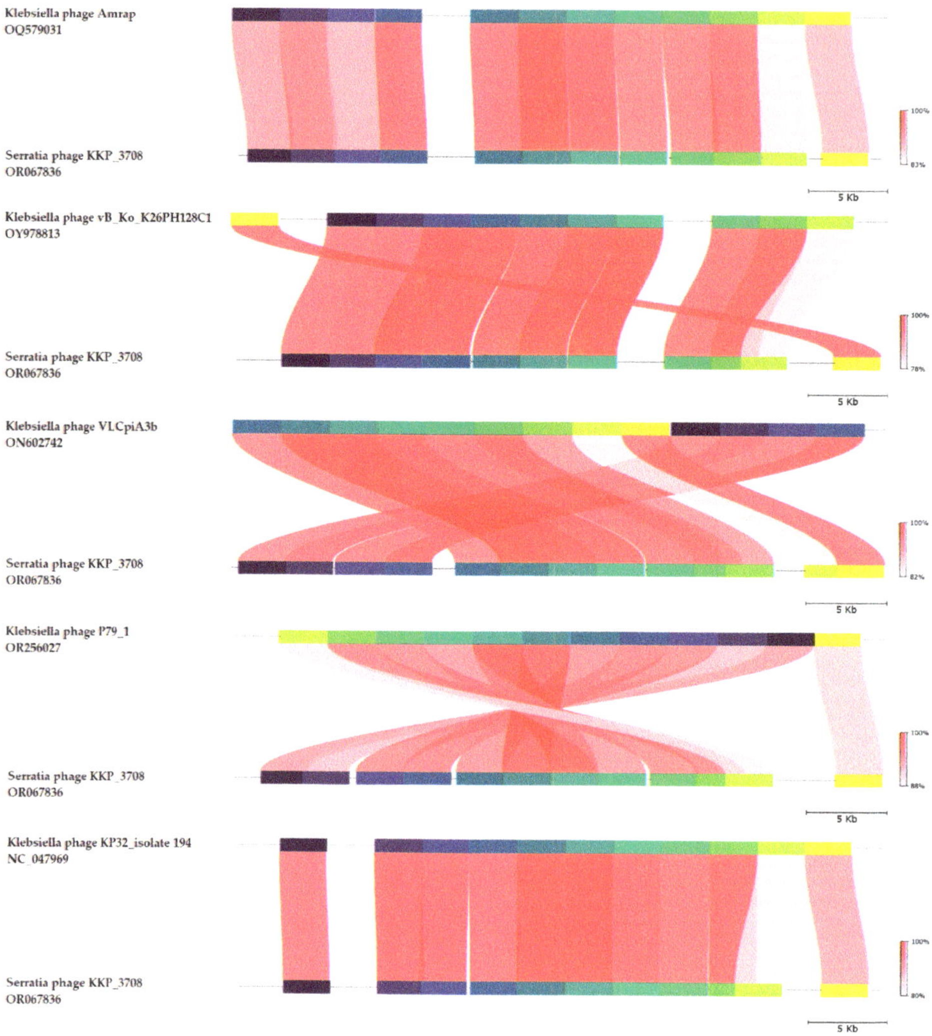

Figure 6. Genome sequence comparison of the Serratia phage KKP_3708 with five other related phage genomes exhibiting co-linearity detected by TBLASTX, using FastANI v1.3.3 software [78] and Proksee software [64]. Homologous regions detected by a TBLASTX search are connected by segments colored based on orthologous matches from query sequence fragments. The color bar shows the percentage identity of TBLASTX.

The nucleotide-based virus overall nucleotide sequence identity between isolated Serratia phage KKP_3708 and its 15 closest relatives was calculated using VIRIDIC v1.1 (Figure 7). It was demonstrated that the highest intergenomic similarity (82.9%) with our phage was to Klebsiella phage Amrap (GenBank Acc. No. OQ579031.1). Moreover, Serratia phage KKP_3708 was compared with other bacterial viruses reported in the PhageAI database. Phage similarity on a 2D scatter plot computationally predicted and rendered through PhageAI v1.0.2 software is shown in the Supplementary Materials (Figure S1).

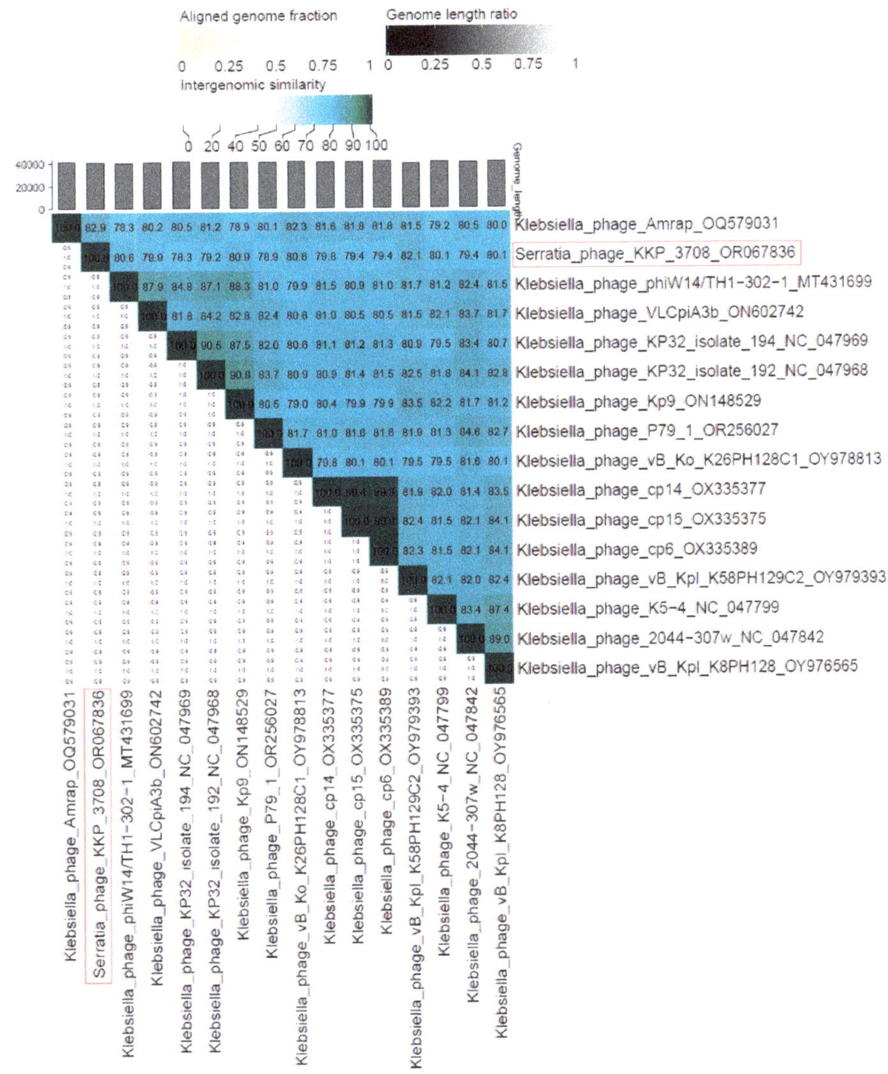

Figure 7. The whole-genome comparison and clustering of Serratia phage KKP_3708 (marked in a red frame), and its 15 closest relatives. The comparison and clustering were performed with the use of VIRIDIC v1.1 [79]. Different shades of blue in the right half of the heatmap represent different intergenomic similarities (%) between the genomes of each pair compared, as indicated above the heatmap and specified by numbers. The left half of the heatmap shows three indicator values for each genome pair: aligned fraction of genome one for the genome in this row (top value), genome length ratio for the two genomes in this pair (middle value), and aligned fraction of genome two for the genome in this column (bottom value).

The proteomic tree of Serratia phage KKP_3709 generated by the BIONJ program-based TBLASTX genomic sequence comparisons of other phage genomes deposited in the Virus-Host DB [62] is presented in Figure 8A. The phylogenetic relationship of Serratia phage KKP_3709 to other described prokaryotic viruses on a genome-wide basis was analyzed using ViPTree v4.0 server [63].

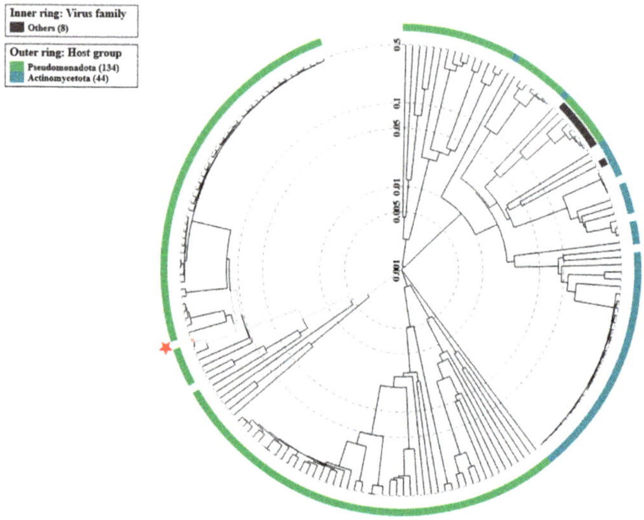

(A)

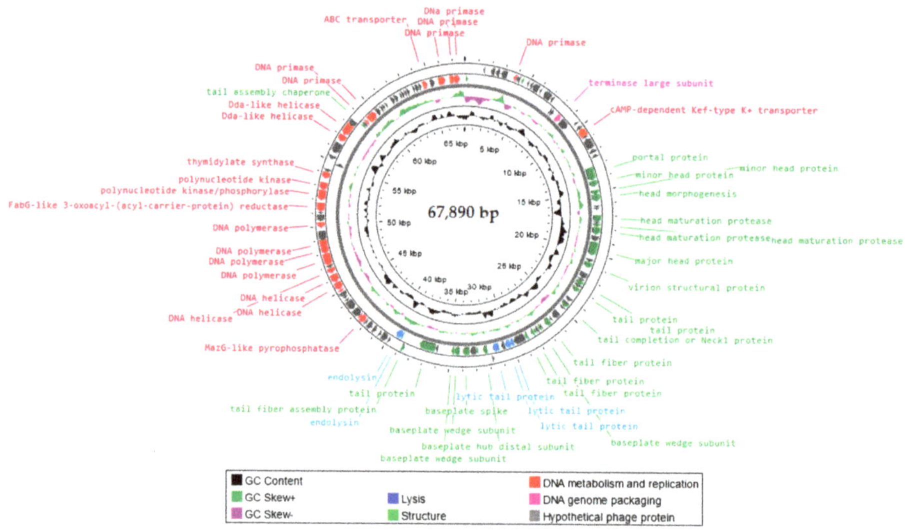

(B)

Figure 8. Genome analysis of Serratia phage KKP_3709: (**A**) The viral proteomic tree of Serratia phage KKP_3709 and other phage genomes is shown in a circular view. The branch represented studied phage is marked by an asterisk. Colored rings indicate virus families (inner ring) and host groups (at phylum level; outer ring). The tree was calculated by BIONJ based on genomic distance matrixes and mid-point rooted. Branch lengths are log-scaled. The sequence and taxonomic data were based on Virus-Host DB [62]. The tree shown was generated using the ViPTree v4.0 server [63]. (**B**) Map of the genomic organization of Serratia phage KKP_3709 generated using the Proksee program [64].

Based on the obtained phylogenetic tree, Serratia phage KKP_3709 was classified as a member from the *Myosmarvirus* genus (Figure 8A). WGS revealed that the Serratia phage KKP_3709 genome consists of 67,890 bp linear dsDNA with a total G+C content of 49.8% (Figure 8B). The described functional proteins have been divided into several groups depending on their functions: related to lysis, structure, DNA metabolisms and replication, and DNA genome packaging (Figure 8B). Out of the 194 predicted ORFs, 64 ORFs are associated with genes encoding proteins with known functions and 130 ORFs encode hypothetical proteins with unknown functions. Among the lytic proteins, endolysin and lytic tail proteins were annotated. Among the proteins responsible for packaging, only a large terminase subunit was found. Structural proteins detected in the Serratia phage KKP_3709 genome included: tail protein, baseplate spike, baseplate wedge subunit, tail fiber protein, virion structural protein, major head protein, head maturation protease, portal protein, tail assembly chaperon and minor head protein. Among the proteins responsible for DNA metabolisms and replication, DNA helicase, DNA polymerase, DNA primase, ABC transporter, thymidylate synthase, and polynucleotide kinase were found (Figure 8B and Table S1B). Importantly, no toxin or antibiotic resistance genes were detected in the Serratia phage KKP_3709 genome.

BLASTn similarity searches performed for Serratia phage KKP_3709 and related phages deposited in GenBank database revealed a 98.36% nucleotide similarity with 97% query coverage to Serratia phage vB_SmaM_Hera (GenBank Acc. No. MW021759.1), 97.32% nucleotide similarity with 95% query coverage to Serratia phage MTx (GenBank Acc. No. NC_048759.1), 82.25% nucleotide similarity with 83% query coverage to Erwinia phage vB_EamM_TropicalSun (GenBank Acc. No. MN013090.1), 83.18% nucleotide similarity with 81% query coverage to Serratia phage MyoSmar (GenBank Acc. No. NC_048800.1), and 82.13% nucleotide similarity with 79% query coverage to Serratia phage SMP (GenBank Acc. No. OP490597.1) (Figure 9 and Table S2).

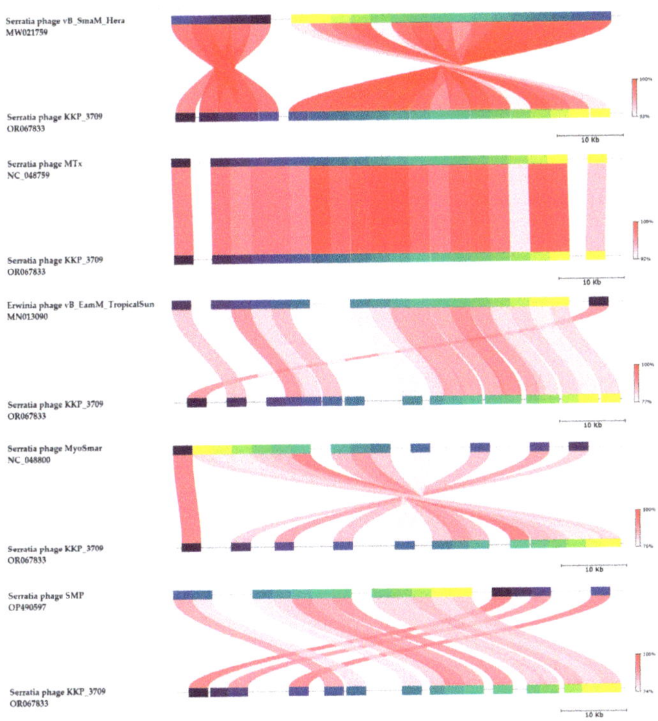

Figure 9. Genome sequence comparison of the Serratia phage KKP_3709 with five other related phage genomes exhibiting co-linearity detected by TBLASTX, using FastANI v1.3.3 software [78] and Proksee software [64]. Homologous regions detected by a TBLASTX search are connected by segments colored based on orthologous matches from query sequence fragments. The color bar shows the percentage identity of TBLASTX.

The nucleotide-based virus overall nucleotide sequence identity between isolated Serratia phage KKP_3709 and its 15 closest relatives was calculated using VIRIDIC v1.1 (Figure 10). Our phage showed the highest intergenomic similarity (95.7%) to Serratia phage vB_SmaM_Hera (GenBank Acc. No. MW021759.1). Furthermore, Serratia phage KKP_3709 was compared with other bacterial viruses reported in the PhageAI database. Phage similarity on a 2D scatter plot computationally predicted and rendered through PhageAI v1.0.2 software is shown in the Supplementary Materials (Figure S1).

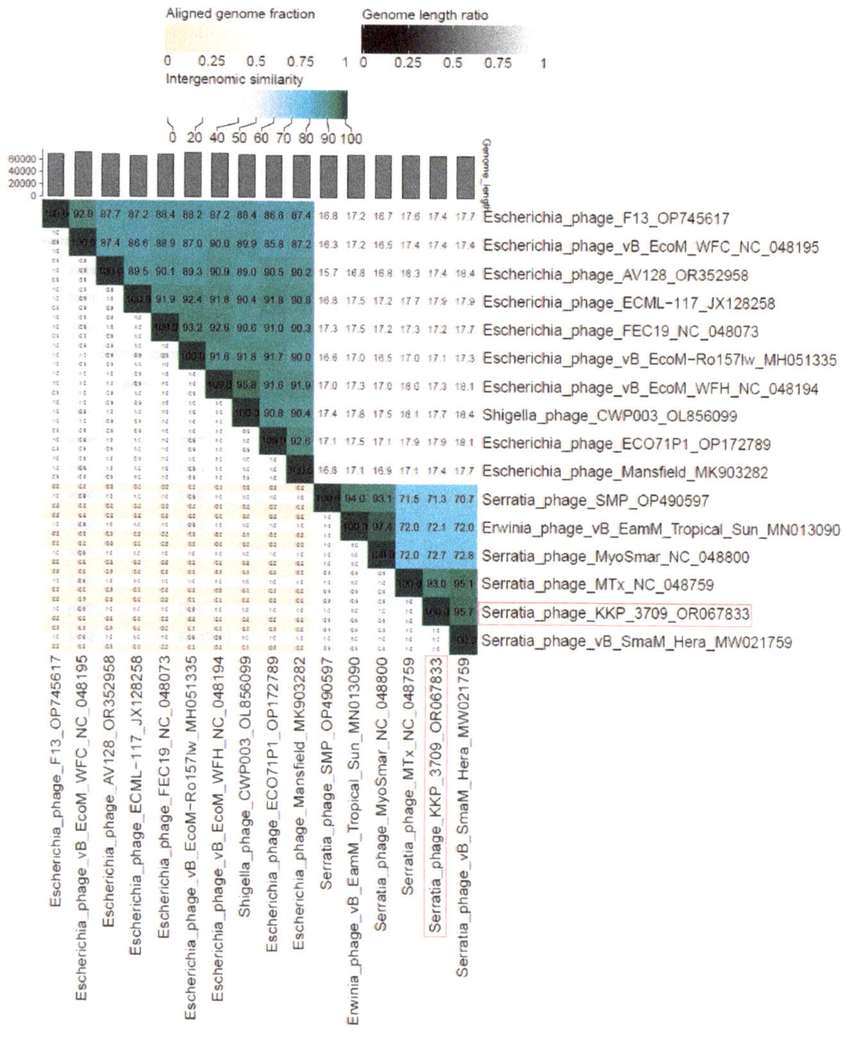

Figure 10. The whole-genome comparison and clustering of Serratia phage KKP_3709 (marked in a red frame) and its 15 closest relatives. The comparison and clustering were performed with the use of VIRIDIC v1.1 [79]. Different shades of blue in the right half of the heatmap represent different intergenomic similarities (%) between the genomes of each pair compared, as indicated above the heatmap and specified by numbers. The left half of the heatmap shows three indicator values for each genome pair: aligned fraction of genome one for the genome in this row (top value), genome length ratio for the two genomes in this pair (middle value), and aligned fraction of genome two for the genome in this column (bottom value).

The proteomic tree of Enterobacter phage KKP_3711 generated by the BIONJ program-based TBLASTX genomic sequence comparisons of other phage genomes deposited in the Virus-Host DB [62] is presented in Figure 11A. The phylogenetic relationship of Enterobacter phage KKP_3711 to other described bacterial viruses on a genome-wide basis was analyzed using ViPTree v4.0 server [63].

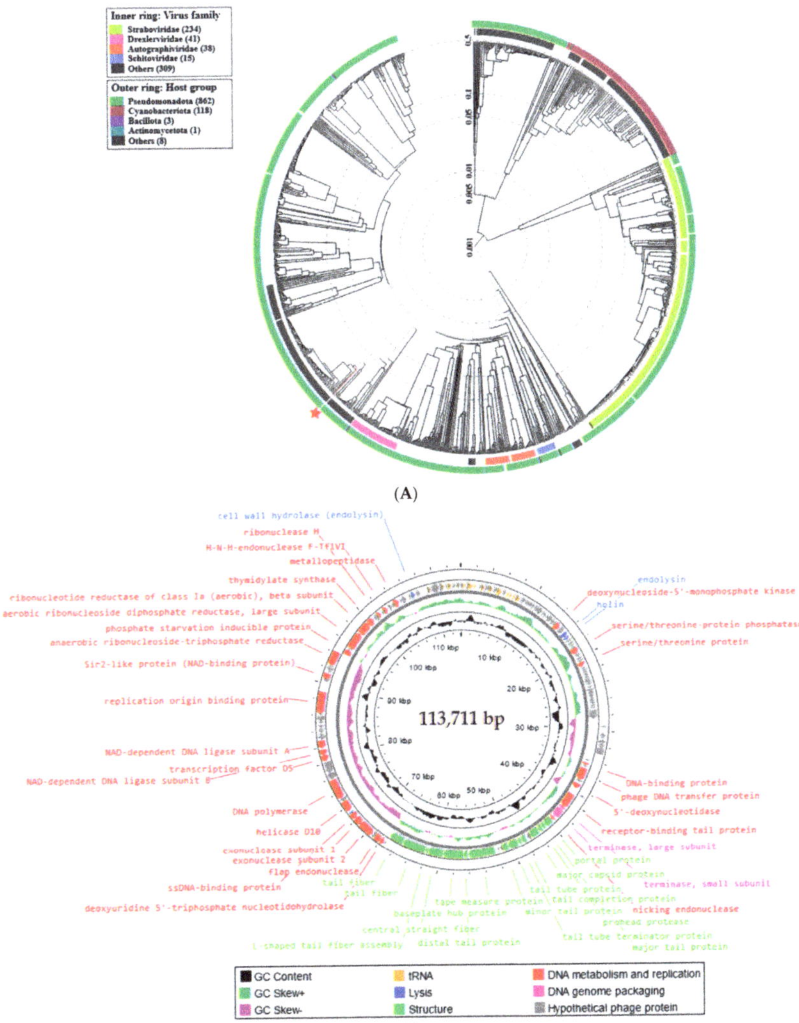

Figure 11. Genome analysis of Enterobacter phage KKP_3711: (**A**) The viral proteomic tree Enterobacter phage KKP_3711 and other phage genomes is shown in a circular view. The branch represented studied phage is marked by an asterisk. Colored rings indicate virus families (inner ring) and host groups (at phylum level; outer ring). The tree was calculated by BIONJ based on genomic distance matrixes and mid-point rooted. Branch lengths are log-scaled. The sequence and taxonomic data were based on Virus-Host DB [62]. The tree shown was generated using the ViPTree v4.0 server [63]. (**B**) Map of the genomic organization of Enterobacter phage KKP_3711 generated using the Proksee program [64].

Based on the obtained phylogenetic tree, the Enterobacter phage KKP_3711 shows the highest similarity to bacteriophages from the *Demerecviridae* family and *Markadamsvirinae* subfamily (Figure 11A). According to the Bacterial Viruses Subcommittee (BVS) of the International Committee on Taxonomy of Viruses (ICTV), two phages are assigned to the same species if their genomes are more than 95% identical, while a genus is described as a cohesive group of viruses sharing a high degree (>70%) of nucleotide identity of the full genome length [80]. Following this and based on the results of the comparative genome sequence analysis performed during this study, we considered that Enterobacter phage KKP_3711 represents a new genus of tailed phages from the *Caudoviricetes* class. Moreover, Serratia phage KKP_3708 could not be classified as any species currently recognized by ICTV, and likely represents a new species of tailed phages of siphovirus morphology from the *Caudoviricetes* class. WGS revealed that the Enterobacter phage KKP_3711 genome consists of 113,711 bp linear dsDNA with a total G+C content of 37.8% (Figure 11B). The described functional proteins have been divided into several groups depending on their functions: related to lysis, structure, DNA metabolisms and replication, and DNA genome packaging. Moreover, unlike the Serratia phage KKP_3708 and Serratia phage KKP_3709 genomes, in the Enterobacter phage KKP_3711 genome 27 tRNA regions were located (Figure 11B). The lack of potential tRNA genes suggests that Serratia phage KKP_3708 and Serratia phage KKP_3709 use their host's tRNA to synthesize their proteins. The lack of tRNA genes may be due to the compressed structure of phage genomes, in which the sequences of translation-related genes are not identified, which may be related to the exclusion of irrelevant information during replication [81]. Out of the 229 predicted ORFs in Enterobacter phage KKP_3711 genome, 49 ORFs are associated with genes encoding proteins with known functions and 120 ORFs encode hypothetical proteins with unknown functions. Proteins related to the phage structure, such as tail fiber, tail protein, capsid protein, portal protein, and baseplate hub protein were detected in the Enterobacter phage KKP_3711 genome. Highlighted proteins are responsible for the adsorption and specificity of the phage (tail fiber) [82]. Other annotated proteins such as holin and endolysin present lytic properties. Large and small subunits of terminase were identified among genes responsible for DNA genome packing. Among the proteins responsible for metabolism and replication, DNA polymerase, exonucleases, endonuclease, helicases, ribonuclease, ligase, and serine/threonine protein were found (Figure 11B and Table S1C). No antibiotic resistance genes, genes encoding virulence factors, integrases, recombinases, or repressors, which are markers of temperate bacteriophages, were identified in the genomes of all three phages.

BLASTn similarity searches performed for Enterobacter phage KKP_3711 and related phages deposited in GenBank database revealed a 71.54% nucleotide similarity with 11% query coverage to Salmonella phage SP33 (GenBank Acc. No. OR862218.1), 72.05% nucleotide similarity with 9% query coverage to Klebsiella phage vB_Kpn_3 (GenBank Acc. No. MZ079855.1), 71.75% nucleotide similarity with 9% query coverage to Escherichia phage EC142 (GenBank Acc. No. ON185584.1), 71.75% nucleotide similarity with 9% query coverage to Escherichia phage EC104 (GenBank Acc. No. ON185581.1), and 71.75% nucleotide similarity with 9% query coverage to Escherichia phage EC122 (GenBank Acc. No. ON185583.1) (Figure 12 and Table S2).

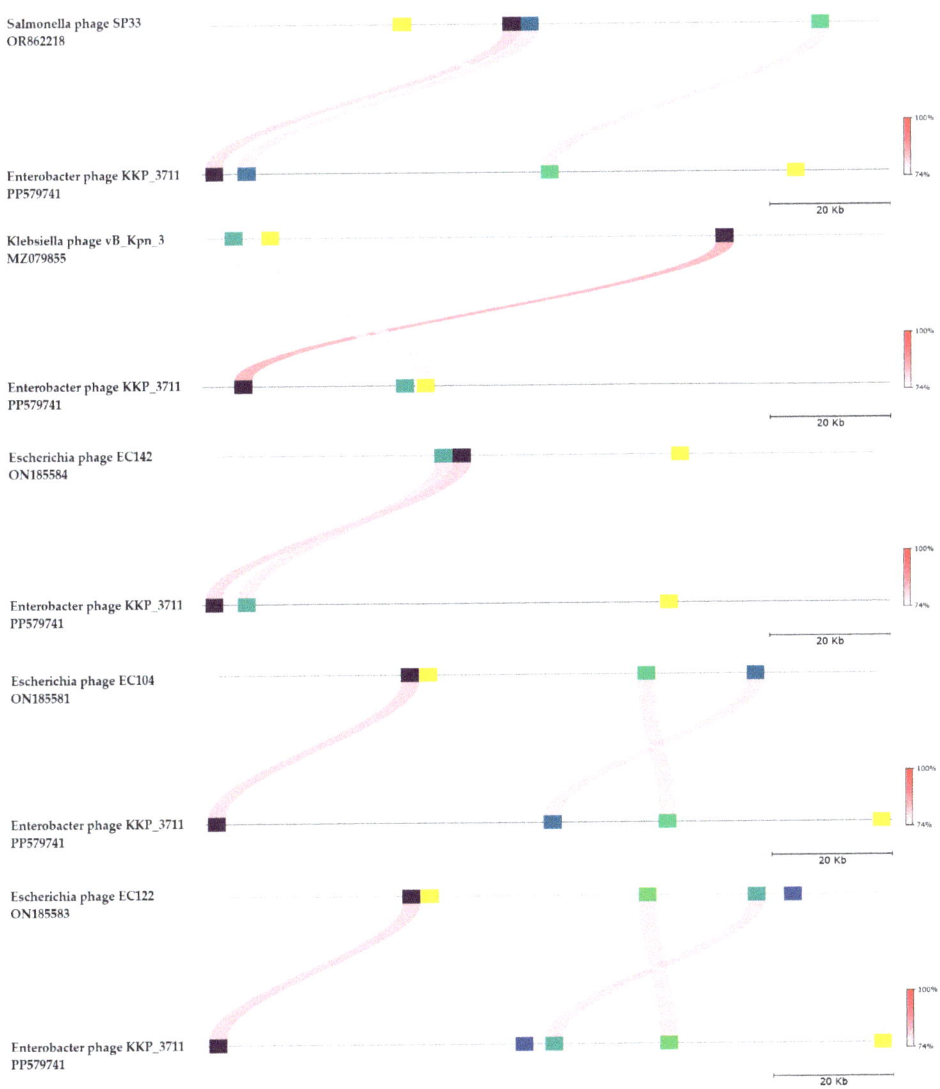

Figure 12. Genome sequence comparison of the Enterobacter phage KKP_3711 with five other related phage genomes exhibiting co-linearity detected by TBLASTX, using FastANI v1.3.3 software [78] and Proksee software [64]. Homologous regions detected by a TBLASTX search are connected by segments colored based on orthologous matches from query sequence fragments. The color bar shows the percentage identity of TBLASTX.

The nucleotide-based virus overall nucleotide sequence identity between isolated Enterobacter phage KKP_3711 and its 15 closest relatives was calculated using VIRIDIC v1.1 (Figure 13). It was demonstrated that the highest intergenomic similarity (32.6%) with our phage was to Salmonella phage vB_SalS_ABTNLsp4 (GenBank Acc. No. MW149273.1). Moreover, Enterobacter phage KKP_3711 was compared with other bacterial viruses reported in the PhageAI database. Phage similarity on a 2D scatter plot computationally predicted and rendered through PhageAI v1.0.2 software is shown in the Supplementary Materials (Figure S1).

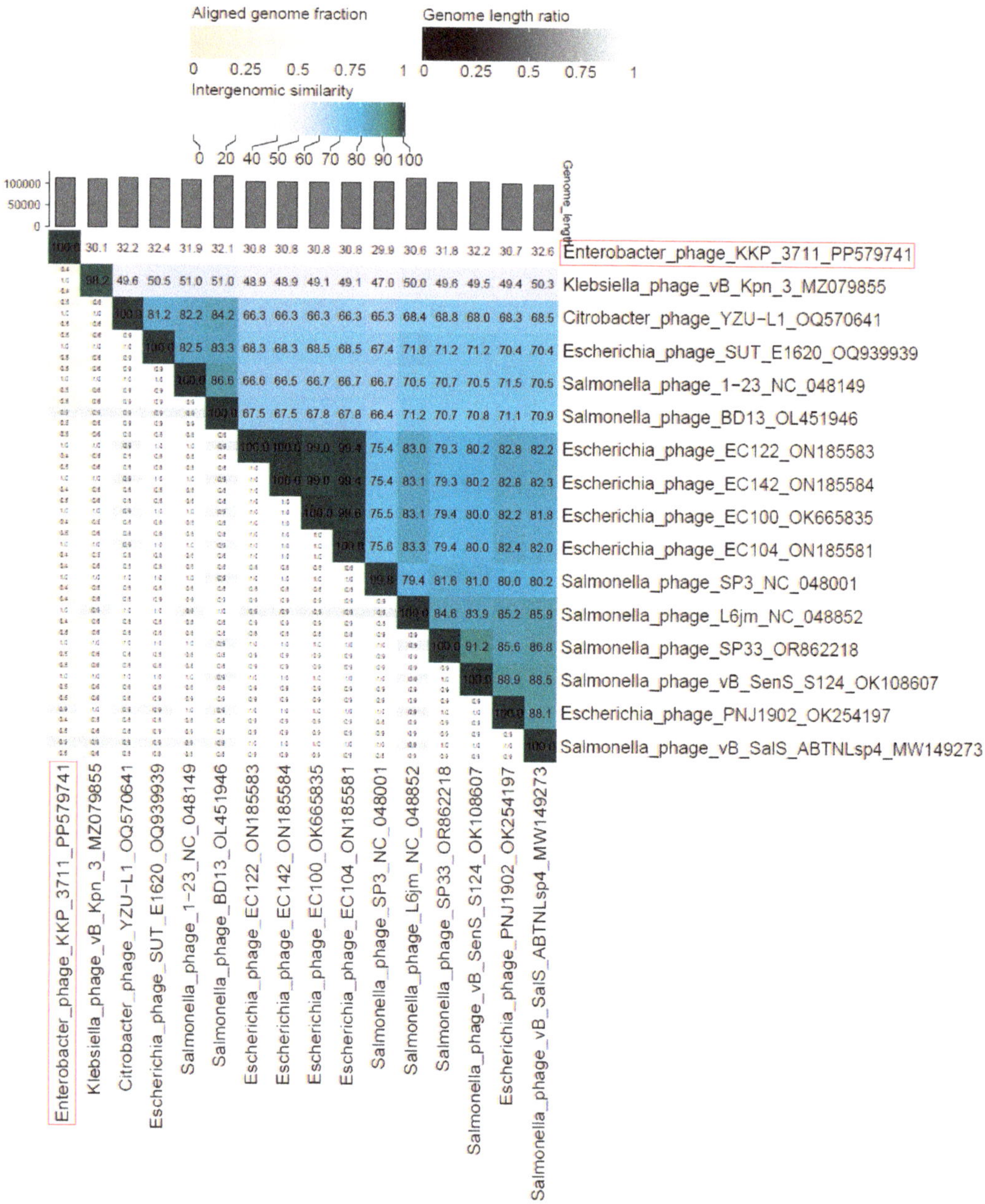

Figure 13. The whole-genome comparison and clustering of Enterobacter phage KKP_3711 (marked in a red frame), and its 15 closest relatives. The comparison and clustering were performed with the use of VIRIDIC v1.1 [79]. Different shades of blue in the right half of the heatmap represent different intergenomic similarities (%) between the genomes of each pair compared, as indicated above the heatmap and specified by numbers. The left half of the heatmap shows three indicator values for each genome pair: aligned fraction of genome one for the genome in this row (top value), genome length ratio for the two genomes in this pair (middle value), and aligned fraction of genome two for the genome in this column (bottom value).

Based on the genomic analyses performed, it was concluded that all three newly isolated bacteriophages, namely, Serratia phage KKP_3708, Serratia phage KKP_3709, and Enterobacter phage KKP_3711 are virulent and do not encode toxin, antibiotic resistance genes or other lysogenic markers, which makes them potential candidates for food preservation.

3. Materials and Methods

3.1. Bacterial Host Strains

In this research, twenty-five saprophytic bacterial strains from the *Enterobacterales* order deposited in the Culture Collection of Industrial Microorganisms—Microbiological Resource Center of the Department of Microbiology of the Prof. Wacław Dąbrowski Institute of Agricultural and Food Biotechnology—State Research Institute (IAFB; Warsaw, Poland) were used. All strains were isolated from the plant-based minimally processed food products, i.e., rucola, salad mix with carrot, salad mix of beetroot, washed spinach, and unwashed spinach as part of previously conducted studies. Identification of strains was carried out by amplification of the *16S* rRNA gene region and sequencing was outsourced to Genomed S.A. company (Warsaw, Poland). The raw sequences were analyzed using BLASTn (NCBI) and deposited in the GenBank database.

3.2. Isolation of Bacteriophages, Propagation and Purification of Phage Particles

Bacteriophages were isolated from municipal sewage (sewage plant treatment "Mokre Łąki", Izabelin, Poland). For this purpose, the wastewater sample was centrifuged at 8000 rpm for 10 min (Sorvall LYNX 6000 ultracentrifuge, Thermo Fisher Scientific, Watertown, MA, USA) to separate bacteria and organic debris. The supernatant was then filtered through a 0.45 μm syringe filter (Minisart® NML Cellulose Acetate; Sartorius, Goettingen, Germany). To isolate phages specific to selected bacteria, 20 mL of filtered supernatant, 20 mL of double-concentrated Luria–Bertani broth (composition: 10.0 g L^{-1} of peptone (BLT, Łódź, Poland), 10.0 g L^{-1} of sodium chloride (Chempur, Piekary Śląskie, Poland), and 5.0 g L^{-1} of yeast extract (BLT, Łódź, Poland)) and 1 mL of overnight bacterial culture in Luria–Bertani broth were added to a 50 mL falcon. The prepared culture was incubated at 37 °C for 24 h, then centrifuged at 8000 rpm for 10 min. The supernatant was filtered through a 0.45 μm syringe filter [38]. Phage titers (in PFU mL^{-1}) were determined by the double-layer agar plate method [83]. For this purpose, a series of dilutions of the obtained phage lysate were performed. To 500 μL of the corresponding dilution, 100 μL of overnight bacterial culture in Luria–Bertani broth was added. After gentle mixing, the suspensions were allowed to stand at room temperature for 20 min. After incubation, the suspensions were poured onto a nutrient agar plate and poured over 4 mL of liquid-chilled Top agar (Luria–Bertani broth with 0.75% agar-agar, pH 7.0–7.2). The liquid mixture was carefully spilled on the surface of the agar plate and allowed to solidify. After drying, the dishes were stored upside down overnight at 37 °C. After incubation, plaques that formed on the bacterial lawn were computed, considering the dilution factor, and expressed as phage titer.

Single-phage plaques were cut with a scalpel and purified in SM (saline magnesium) buffer [84]. Purification was performed in four rounds of single-plaque passage to ensure that the isolate represented the clonal phage population.

3.3. Electron Micrographs of Phage Particles

Transmission electron microscopy (TEM) was used to determine phage morphology according to the protocol described in previous articles [38,52]. Briefly, phage lysates were concentrated and purified by centrifugation at 4 °C and 14,500 rpm for 40 min (centrifuge MiniSpin® plus, Eppendorf, Hamburg, Germany). The supernatant was removed and the pellet was suspended in 2 mL of 100 mM cold ammonium acetate (Chempur, Piekary Śląskie, Poland) filtered through a 0.22 μm syringe filter. The pellet was carefully pipetted and centrifuged again using the previously described parameters. The entire procedure was repeated four times. After the final centrifugation, the pellet was suspended in 50 µL of ammonium acetate. The obtained phage suspension in an amount of 2 µL was applied to carbon-sputtered copper–wolfram mesh grids and left to dry at room temperature. The sample was then stained with 2% uranyl acetate (Warchem, Zakręt, Poland) solution for 1 min. The prepared samples were dried for 12 h under sterile conditions and visualized in a JEM-1220 transmission microscope (JEOL, Tokyo, Japan) at 100,000–200,000× magnification at a voltage of 80 kV.

3.4. Determination of the Bacterial Host Range for the Tested Phages

The ability of the strains to be infected by a specific bacteriophage was determined by a drop test. For this purpose, on nutrient agar plates were added 100 µL of overnight bacterial culture and equimoniously spread using 4 mL of 0.75% Top-agar. Then, 5 µL of each lysate of a predetermined titer (in PFU mL^{-1}) was applied to the bacterial lawns. Plates were incubated for 24 h at 37 °C. After incubation, the results were interpreted according to the assays: "++"—transparent plaques (sensitive bacterial strain); "+"—cloudy (opaque) plaques; "−"—no plaques (insensitive bacterial strain). The efficiency of plating (EOP) in the case of strains sensitive to phage infection was determined using the formula:

$$EOP = \frac{\text{phage titer on test strain}}{\text{phage titer on host strain}} \quad (1)$$

A series of dilutions with seeding by the double-layer plate method was used to determine phage titers, according to the procedure described earlier. Depending on the average EOP value for a specific phage–bacteria combination, the phages were classified as: EOP $\geq$ 0.5—"high production", infection of at least 50% compared to the primary host; EOP $\geq$ 0.1 but < 0.5—"average production"; EOP > 0.001 but < 0.1 for "low production" efficiency; and EOP $\leq$ 0.001—"ineffective" [43]. According to another division, phages were classified as: EOP < 0.001—the phage is not effective against the target bacteria or is avirulent; EOP $\geq$ 0.001 but $\leq$0.099—moderately virulent phages, while EOP $\geq$ 0.1 to 1.0—high virulence of bacteriophages, indicating strong effectiveness of the phage against the target bacteria [45]. All experiments were performed in triplicate.

3.5. Changes in the Growth Kinetics of Bacterial Hosts after Phage Infection at Different MOIs

Growth curves were performed for each bacterial host (unpublished data). For this purpose, a bacterial strain was inoculated into the Luria–Bertani broth and incubated at 37 °C for 24 h. During the incubation, the dependence of the optical density (DU® 640 spectrophotometer, Beckman Instruments, Inc., Fullerton, CA, USA) on the bacterial concentration [CFU mL^{-1}] was performed in triplicate. After plotting bacterial growth curves, the growth kinetics of the bacterial culture after phage infection versus the control was measured using a Bioscreen C Pro automatic growth analyzer (Yo AB Ltd., Growth Curves, Helsinki, Finland). For this purpose, fresh bacterial culture was diluted with Luria–Bertani broth to obtain the desired optical density/number of bacterial cells depending on the phage titer. Then, 180 µL of Luria–Bertani broth, 10 µL of an appropriately diluted bacterial culture, and 10 µL of an appropriately diluted phage lysate were pipetted into each well on a multi-well plate to obtain the desired MOI. MOI values of 1000, 100, 10, 1.0, 0.1, 0.01, 0.001, and 0.0001 were used in this study. Based on the analysis of the growth curves of bacterial hosts, the minimum inhibitory multiplicity of infection (miMOI) of bacteriophages was also determined. All experiments were performed with ten repetitions.

3.6. Phage Stability after Exposure to Selected Factors

Bacteriophage suspensions of known concentration were incubated for 1 h at various temperatures (−20 °C, 4 °C, 20 °C, 30 °C, 40 °C, 50 °C, 60 °C, 70 °C, 80 °C) and pH values from 3 to 12. Then, the phage titer was determined using the double-layer plate method as mentioned earlier [85].

3.7. Extraction of Bacteriophage Genomic DNA

Phage genomic DNA was isolated using the PureLink™ RNA/DNA Mini Kit (Thermo Fisher Scientific Inc., Carlsbad, CA, USA) according to the manufacturer's protocol, with modifications by Wójcicki et al. [38]. Briefly, 8 mL of precipitation solution (PEG-NaCl: 2.5 M NaCl, 20% PEG 8000) was added to 40 mL of phage lysates and incubated overnight on ice. After incubation, samples were concentrated by centrifugation at 27,000 rpm for 1.5 h at 4 °C (Sorvall LYNX 6000 ultracentrifuge, Thermo Fisher Scientific, Watertown, MA, USA). The pellet was suspended in 400 µL of lysis buffer (containing 5.6 µg of carrier RNA) and vortexed for 15 s. Then, 50 µL of proteinase K was added and incubated for 1 h at 56 °C with shaking at 900 rpm (ThermoMixer C, Eppendorf, Hamburg, Germany). After incubation, the tubes were briefly vortexed and 300 µL of 100% ice-cold molecular biology-grade ethanol was added. Samples were vortexed and incubated at 20 °C for 5 min. After incubation, the tubes were briefly centrifuged to remove any drops from the inside of the lids. A total of 675 µL of each sample was transferred to the viral spin column and centrifuged for 1 min at 10,000 rpm (centrifuge MiniSpin® plus, Eppendorf, Hamburg, Germany). Columns were transferred to new wash tubes, and 500 µL of wash buffer was added and centrifuged for 1 min at 10,000 rpm. The procedure was repeated triplicate, with the last centrifugation for 3 min at 14,500 rpm. To elute the genetic material, the viral spin column was transferred to new tubes and 20 µL of RNase-free water was added. After a 1 min incubation, the samples were centrifuged twice for 1 min at 14,500 rpm. DNA purity was measured by the Nanodrop ND-1000 Spectrophotometer (Thermo Fisher Scientific, Watertown, MA, USA), and DNA concentration was quantified by a Qubit 4.0 Fluorometer using the Qubit dsDNA BR Assay Kit (Invitrogen, Carlsbad, CA, USA). DNA samples were stored at 4 °C until further processing for WGS analysis.

3.8. Genome Sequencing and Bioinformatic Analysis

The isolated genomic phage DNA was sent to genXone SA (Złotniki, Poland) for next-generation sequencing (NGS) in the WGS application. DNA libraries were prepared using Rapid Barcoding Kit reagents (Oxford Nanopore Technologies, Oxford, UK) according to the manufacturer's protocol. A sequencing depth of at least 50× genome coverage was assumed. NGS sequencing was performed by nanopore technology on the GridION X5 sequencing device (Oxford Nanopore Technologies, Oxford, UK) under the control of MinKnow v22.10.5. Bases were called with Guppy v6.3.8 Basecaller (Oxford Nanopore Technologies, Oxford, UK), followed by barcode demultiplexing, also using Guppy Barcoder v6.3.8 (Oxford Nanopore Technologies, Oxford, UK), generating a.fastq file for each barcode. De novo assembly of genomes was performed in Flye v2.8.1 software [86] and annotation of phage genomes in Phanotate v1.5.0 [87] and PhaGAA software (available online: http://phage.xialab.info/home (accessed on 24 April 2024)) [88]. Proksee software (available online: https://proksee.ca/ (accessed on 24 April 2024)) [64] was used to visualize phage genomes. Viral proteomic trees of phage genomes were calculated by BIONJ based on genomic distance matrixes and mid-point rooted and were represented in the circular view. Branch lengths were log-scaled. The sequence and taxonomic data were based on Virus-Host DB [62]. The trees were generated using the ViPTree v4.0 server [63]. Genome sequence comparison of newly isolated phages with five other related phage genomes exhibiting co-linearity was detected by TBLASTX using FastANI v1.3.3 software [78] and Proksee software [64]. The overall isolated phages nucleotide sequence identity with the other 15 phages deposited in the GenBank database was calculated using VIRIDIC v1.1 (intergenomic distance calculator) [79]. Phage similarity was computationally predicted on a 2D scatter plot and rendered through PhageAI v1.0.2 software [89]. The phage genomes were deposited in the GenBank database.

3.9. Statistical Analysis

All the experiments were repeated at least three times. The graphical data presented were statistically analyzed using Graph Prism v8.0.2 (GraphPad Software Inc., San Diego,

CA, USA). One-way ANOVA followed by Tukey's test with a 95% confidence interval ($\alpha = 0.05$) was used to assess the impact of selected physical and chemical factors on phage stability. Error bars shown in Bioscreen C Pro plots represent the standard error of the mean (±SEM).

4. Conclusions

Due to increasing antibiotic resistance and the difficulty in eliminating bacteria that form biofilms, scientists are turning to lytic bacteriophages and their enzymes. Therefore, in this study, we present the biological and genomic characterization of three newly isolated phages, namely, Serratia phage KKP_3708, Serratia phage KKP_3709, and Enterobacter phage KKP_3711. Based on genome-wide analyses, we could assign a taxonomic classification. The newly isolated phages belong to the tailed dsDNA bacteriophages from the *Caudoviricetes* class. Bacteriophages were characterized as strictly lytic, lacking toxin, antibiotic resistance genes, and other lysogenic markers. *Serratia*- and *Enterobacter*-specific phages have broad-spectrum activity and show a tolerance to a wide range of temperatures (from −20 °C to 60 °C) and active acidity values (pH from 3 to 12).

In future research, a reasonable approach would be to use strictly lytic (virulent) bacteriophages in combination with other conventional (chemical and/or physical) food preservation methods, including spraying with solutions of weak organic acids or hot water. The features of our newly isolated and characterized lytic bacteriophages indicate their potential for application in food biocontrol, for example, as one of the factors in hurdle technology.

Supplementary Materials: The following supporting information can be downloaded at: https://www.mdpi.com/article/10.3390/ijms25115944/s1.

Author Contributions: Conceptualization, D.S.; methodology, D.S. and M.W.; formal analysis, D.S., P.Ś. and M.W.; investigation, D.S., S.B., P.Ś., H.C., A.O., B.S. and M.W.; writing—original draft preparation, D.S.; writing—review and editing, M.W.; visualization, D.S., A.O. and M.W.; supervision, S.B., B.S. and M.W. All authors have read and agreed to the published version of the manuscript.

Funding: This research, under the name "The use of lytic bacteriophages in the eradication of saprophytic bacteria in food processing factories," (grant 11/2021) was carried out as part of the Innovation Incubator 4.0 program and was funded by the Ministry of Education and Science granted under contract MNiSW/2020/318/DIR.

Institutional Review Board Statement: Not applicable.

Informed Consent Statement: Not applicable.

Data Availability Statement: Data are contained within the article and Supplementary Materials.

Acknowledgments: We thank Grzegorz Nowicki and Jakub Grabowski from the genXone SA for helping in the sequence analysis of whole bacterial and phage genomes.

Conflicts of Interest: The authors declare no conflicts of interest.

References

1. Głąbska, D.; Guzek, D.; Groele, B.; Gutkowska, K. Fruit and Vegetable Intake and Mental Health in Adults: A Systematic Review. *Nutrients* **2020**, *12*, 115. [CrossRef] [PubMed]
2. Coppens, P. The Importance of Food Supplements for Public Health and Well-Being. *World Rev. Nutr. Diet.* **2020**, *121*, 66–72. [PubMed]
3. Cena, H.; Calder, P.C. Defining a Healthy Diet: Evidence for the Role of Contemporary Dietary Patterns in Health and Disease. *Nutrients* **2020**, *12*, 334. [CrossRef] [PubMed]
4. Apicella, A.; Carlucci, E.; Leneveu-Jenvrin, C.; Remize, F.; Scarfato, P. Active Coated Films to Decrease Browning of Minimally Processed Banana Fruit. *Macromol. Symp.* **2022**, *405*, 2100256. [CrossRef]
5. Collado-López, S.; Betanzos-Robledo, L.; Téllez-Rojo, M.M.; Lamadrid-Figueroa, H.; Reyes, M.; Ríos, C.; Cantoral, A. Heavy Metals in Unprocessed or Minimally Processed Foods Consumed by Humans Worldwide: A Scoping Review. *Int. J. Environ. Res. Public Health* **2022**, *19*, 8651. [CrossRef] [PubMed]

6. Alegbeleye, O.; Odeyemi, O.A.; Strateva, M.; Stratev, D. Microbial Spoilage of Vegetables, Fruits and Cereals. *Appl. Food Res.* **2022**, *2*, 100122. [CrossRef]
7. Morya, S.; Amoah, A.E.D.D.; Snaebjornsson, S.O. Food Poisoning Hazards and Their Consequences over Food Safety. In *Microorganisms for Sustainable Environment and Health*; Chowdhary, P., Raj, A., Verma, D., Akhter, Y., Eds.; Elsevier Inc.: Amsterdam, The Netherlands, 2020; pp. 383–400.
8. European Food Safety Authority; European Centre for Disease Prevention and Control. The European Union One Health 2022 Zoonoses Report. *EFSA J.* **2023**, *21*, e8442.
9. Dorado-García, A.; Smid, J.H.; Van Pelt, W.; Bonten, M.J.M.; Fluit, A.C.; Van Den Bunt, G.; Wagenaar, J.A.; Hordijk, J.; Dierikx, C.M.; Veldman, K.T.; et al. Molecular Relatedness of ESBL/AmpC-Producing *Escherichia coli* from Humans, Animals, Food and the Environment: A Pooled Analysis. *J. Antimicrob. Chemother.* **2018**, *73*, 339–347. [CrossRef] [PubMed]
10. Chelaghma, W.; Loucif, L.; Bendahou, M.; Rolain, J.M. Vegetables and Fruit as a Reservoir of β-Lactam and Colistin-Resistant Gram-Negative Bacteria: A Review. *Microorganisms* **2021**, *9*, 2534. [CrossRef]
11. Formenti, N.; Calò, S.; Parisio, G.; Guarneri, F.; Birbes, L.; Pitozzi, A.; Scali, F.; Tonni, M.; Guadagno, F.; Giovannini, S.; et al. Esbl/Ampc-Producing *Escherichia coli* in Wild Boar: Epidemiology and Risk Factors. *Animals* **2021**, *11*, 1855. [CrossRef]
12. Begrem, S.; Jérôme, M.; Leroi, F.; Delbarre-Ladrat, C.; Grovel, O.; Passerini, D. Genomic Diversity of *Serratia proteamaculans* and *Serratia liquefaciens* Predominant in Seafood Products and Spoilage Potential Analyses. *Int. J. Food Microbiol.* **2021**, *354*, 109326. [CrossRef] [PubMed]
13. Allen, J.L.; Doidge, N.P.; Bushell, R.N.; Browning, G.F.; Marenda, M.S. Healthcare-Associated Infections Caused by Chlorhexidine-Tolerant *Serratia marcescens* Carrying a Promiscuous IncHI2 Multi-Drug Resistance Plasmid in a Veterinary Hospital. *PLoS ONE* **2022**, *17*, e0264848. [CrossRef] [PubMed]
14. Labbate, M.; Queck, S.Y.; Koh, K.S.; Rice, S.A.; Givskov, M.; Kjelleberg, S. Quorum Sensing-Controlled Biofilm Development in *Serratia liquefaciens* MG1. *J. Bacteriol.* **2004**, *186*, 692–698. [CrossRef] [PubMed]
15. Salgado, C.A.; Baglinière, F.; Vanetti, M.C.D. Spoilage Potential of a Heat-Stable Lipase Produced by *Serratia liquefaciens* Isolated from Cold Raw Milk. *LWT* **2020**, *126*, 109289. [CrossRef]
16. Lavizzari, T.; Breccia, M.; Bover-Cid, S.; Vidal-Carou, M.C.; Veciana-Nogués, M.T. Histamine, Cadaverine, and Putrescine Produced in Vitro by *Enterobacteriaceae* and *Pseudomonadaceae* Isolated from Spinach. *J. Food Prot.* **2010**, *73*, 385–389. [CrossRef]
17. Özkütük, A.S. Antimicrobial Effects of Carnosic Acid, Kaempferol and Luteolin on Biogenic Amine Production by Spoilage and Food-Borne Pathogenic Bacteria. *Food Biosci.* **2022**, *46*, 101588. [CrossRef]
18. Zhang, J.; Zhao, G.; Liu, X.; Wang, J. Genomic Analysis and Biological Properties of the Novel *Serratia liquefaciens* Phage VB_SlqM_MQ-4. *Arch. Virol.* **2023**, *168*, 38. [CrossRef] [PubMed]
19. Mohsina, K.; Ratkowsky, D.A.; Bowman, J.P.; Powell, S.; Kaur, M.; Tamplin, M.L. Effect of Glucose, PH and Lactic Acid on *Carnobacterium maltaromaticum*, *Brochothrix thermosphacta* and *Serratia liquefaciens* within a Commercial Heat-Shrunk Vacuum-Package Film. *Food Microbiol.* **2020**, *91*, 103515. [CrossRef]
20. Khayat, M.T.; Elbaramawi, S.S.; Nazeih, S.I.; Safo, M.K.; Khafagy, E.S.; Ali, M.A.M.; Abbas, H.A.; Hegazy, W.A.H.; Seleem, N.M. Diminishing the Pathogenesis of the Food-Borne Pathogen *Serratia marcescens* by Low Doses of Sodium Citrate. *Biology* **2023**, *12*, 504. [CrossRef]
21. Liu, J.; Yu, S.; Han, B.; Chen, J. Effects of Benzalkonium Chloride and Ethanol on Dual-Species Biofilms of *Serratia liquefaciens* S1 and *Shewanella putrefaciens* S4. *Food Control* **2017**, *78*, 196–202. [CrossRef]
22. Iguchi, A.; Nagaya, Y.; Pradel, E.; Ooka, T.; Ogura, Y.; Katsura, K.; Kurokawa, K.; Oshima, K.; Hattori, M.; Parkhill, J.; et al. Genome Evolution and Plasticity of *Serratia marcescens*, an Important Multidrug-Resistant Nosocomial Pathogen. *Genome Biol. Evol.* **2014**, *6*, 2096–2110. [CrossRef]
23. Åttman, E.; Korhonen, P.; Tammela, O.; Vuento, R.; Aittoniemi, J.; Syrjänen, J.; Mattila, E.; Österblad, M.; Huttunen, R.A. *Serratia marcescens* Outbreak in a Neonatal Intensive Care Unit Was Successfully Managed by Rapid Hospital Hygiene Interventions and Screening. *Acta Paediatr.* **2018**, *107*, 425–429. [CrossRef] [PubMed]
24. Rao, W.; Fang, Z.; Chen, Z.; Wu, J.; Fang, X. Antibacterial Mechanism of Metabolites of *Leuconostoc mesenteroides* against *Serratia liquefaciens*. *LWT* **2023**, *187*, 115335. [CrossRef]
25. Mladenović, K.G.; Grujović, M.; Kiš, M.; Furmeg, S.; Tkalec, V.J.; Stefanović, O.D.; Kocić-Tanackov, S.D. *Enterobacteriaceae* in Food Safety with an Emphasis on Raw Milk and Meat. *Appl. Microbiol. Biotechnol.* **2021**, *105*, 8615–8627. [CrossRef]
26. Liu, Y.M.; Chen, Y.S.; Toh, H.S.; Huang, C.C.; Lee, Y.L.; Ho, C.M.; Lu, P.L.; Ko, W.C.; Chen, Y.H.; Wang, J.H.; et al. In Vitro Susceptibilities of Non-Enterobacteriaceae Isolates from Patients with Intra-Abdominal Infections in the Asia-Pacific Region from 2003 to 2010: Results from the Study for Monitoring Antimicrobial Resistance Trends (SMART). *Int. J. Antimicrob. Agents* **2012**, *40*, S11–S17. [CrossRef] [PubMed]
27. Paterson, D.L.; Rossi, F.; Baquero, F.; Hsueh, P.R.; Woods, G.L.; Satishchandran, V.; Snyder, T.A.; Harvey, C.M.; Teppler, H.; DiNubile, M.J.; et al. In Vitro Susceptibilities of Aerobic and Facultative Gram-Negative Bacilli Isolated from Patients with Intra-Abdominal Infections Worldwide: The 2003 Study for Monitoring Antimicrobial Resistance Trends (SMART). *J. Antimicrob. Chemother.* **2005**, *55*, 965–973. [CrossRef] [PubMed]
28. Budiarso, T.Y.; Amarantini, C.; Restiani, R.; Sitanggang, P.K.; Samodra, E.M.A.; Mamoto, A.V.; Puteri, Y.; Kindangen, V.; Wattimury, A.; Yokebeth, I.P. Isolation and Biochemical Characterization of *Enterobacter cloacae* Isolates from Ready-to-Eat Foods Using API 20E. *AIP Conf. Proc.* **2023**, *2556*, 030006.

29. Nyenje, M.E.; Odjadjare, C.E.; Tanih, N.F.; Green, E.; Ndip, R.N. Foodborne Pathogens Recovered from Ready-to-Eat Foods from Roadside Cafeterias and Retail Outlets in Alice, Eastern Cape Province, South Africa: Public Health Implications. *Int. J. Environ. Res. Public Health* **2012**, *9*, 2608–2619. [CrossRef]
30. Bouige, A.; Fourcade, C.; Bicart-See, A.; Félicé, M.P.; Gautie, L.; Krin, G.; Lourtet-Hascoet, J.; Marlin, P.; Giordano, G.; Bonnet, E. Characteristics of *Enterobacter cloacae* Prosthetic Joint Infections. *Med. Mal. Infect.* **2019**, *49*, 511–518. [CrossRef]
31. Yang, H.F.; Pan, A.J.; Hu, L.F.; Liu, Y.Y.; Cheng, J.; Ye, Y.; Li, J.B. *Galleria Mellonella* as an *in vivo* Model for Assessing the Efficacy of Antimicrobial Agents against *Enterobacter cloacae* Infection. *J. Microbiol. Immunol. Infect.* **2017**, *50*, 55–61. [CrossRef]
32. Ganbold, M.; Seo, J.; Wi, Y.M.; Kwon, K.T.; Ko, K.S. Species Identification, Antibiotic Resistance, and Virulence in *Enterobacter cloacae* Complex Clinical Isolates from South Korea. *Front. Microbiol.* **2023**, *14*, 1122691. [CrossRef] [PubMed]
33. Mezzatesta, M.L.; Gona, F.; Stefani, S. *Enterobacter cloacae* Complex: Clinical Impact and Emerging Antibiotic Resistance. *Future Microbiol.* **2012**, *7*, 887–902. [CrossRef] [PubMed]
34. Hryniewicz, W.; Żabicka, D. Stanowisko Zespołu Roboczego ds. Oznaczania Lekowrażliwości Zgodnie z Zaleceniami EUCAST w Sprawie Najczęściej Zgłaszanych pytań Dotyczących Stosowania Rekomendacji EUCAST–wersja 6.0 z dnia 30 maja 2023 [Position of the Working Group on Antimicrobial Susceptibility Testing in Accordance with EUCAST Recommendations on the Most Frequently Reported Questions Regarding the Application of EUCAST Recommendations-version 6.0 of May 30, 2023]. Available online: https://www.google.com/url?sa=t&source=web&rct=j&opi=89978449&url=https://korld.nil.gov.pl/wp-content/uploads/2023/07/Zalecenia-Zespolu-wersja-6_30-05-2023.pdf&ved=2ahUKEwijkNDh5uaFAxUrFBAIHV5XDygQFnoECA8QAQ&usg=AOvVaw0Ho3V3wKxxD25Fxh3DibJM (accessed on 23 April 2024).
35. Li, X.; Xu, L.; Cui, Y.; Pang, M.; Wang, F.; Qi, J. Anti-Bacteria Effect of Active Ingredients of Cacumen Platycladi on the Spoilage Bacteria of Sauced Pork Head Meat. *IOP Conf. Ser. Mater. Sci. Eng.* **2018**, *275*, 012013. [CrossRef]
36. Liu, F.; Jin, P.; Sun, Z.; Du, L.; Wang, D.; Zhao, T.; Doyle, M.P. Carvacrol Oil Inhibits Biofilm Formation and Exopolysaccharide Production of *Enterobacter cloacae*. *Food Control* **2021**, *119*, 107473. [CrossRef]
37. Shymialevich, D.; Wójcicki, M.; Świder, O.; Średnicka, P.; Sokołowska, B. Characterization and Genome Study of a Newly Isolated Temperate Phage Belonging to a New Genus Targeting *Alicyclobacillus acidoterrestris*. *Genes* **2023**, *14*, 1303. [CrossRef] [PubMed]
38. Wójcicki, M.; Świder, O.; Średnicka, P.; Shymialevich, D.; Ilczuk, T.; Koperski, Ł.; Cieślak, H.; Sokołowska, B.; Juszczuk-Kubiak, E. Newly Isolated Virulent Salmophages for Biocontrol of Multidrug-Resistant *Salmonella* in Ready-to-Eat Plant-Based Food. *Int. J. Mol. Sci.* **2023**, *24*, 10134. [CrossRef] [PubMed]
39. Wójcicki, M.; Świder, O.; Gientka, I.; Błażejak, S.; Średnicka, P.; Shymialevich, D.; Cieślak, H.; Wardaszka, A.; Emanowicz, P.; Sokołowska, B.; et al. Effectiveness of a Phage Cocktail as a Potential Biocontrol Agent against Saprophytic Bacteria in Ready-To-Eat Plant-Based Food. *Viruses* **2023**, *15*, 172. [CrossRef] [PubMed]
40. Dakheel, K.H.; Rahim, R.A.; Neela, V.K.; Al-Obaidi, J.R.; Hun, T.G.; Isa, M.N.M.; Yusoff, K. Genomic Analyses of Two Novel Biofilm-Degrading Methicillin-Resistant *Staphylococcus aureus* Phages. *BMC Microbiol.* **2019**, *19*, 114. [CrossRef] [PubMed]
41. Shymialevich, D.; Wójcicki, M.; Wardaszka, A.; Świder, O.; Sokołowska, B.; Błażejak, S. Application of Lytic Bacteriophages and Their Enzymes to Reduce Saprophytic Bacteria Isolated from Minimally Processed Plant-Based Food Products-In Vitro Studies. *Viruses* **2022**, *15*, 9. [CrossRef]
42. Dion, M.B.; Oechslin, F.; Moineau, S. Phage Diversity, Genomics and Phylogeny. *Nat. Rev. Microbiol.* **2020**, *18*, 125–138. [CrossRef]
43. Ribeiro, J.M.; Pereira, G.N.; Durli, I.; Teixeira, G.M.; Bertozzi, M.M.; Verri, W.A.; Kobayashi, R.K.T.; Nakazato, G. Comparative Analysis of Effectiveness for Phage Cocktail Development against Multiple *Salmonella* Serovars and Its Biofilm Control Activity. *Sci. Rep.* **2023**, *13*, 13054. [CrossRef]
44. Green, S.I.; Kaelber, J.T.; Ma, L.; Trautner, B.W.; Ramig, R.F.; Maresso, A.W. Bacteriophages from ExPEC Reservoirs Kill Pandemic Multidrug-Resistant Strains of Clonal Group ST131 in Animal Models of Bacteremia. *Sci. Rep.* **2017**, *7*, 46151. [CrossRef]
45. Gibson, S.B.; Green, S.L.; Liu, C.G.; Salazar, K.C.; Clark, J.R.; Terwiliger, A.L.; Kaplan, H.B.; Maresso, A.W.; Trautner, B.W.; Ramig, R.F. Constructing and characterizing bacteriophages libraries for phage therapy of human infections. *Front. Microbiol.* **2019**, *10*, 2537. [CrossRef] [PubMed]
46. Lukman, C.; Yonathan, C.; Magdalena, S.; Waturangi, D.E. Isolation and characterization of pathogenic *Escherichia coli* bacteriophages from chicken and beef offal. *BMC Res. Notes* **2020**, *13*, 8. [CrossRef]
47. Łobocka, M.; Hejnowicz, M.S.; Gągała, U.; Weber-Dąbrowska, B.; Węgrzyn, G.; Dadlez, M. The First Step to Bacteriophage Therapy: How to Choose the Correct Phage. In *Phage Therapy: Current Research and Applications*; Borysowski, J., Międzybrodzki, R., Górski, A., Eds.; Caister Academic Press: Norfalk, UK, 2014; pp. 23–67.
48. Guerrero-Bustamante, C.A.; Dedrick, R.M.; Garlena, R.A.; Russell, D.A.; Hatfull, G.F. Toward a Phage Cocktail for Tuberculosis: Susceptibility and Tuberculocidal Action of Mycobacteriophages against Diverse *Mycobacterium tuberculosis* Strains. *MBio* **2021**, *12*, e00973-21. [CrossRef] [PubMed]
49. Mohammed, M.; Casjens, S.R.; Millard, A.D.; Harrison, C.; Gannon, L.; Chattaway, M.A. Genomic analysis of Anderson typing phages of *Salmonella* Typhimrium: Towards understanding the basis of bacteria-phage interaction. *Sci. Rep.* **2023**, *13*, 10484. [CrossRef] [PubMed]
50. Vipra, A.; Desay, S.N.; Junjappa, R.P.; Roy, P.; Poonacha, N.; Ravinder, P.; Sriram, B.; Padmanabhan, S. Determining the minimum inhibitory concentration of bacteriophages: Potential advantages. *Adv. Microbiol.* **2013**, *3*, 32359. [CrossRef]
51. Cieślik, M.; Harhala, M.; Orwat, F.; Dąbrowska, K.; Górski, A.; Jończyk-Matysiak, E. Two Newly Isolated Enterobacter-Specific Bacteriophages: Biological Properties and Stability Studies. *Viruses* **2022**, *14*, 1518. [CrossRef] [PubMed]

52. Wójcicki, M.; Średnicka, P.; Błażejak, S.; Gientka, I.; Kowalczyk, M.; Emanowicz, P.; Świder, O.; Sokołowska, B.; Juszczuk-Kubiak, E. Characterization and Genome Study of Novel Lytic Bacteriophages against Prevailing Saprophytic Bacterial Microflora of Minimally Processed Plant-Based Food Products. *Int. J. Mol. Sci.* **2021**, *22*, 12460. [CrossRef] [PubMed]
53. Shang, Y.; Sun, Q.; Chen, H.; Wu, Q.; Chen, M.; Yang, S.; Du, M.; Zha, F.; Ye, Q.; Zhang, J. Isolation and Characterization of a Novel *Salmonella* Phage VB_SalP_TR2. *Front. Microbiol.* **2021**, *12*, 664810. [CrossRef]
54. Egido, J.E.; Costa, A.R.; Aparicio-Maldonado, C.; Haas, P.J.; Brouns, S.J.J. Mechanisms and Clinical Importance of Bacteriophage Resistance. *FEMS Microbiol. Rev.* **2022**, *46*, fuab048. [CrossRef] [PubMed]
55. Kim, J.W.; Dutta, V.; Elhanafi, D.; Lee, S.; Osborne, J.A.; Kathariou, S. A Novel Restriction-Modification System Is Responsible for Temperature-Dependent Phage Resistance in *Listeria monocytogenes* ECII. *Appl. Environ. Microbiol.* **2012**, *78*, 1995–2004. [CrossRef] [PubMed]
56. Christensen, L.L.; Josephsen, J. The Methyltransferase from the LlaDII Restriction-Modification System Influences the Level of Expression of Its Own Gene. *J. Bacteriol.* **2004**, *186*, 287–295. [CrossRef] [PubMed]
57. Dupuis, M.È.; Villion, M.; Magadán, A.H.; Moineau, S. CRISPR-Cas and Restriction-Modification Systems Are Compatible and Increase Phage Resistance. *Nat. Commun.* **2013**, *4*, 2087. [CrossRef] [PubMed]
58. Morozova, V.; Bokovaya, O.; Kozlova, Y.; Kurilshikov, A.; Babkin, I.; Tupikin, A.; Bondar, A.; Ryabchikova, E.; Brayanskaya, A.; Peltek, S.; et al. A Novel Thermophilic *Aeribacillus* Bacteriophage AP45 Isolated from the Valley of Geysers, Kamchatka: Genome Analysis Suggests the Existence of a New Genus within the *Siphoviridae* Family. *Extremophiles* **2019**, *23*, 599–612. [CrossRef] [PubMed]
59. Estilo, E.E.C.; Nakano, H. Effects of Diluents, Temperature and PH on the Enumeration and Growth Kinetics of *Alicyclobacillus acidoterrestris* in Standard Growth Media. *LWT* **2019**, *104*, 148–158. [CrossRef]
60. Shang, C.; Zhang, T.; Xu, J.; Zhao, N.; Zhang, W.; Fan, M. Exploring the Growth Characteristics of *Alicyclobacillus acidoterrestris* for Controlling Juice Spoilage with Zero Additives. *Food Chem. X* **2023**, *19*, 100790. [CrossRef] [PubMed]
61. Turner, D.; Shkoporov, A.N.; Lood, C.; Millard, A.D.; Dutilh, B.E.; Alfenas-Zerbini, P.; van Zyl, L.J.; Aziz, R.K.; Oksanen, H.M.; Poranen, M.M.; et al. Abolishment of morphology-based taxa and change to binomial species names: 2022 taxonomy update of the ICTV bacterial viruses subcommittee. *Arch. Virol.* **2023**, *168*, 74. [CrossRef] [PubMed]
62. Mihara, T.; Nishimura, Y.; Shimizu, Y.; Nishiyama, H.; Yoshikawa, G.; Uehara, H.; Hingamp, P.; Goto, S.; Ogata, H. Linking Virus Genomes with Host Taxonomy. *Viruses* **2016**, *8*, 66. [CrossRef]
63. Nishimura, Y.; Yoshida, T.; Kuronishi, M.; Uehara, H.; Ogata, H.; Goto, S. ViPTree: The viral proteomic tree server. *Bioinformatics* **2017**, *33*, 2379–2380. [CrossRef]
64. Grant, J.R.; Enns, E.; Marinier, E.; Mandal, A.; Herman, E.K.; Chen, C.; Graham, M.; Van Domselaar, G.; Stothard, P. Proksee: In-depth characterization and visualization of bacterial genomes. *Nucleic Acids Res.* **2023**, *51*, 484–492. [CrossRef] [PubMed]
65. Li, X.; Zhang, C.; Wei, F.; Yu, F.; Zhao, Z. Bactericidal Activity of a Holin-Endolysin System Derived from *Vibrio alginolyticus* Phage HH109. *Microb. Pathog.* **2021**, *159*, 105135. [CrossRef] [PubMed]
66. Grabowski, Ł.; Łepek, K.; Stasiłojć, M.; Kosznik-Kwaśnicka, K.; Zdrojewska, K.; Maciąg-Dorszyńska, M.; Węgrzyn, G.; Węgrzyn, A. Bacteriophage-Encoded Enzymes Destroying Bacterial Cell Membranes and Walls, and Their Potential Use as Antimicrobial Agents. *Microbiol. Res.* **2021**, *248*, 126746. [CrossRef] [PubMed]
67. Broendum, S.S.; Buckle, A.M.; McGowan, S. Catalytic Diversity and Cell Wall Binding Repeats in the Phage-Encoded Endolysins. *Mol. Microbiol.* **2018**, *110*, 879–896. [CrossRef] [PubMed]
68. Jado, I.; López, R.; García, E.; Fenoll, A.; Casal, J.; García, P.; Pallares, R.; de la Campa, A.G.; Bouza, E.; Baquero, F.; et al. Phage Lytic Enzymes as Therapy for Antibiotic-Resistant Streptococcus Pneumoniae Infection in a Murine Sepsis Model. *J. Antimicrob. Chemother.* **2003**, *52*, 967–973. [CrossRef] [PubMed]
69. Chamakura, K.; Young, R. Phage Single-Gene Lysis: Finding the Weak Spot in the Bacterial Cell Wall. *J. Biol. Chem.* **2019**, *294*, 3350–3358. [CrossRef] [PubMed]
70. Leon-Velarde, C.G.; Kropinski, A.M.; Chen, S.; Abbasifar, A.; Griffiths, M.W.; Odumeru, J.A. Complete Genome Sequence of Bacteriophage VB-YenP-AP5 Which Infects *Yersinia enterocolitica* of Serotype O:3. *Virol. J.* **2014**, *11*, 188. [CrossRef] [PubMed]
71. Rao, V.B.; Feiss, M. The Bacteriophage DNA Packaging Motor. *Annu. Rev. Genet.* **2008**, *42*, 647–681. [CrossRef] [PubMed]
72. Shen, X.; Li, M.; Zeng, Y.; Hu, X.; Tan, Y.; Rao, X.; Jin, X.; Li, S.; Zhu, J.; Zhang, K.; et al. Functional Identification of the DNA Packaging Terminase from *Pseudomonas aeruginosa* Phage PaP3. *Arch. Virol.* **2012**, *157*, 2133–2141. [CrossRef]
73. Huiting, E.; Bondy-Denomy, J. Defining the expanding mechanisms of phage-mediated activation of bacterial immunity. *Curr. Opin. Microbiol.* **2023**, *74*, 102325. [CrossRef]
74. Gao, Y.; Cao, D.; Zhu, J.; Feng, H.; Luo, X.; Liu, S.; Yan, X.X.; Zhang, X.; Gao, P. Structural Insights into Assembly, Operation and Inhibition of a Type I Restriction–Modification System. *Nat. Microbiol.* **2020**, *5*, 1107–1118. [CrossRef] [PubMed]
75. Atanasiu, C.; Su, T.J.; Sturrock, S.S.; Dryden, D.T.F. Interaction of the Ocr Gene 0.3 Protein of Bacteriophage T7 with EcoKI Restriction/Modification Enzyme. *Nucleic Acids Res.* **2002**, *30*, 3936–3944. [CrossRef] [PubMed]
76. Klemm, B.P.; Singh, D.; Smith, C.E.; Hsu, A.L.; Dillard, L.B.; Krahn, J.M.; London, R.E.; Mueller, G.A.; Borgnia, M.J.; Schaaper, R.M. Mechanism by Which T7 Bacteriophage Protein Gp1.2 Inhibits *Escherichia coli* DGTPase. *Proc. Natl. Acad. Sci. USA* **2022**, *119*, e2123092119. [CrossRef] [PubMed]

77. Zhang, L.; Xu, D.; Huang, Y.; Zhu, X.; Rui, M.; Wan, T.; Zheng, X.; Shen, Y.; Chen, X.; Ma, K.; et al. Structural and Functional Characterization of Deep-Sea Thermophilic Bacteriophage GVE2 HNH Endonuclease. *Sci. Rep.* **2017**, *7*, 42542. [CrossRef] [PubMed]
78. Jain, C.; Rodriguez, R.L.M.; Phillippy, A.M.; Konstantinidis, K.T.; Aluru, S. High throughput ANI analysis of 90K prokaryotic genomes reveals clear species boundaries. *Nat. Commun.* **2018**, *9*, 5114. [CrossRef] [PubMed]
79. Moraru, C.; Varsani, A.; Kropinski, A.M. VIRIDIC-a novel tool to calculate the intergenomic similarities of prokaryote-infecting viruses. *Viruses* **2020**, *12*, 1268. [CrossRef] [PubMed]
80. Turner, D.; Kropinski, A.M.; Adriaenssens, E.M. A roadmap for genome-based phage taxonomy. *Viruses* **2021**, *13*, 506. [CrossRef] [PubMed]
81. Bailly-Bechet, M.; Vergassola, M.; Rocha, E. Causes for the intriguing presence of tRNAs in phages. *Genome Res.* **2007**, *17*, 1486–1495. [CrossRef] [PubMed]
82. Kemp, P.; Garcia, L.R.; Molineux, I.J. Changes in Bacteriophage T7 Virion Structure at the Initiation of Infection. *Virology* **2005**, *340*, 307–317. [CrossRef]
83. Jamal, M.; Andleeb, S.; Jalil, F.; Imran, M.; Nawaz, M.A.; Hussain, T.; Ali, M.; Das, C.R. Isolation and Characterization of a Bacteriophage and Its Utilization against Multi-Drug Resistant *Pseudomonas aeruginosa*-2995. *Life Sci.* **2017**, *190*, 21–28. [CrossRef]
84. Mirzaei, K.; Nilsson, A.S. Isolation of phage for phage therapy: A comparison of spot tests and efficiency of plating analyses for determination of host range and efficacy. *PLoS ONE* **2015**, *10*, e0118557. [CrossRef] [PubMed]
85. Sváb, D.; Falgenhauer, L.; Rohde, M.; Szabó, J.; Chakraborty, T.; Tóth, I. Identification and characterization of T5-like bacteriophages representing two novel subgroups from food products. *Front. Microbiol.* **2018**, *9*, 222. [CrossRef]
86. Kolmogorov, M.; Yuan, J.; Lin, Y.; Pevzner, P.A. Assembly of Long, Error-Prone Reads Using Repeat Graphs. *Nat. Biotechnol.* **2019**, *37*, 540–546. [CrossRef] [PubMed]
87. Mcnair, K.; Zhou, C.; Dinsdale, E.A.; Souza, B.; Edwards, R.A. PHANOTATE: A Novel Approach to Gene Identification in Phage Genomes. *Bioinformatics* **2019**, *35*, 4537–4542. [CrossRef] [PubMed]
88. Wu, J.; Liu, Q.; Li, M.; Xu, J.; Wang, C.; Zhang, J.; Xiao, M.; Bin, Y.; Xia, J. PhaGAA: An Integrated Web Server Platform for Phage Genome Annotation and Analysis. *Bioinformatics* **2023**, *39*, btad120. [CrossRef]
89. Tynecki, P.; Guziński, A.; Kazimierczak, J.; Jadczuk, M.; Dastych, J.; Onisko, A. PhageAI-Bacteriophage Life Cycle Recognition with Machine Learning and Natural Language Processing. *bioRXiv* **2020**, 198606.

International Journal of
Molecular Sciences

MDPI

Article

Maize Endophytic Plant Growth-Promoting Bacteria *Peribacillus simplex* Can Alleviate Plant Saline and Alkaline Stress

Guoliang Li [1], Miaoxin Shi [1,2], Wenhao Wan [2], Zongying Wang [2], Shangwei Ji [2], Fengshan Yang [3,4,5], Shumei Jin [2,*] and Jianguo Zhang [1,*]

1 Heilongjiang Academy of Agricultural Sciences, Harbin 150086, China
2 Key Laboratory of Saline-Alkali Vegetation Ecology Restoration, Ministry of Education, College of Life Sciences, Northeast Forestry University, Harbin 150040, China
3 Engineering Research Center of Agricultural Microbiology Technology, Ministry of Education, Heilongjiang University, Harbin 150080, China
4 Heilongjiang Provincial Key Laboratory of Ecological Restoration and Resource Utilization for Cold Region, Heilongjiang University, Harbin 150080, China
5 Key Laboratory of Molecular Biology, College of Heilongjiang Province, College of Life Sciences, Heilongjiang University, Harbin 150080, China
* Correspondence: jinshumei1972@nefu.edu.cn (S.J.); zhangjianguo72@163.com (J.Z.)

Abstract: Soil salinization is currently one of the main abiotic stresses that restrict plant growth. Plant endophytic bacteria can alleviate abiotic stress. The aim of the current study was to isolate, characterize, and assess the plant growth-promoting and saline and alkaline stress-alleviating traits of *Peribacillus simplex M1* (*P. simplex M1*) isolates from maize. One endophytic bacterial isolate, named *P. simplex M1*, was selected from the roots of maize grown in saline–alkali soil. The *P. simplex M1* genome sequence analysis of the bacteria with a length of 5.8 Mbp includes about 700 genes that promote growth and 16 antioxidant activity genes that alleviate saline and alkaline stress. *P. simplex M1* can grow below 400 mM $NaHCO_3$ on the LB culture medium; The isolate displayed multiple plant growth-stimulating features, such as nitrogen fixation, produced indole-3-acetic acid (IAA), and siderophore production. This isolate had a positive effect on the resistance to salt of maize in addition to the growth. *P. simplex M1* significantly promoted seed germination by enhancing seed vigor in maize whether under normal growth or $NaHCO_3$ stress conditions. The seeds with $NaHCO_3$ treatment exhibited higher reactive oxygen species (ROS) levels than the maize in *P. simplex M1* inoculant on maize. *P. simplex M1* can colonize the roots of maize. The *P. simplex M1* inoculant plant increased chlorophyll in leaves, stimulated root and leaf growth, increased the number of lateral roots and root dry weight, increased the length and width of the blades, and dry weight of the blades. The application of inoculants can significantly reduce the content of malondialdehyde (MDA) and increase the activity of plant antioxidant enzymes (Catalase (CAT), Superoxide Dismutase (SOD), and Peroxidase (POD)), which may thereby improve maize resistance to saline and alkaline stress. Conclusion: *P. simplex M1* isolate belongs to plant growth-promoting bacteria by having high nitrogen concentration, indoleacetic acid (IAA), and siderophore, and reducing the content of ROS through the antioxidant system to alleviate salt alkali stress. This study presents the potential application of *P. simplex M1* as a biological inoculant to promote plant growth and mitigate the saline and alkaline effects of maize and other crops.

Keywords: endophytic bacteria; maize; *Peribacillus simplex*; plant growth promote; saline and alkaline stress

1. Introduction

There are data indicating that over 800 million hectares of arable land worldwide are affected by salinization, and in China, the area of saline–alkali land accounts for 25% of the total farmland area [1]; sodic soils are characterized by highly alkaline pH (i.e., >pH 8) with a high level of exchangeable sodium [2]. By 2050, it is expected that 50% of the agricultural land will be lost due to soil salinization [3]. How to utilize saline–alkali land and improve the ability of plants to grow on saline–alkali land is an urgent problem to be solved.

In 2018, China's maize production ranked second only to the United States [4,5]. Maize is not only the main grain crop, but also an important industrial sector, and ensuring the sustained high yields of maize as raw materials is beneficial for maintaining China's food security [6]. Maize is a moderately salt-sensitive crop, so its growth and development status can be severely affected by salt stress [7]. Maize is affected by salt stress, which can result in a significant decrease in the photosynthetic rate, the growth rate, and dry matter accumulation, and the decline leads to a significant slowdown in maize growth and development [8]. Therefore, increasing the growth and development of maize under saline and alkaline stress is crucially important and meaningful.

Plant growth promotion (PGP) represents a promising solution to strengthen agricultural productivity. Plant-growth-promoting endophytic bacteria (PGPB) and plant-growth-promoting rhizobacteria (PGPR) are essential parts of the microbiome. They are expected to provide new solutions to improve the salt tolerance characteristics of plants [9,10]. PGPR can establish mutualistic relationships with plants through endophytic bacteria, thereby promoting plant growth and enhancing plant resistance to biotic/abiotic stress [11]. PGPR can absorb nutrients such as nitrogen, phosphorus, and iron from the soil and transport them to the rhizosphere for plant absorption to promote plant growth and development [12]. PGPB can alleviate both biotic and abiotic stress on plants by regulating and controlling the photosynthesis and hormone concentration of symbiotic plants, thereby improving plant stress resistance [13,14]. In some cases, microbiome growth stimulation can notably reach levels similar to chemical fertilizers, making the bacterium a sustainable alternative to potentially harmful chemicals in food production [15,16]. Therefore, in the past few decades, products based on PGPB and PGPR have been used in agriculture, but current research on the mechanism of improving plant growth and stress resistance is still very limited.

A number of studies have highlighted *P. simplex*'s potential to act as a PGP microorganism; *Bacillus simplex* was named by Meyer and Gottheil in 1901 and amended by Priest et al. in 1989 [17,18]. *Peribacillus simplex* was the type species of this genus [19], the isolates matching with *B. simplex*, which is currently named *Peribacillus simplex* [20]. *P. simplex* demonstrates a broad range of activity, stimulating growth in a large variety of commercially relevant crops; *P. simplex* isolates stimulate primary root growth and lateral root development, increased shoot and root weight in plants [20–22], increased IAA production and high phosphate solubilization [23], increased soil nutrient concentrations (phosphate, magnesium, manganese, and sulfur), and the grain yield and straw weight [24]. Plants inoculated with *P. simplex* showed a change in root architecture due to the emergence of more lateral roots [25]; the *P. simplex* PHYB1 had the highest impact in the highest seed production of *Black cumin* [26]. *P.simplex* isolates enhanced germination, root growth, a high phosphate, siderophore, and zinc solubilization index in wheat [27]; *Peribacillus simplex* had a positive effect on the growth of cultivated rice, and the contents of total nitrogen, total phosphorus, IAA, and chlorophyll in plant leaves were increased [22]. Increased shoot and root weight, IAA production, and high phosphate solubilization were detected in tomato plants treated with *P. simplex* [28]. Bacillus simplex significantly improved photosynthesis yield, potato transpiration rate, water use efficiency, iron uptake, and overall yield [29].

Previous reports suggest that *Bacillus* spp. can increase maize growth and confer salt stress amelioration, and the *P. Simplex* R180 strain coming from wheat increased maize growth significantly compared to the controls [30]. *Bacillus simplex* strain SYM00260 is a recently developed active ingredient for the growth promotion of maize in Brazil. *Bacillus simplex* has sometimes achieved over a quarter of crop yield increase [31,32]. *Bacillus* spp. (*B. safensis*) mitigates NaCl stress and promotes maize growth by modulating ethylene metabolism [31,32]. *P. Simplex M1* was selected from maize roots and its role as an endophytic bacterium in promoting plant growth and alleviating salt alkali stress is worth studying.

With the aim of searching for sustainable plant supplements or alternatives to chemical fertilizers, the use of PGPB has shown great potential, minimizing environmental impacts; thus, the present study was conducted to isolate endophytic isolates of *P. simplex M1* from the maize grown in an alkaline-stressed environment and characterize it for its potential as an endophyte-based bioformulation for enhancing plant growth and alleviating alkaline stress by inoculating maize under salt alkaline stress conditions.

2. Results

2.1. Sequencing of the 16s rRNA Gene

A bacterial isolate was obtained from the NA (Nutrient Agar) solid medium added with 400 mM $NaHCO_3$ (the colony indicated by the arrow in Figure 1A); the colony that can grow on 400 mM $NaHCO_3$ medium indicates that the colony has strong salt alkali resistance. However, whether the colonization of this strain into plants can improve their growth and stress resistance is a problem that needs to be addressed in this study. The 16S rRNA gene was amplified by PCR. The amplified 16S rRNA gene DNA PCR product sequence was shown in NCBI (GenBank accession number: PQ358470); Blast the 16S rRNA sequence using the NCBI database. Even though it is just 99.32% identical to the *P. simplex* type strain, most of the Blast result belongs to the genus of *P. simplex*. The analysis revealed that the bacterial isolate corresponds to *Peribacillus simplex* NBRC 15,720 (NR_112726.1) (standard bacterial strain in the BERGEY'S MANUAL) with 100% identity (Figure 1B).

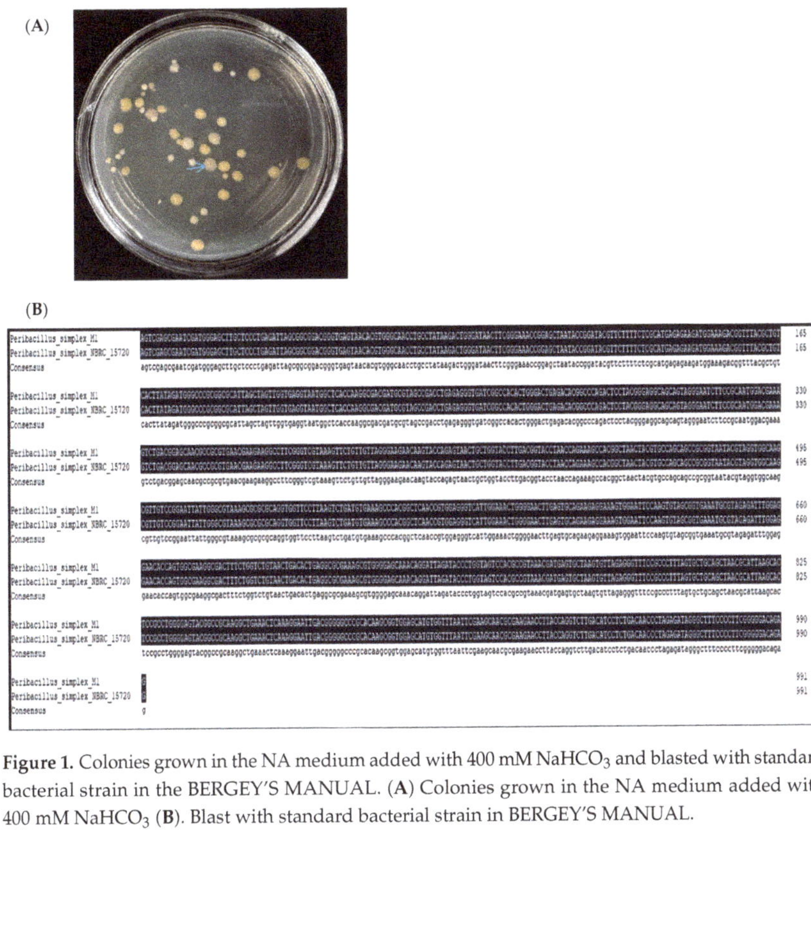

Figure 1. Colonies grown in the NA medium added with 400 mM $NaHCO_3$ and blasted with standard bacterial strain in the BERGEY'S MANUAL. (**A**) Colonies grown in the NA medium added with 400 mM $NaHCO_3$ (**B**). Blast with standard bacterial strain in BERGEY'S MANUAL.

2.2. The Bacterium Genome Sequence Analysis

Average nucleic acid similarity (ANI) is one of the most powerful measurement indicators for identifying the genetic relationships of bacterial genomes. The average value of the comparison of all the orthologous protein sequences based on genome quality inspection reflects the evolutionary distance of genome quality inspection. This bacterium genome (GenBank accession number: SUB14754556) has an ANI value of 95.7661 compared to the reference genome (NR_112726.1, *Peribacillus simplex* NBRC 15,720 = DSM 1321); when ANI >95%, it indicates that the two genomes belong to the same species.

Glimmer 3 is used for genome coding gene prediction in this project [33]. Glimmer is the most widely used prokaryotic gene structure prediction software, with high accuracy in prediction results. The analysis of the genome sequencing results of the bacteria showed that a total of 9,153,766 read fragments were obtained, with a length of 5.8 Mbp and a GC content of 34.91%. A total of 5882 coding genes and 123 non coding genes were predicted. Search for the genes related to promoting plant growth and stress tolerance is shown in Table 1.

Table 1. The genes related to promoting plant growth and stress tolerance.

Gene Function	Gene Number	Genes Name
nitrogen compound metabolic process	515	*ntcA* and *nifS*
phosphorus metabolic process	160	*Pqq*, *pstA*, *pstB*, and *pstS*
IAA production	60	*AdhR*, *aroA*, and *trpA*
Iron carrier synthesis	19	*Fer*, *pksS*, and *rub3*
antioxidant activity	16	*CAT*, *SOD*, and *peroxidase*

2.3. The Bacterium Trait

The morphology of the bacteria is rods and straight, cream-colored, and glossy, with irregular margins. The bubbles appear within half a minute in the catalase characteristic experiment, and it is considered catalase positive (Figure 2A). There were no black precipitates of lead sulfide forming in the hydrogen sulfide test, showing that the bacteria were negative for the hydrogen sulfide test (Figure 2B). We generated a organic compound in the Voges–Proskauer (VP) test (Figure 2C) and produced red acid (Figure 2D) in the methyl red test, showing the ability of the bacteria to produce acidic products using glucose. These features conform to the basic physiological characteristics of *Peribacillus simplex* in the Berger Handbook. Combining 16s RNA, genome sequencing, and physiological index analysis, it is highly likely that this bacterium is *Peribacillus simplex*; this bacterium comes from maize and it is the first time we found this bacterium, so we named this bacterium *P. simplex M1*.

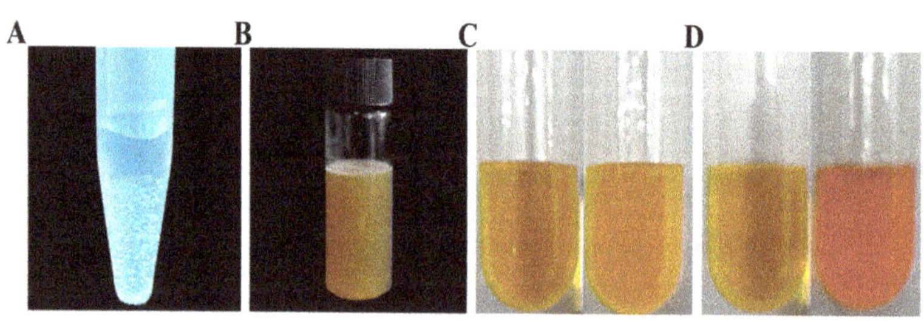

Figure 2. Physiological and biochemical reactions of bacteria. (**A**) Catalase characteristic experiment for catalase positive or negative. (**B**) hydrogen sulfide test for H_2S production. (**C**) Voges–Proskauer (VP) test for producing acid reaction. (**D**) methyl red test for producing acid reaction.

The *P. simplex M1* bacterial isolate can grow normally in an Ashby nitrogen-free culture medium, and this indicates that *P. simplex M1* has nitrogen fixation ability (Figure 3A). A transparent circle around the *P. simplex M1* colony on a NBRIP phosphorus solubilized culture medium (Figure 3B) and a blue to pink color change in the agar under and around a bacterial colony within 30 min of flooding with CAS solution was an indication of siderophore production by the bacterium (Figure 3C), indicating the *P. simplex M1* isolate has the ability to solubilize inorganic phosphate and produce iron carriers. The *P. simplex M1* isolate exhibited the capacity to produce IAA when grown on no added tryptophan (Figure 3D).

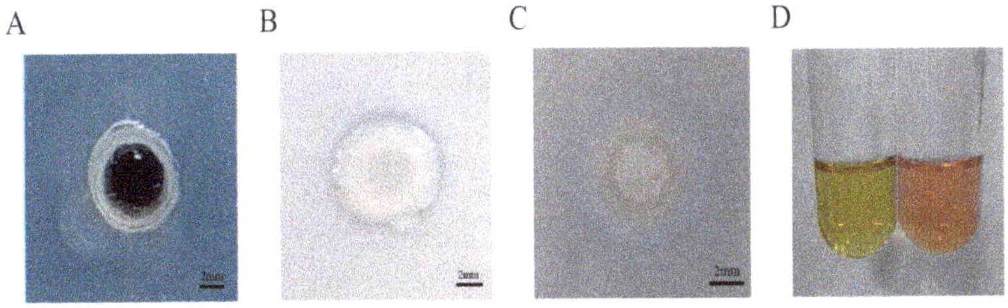

Figure 3. Schematic illustration showing the various tests performed on plates (Petri dishes 9 cm in diameter) for the assessment of the PGP traits of *P. simplex M1*. (**A**) Nitrogen fixation ability on the Nfb culture medium; (**B**) phosphorus solubilization ability on the NBRIP culture medium; (**C**) Siderophore production on the CAS culture medium; (**D**) produce IAA grown on no added tryptophan.

2.4. *Growth of P. simplex M1 under $NaHCO_3$ Stress and Different pH*

2.4.1. Growth of *P. simplex* M1 under Different Concentrations of $NaHCO_3$

The *P. simplex M1* isolates were able to grow at 0–400 mM $NaHCO_3$. When the concentration was increased to 400 mM $NaHCO_3$, which showed reduced growth, the bacteria isolate only had little growth, and the bacterial isolate could not grow at 500 mM $NaHCO_3$ (Figure 4).

Figure 4. Growth of *P. simplex M1* under different concentrations of $NaHCO_3$ stress. A total of 5 μL *P. simplex M1* (OD_{600} = 0.5) were spotted on solid LB media supplemented with the indicated stresses and grew at 30 °C for 3 d. No treatment is a control (CK).

2.4.2. Growth of *P. simplex M1* under Different pH Value

The *P. simplex M1* bacterium can grow at the pH values of 5–9 and be inhibited in growth at pH 3, 4, and 10 in 24 h (Figure 5).

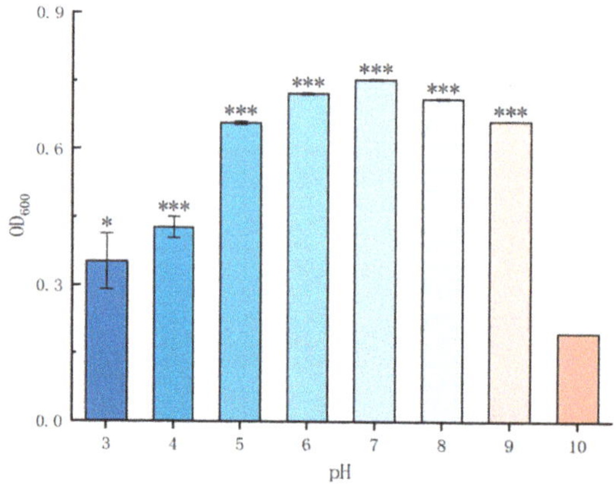

Figure 5. Growth of *P. simplex M1* under different pH conditions. A total of 200 μL *P. simplex M1* (OD_{600} = 0.5) liquid was added to 1 mL sterilized LB liquid medium with different pH values (3–10) at 30 °C with constant shaking at 1300 rpm. The OD_{600} value was measured using an ELISA reader every hour and measured continuously for 24 h. * $p < 0.05$, *** $p < 0.001$, standard error of three biological replicates.

2.5. *Effect of P. simplex M1 Bacterial Isolates on Growth of Maize*

2.5.1. Germination of Maize Seeds under Normal Growth or $NaHCO_3$ Stresses Conditions

The maize seed germination of the bacterial inoculation group with *P. simplex M1* (germination in 3 days, seed germination rate is 100%) on the culture medium was faster than that in the control group (germination in 4 days, seed germination rate is 100%). Under the stress treatment of 10 mM $NaHCO_3$, the germination of maize seeds was inhibited, but the germination rate of the maize seeds in the inoculation group (germination in 5 days, seed germination rate is 100%) was faster than that in the control group (germination in 6 days, seed germination rate is 80%), and the roots and buds also grew longer than the control. So, the maize seeds in the inoculation group had longer roots and shoots than those in the control group in the same period of time, revealing that the bacteria exerted a significant positive influence on early plant growth, namely germination (Figure 6).

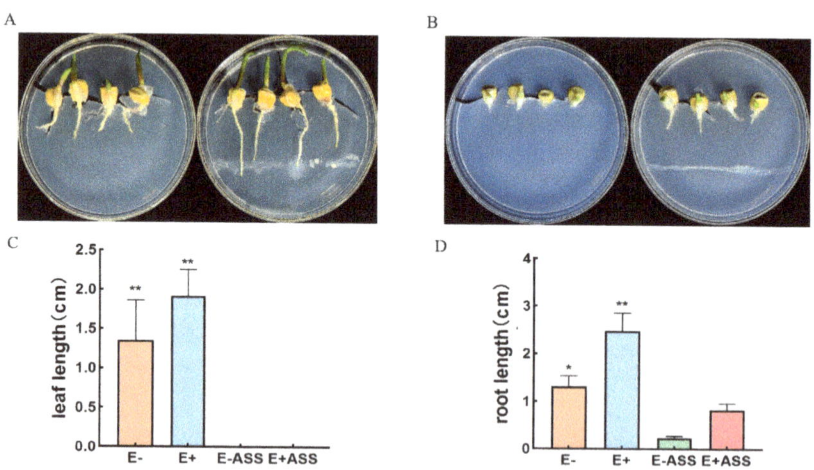

Figure 6. The effect of *P. simplex M1* on maize seed germination. (**A**) Left: the sterile maize seeds were planted on the 1/2 MS; right: the sterile maize seeds in the inoculation with *P. simplex M1* were planted on the 1/2 MS; (**B**) left: the sterile maize seeds were planted on the 1/2 MS + 10 mM $NaHCO_3$; right: the

sterile maize seeds in the inoculation with *P. simplex M1* were planted on the 1/2 MS + 10 mM $NaHCO_3$; (**C**) the leaf length of maize seed germination; (**D**) the root length of maize seed germination for 10 d. The germination percentage of the maize seeds was recorded during 10 d of $NaHCO_3$ treatment. Data show the means ± SE of three replicates. At least 50 seeds in each treatment were measured in each repeat. * $p < 0.05$, ** $p < 0.01$, standard error of three biological replicates.

2.5.2. TTC, DAB, and NBT Staining

For maize seed tissues, the staining results show that living tissues are stained to varying degrees of red, and the maize seed embryos showed stronger TTC (2, 3, 5-triphenyltetrazolium chloride) staining (deep red) than those seed embryos that were treated with 10 mM $NaHCO_3$ (light red), indicating that the seed vitality is affected under the stress. The maize seeds in the inoculation with *P. simplex M1* were stained darker than the control, indicating that *P. simplex M1* is beneficial for increasing seed vitality (Figure 7).

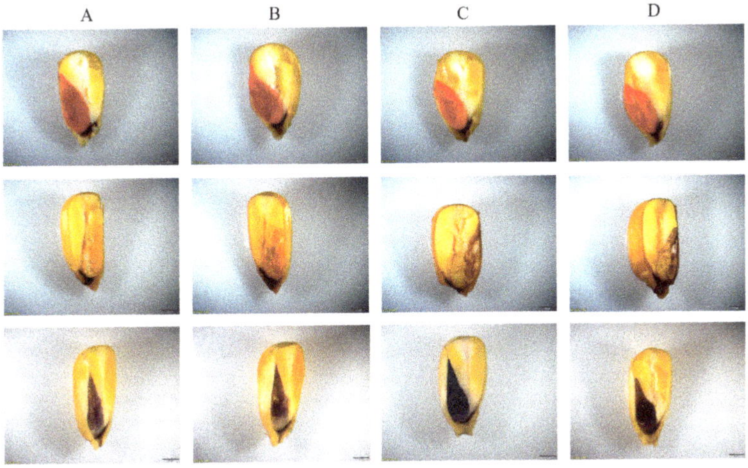

Figure 7. TTC, DAB, and NBT staining of maize seed. (**A**) Each group represents the sterile maize seeds under free salt stress (control); (**B**) the sterile maize seeds in the inoculation with *P. simplex M1* under free salt stress; (**C**) the sterile maize seeds under the 10 mM $NaHCO_3$ stress; (**D**) the sterile maize seeds in the inoculation with *P. simplex M1* under the 10 mM $NaHCO_3$.

The maize seed was observed by DAB (diaminobenzidine) staining (brown pigment) and NBT (nitro blue tetrazolium) staining (blue). The maize seed embryos showed more dark brown or dark blue staining than those seed embryos that were treated with 10 mM $NaHCO_3$; those infected with *P. simplex M1* were lighter in color than those without infection under stress. These results demonstrated that $NaHCO_3$ treatment significantly increased ROS levels during seed germination and *P. simplex M1* can decrease ROS content (Figure 7).

The top group is the TTC staining of maize seed vigor; the seed embryos were stained deep red under free salt stress and light red under the 10 mM $NaHCO_3$ stress, but those infected with *P. simplex M1* were darker in color than those without infection under stress. The middle group is the DAB staining of maize dynamic accumulation of H_2O_2; the seed embryos were stained light brown under free salt stress and darker brown under the 10 mM $NaHCO_3$ stress, but those infected with *P. simplex M1* were lighter in color than those without infection under stress. The bottom group is the NBT staining of maize dynamic accumulation of O_2^-; the seed embryos were stained light blue under free salt stress and darker blue under the 10 mM $NaHCO_3$ stress, but those infected with *P. simplex M1* were lighter in color than those without infection under stress.

2.6. The Effect of *P. simplex M1* on the Maize Seedlings

2.6.1. *P. simplex M1* Bacterial Strains to Colonize Seedling Roots

To assess the ability of the *P. simplex M1* bacterial strains to colonize seedling roots we labeled them with GFP following inoculation. The colonization roots with GFP-expressing *P. simplex M 1* bacterial strains were observed using microscopy (Figure 8).

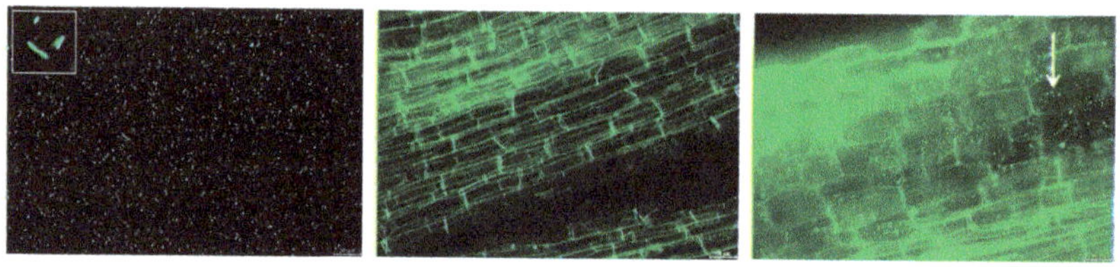

Figure 8. Colonization of *P. simplex M1* on maize seedling root. (**A**) The morphology of GFP-expressing *P. simplex M1*; (**B**) the maize root as a control; (**C**) the colonization of root with GFP-expressing *P. simplex M1*. The white arrow points to a bacterial cluster, scale Bar = 50 μm.

2.6.2. The Effect of *P. simplex M1* on the Maize Seedling

To characterize the effect of *P. simplex M1* on the maize seedling, the *P. simplex M1* strain promoted plant growth in the absence as well as in the presence of salt stress. Both uninoculated and inoculated plants were irrigated with 50 mL water and 50 mL saline solution of 400 mM $NaHCO_3$, respectively. The phenotypic growth parameters of the plants recorded in the pot study are presented in Figure 8. The growth of the plants inoculated with *P. simplex M1* was higher compared to the non-inoculated plants under free salt treatments or salt (400 mM $NaHCO_3$). The magnitude of leaf SPAD (Soil–Plant Analyzer Development) is a relative chlorophyll content reading that reflects the relative content of chlorophyll. We examined the chlorophyll content (38.4, 33.7, 28.6, and 36.5); the healthier the crop, the higher the SAPD index. The plants' inoculation with *P. simplex M1* significantly increased the chlorophyll content index of leaves, and this showed that *P. simplex M1* can promote plant growth.

The positive effects of *P. simplex M1* on the growth of maize seedlings were observed in the root and leave system. The leaf length (6.476, 8.29, 2.341, and 5.27 cm), leaf depth (0.844, 1.504, 0.542, and 0.818), dry weight of the leaf (0.452, 0.601, 0.2245, and 0.4026 g), the root length (7.438, 8.66, 4.792, and 6.416 cm), root number (10.2, 13.6, 5.2, and 7.8), and dry weight of root (0.1326, 0.1793, 0.059, and 0.084 g) of the uninoculated seeding, inoculated seeding, uninoculated seeding treated with 400 mM $NaHCO_3$, and inoculated seeding treated with 400 mM $NaHCO_3$ were measured, respectively. When compared to the uninoculated maizes, the *P. simplex M1* strain improved the leaf and root development under four irrigation conditions (Figure 9).

2.6.3. Antioxidant Enzyme Determination of Maize Seedlings under 400 mM $NaHCO_3$ Stress

The MDA (malondialdehyde) content (4.11876, 3.25508, 17.73624, and 14.90944 mmol/g), the CAT (catalase) activity (64.33, 106.12, 192.12, and 247.93 U/g), the POD (peroxidase) activity (450.51, 524.83, 782.21, and 908.58 U/g), the SOD (superoxide dismutase) activity (6.82, 6.70, 43.14, and 61.38 U/g), O_2^- (0.2135, 0.1723, 0.5526, and 0.4143 μmol/g), and H_2O_2 (0.026, 0.031, 0.072, and 0.053 μmol/g) was measured in the maize seedlings irrigated with water, maize seedlings irrigated with *P. simplex* bacteria, maize seedlings inoculated with *P. simplex* for 5 days and irrigated by 400 mM $NaHCO_3$ stress, and maize seedlings irrigated with *P. simplex* bacteria before $NaHCO_3$ stress, respectively (Figure 10). Under salt-free treatment, the activities of the antioxidant enzymes (MDA, CAT, POD, and SOD) of maize irrigated with water and *P. simplex* bacteria were basically the same, but after $NaHCO_3$ treatment, they were significantly different. The production of MDA can exacerbate membrane damage. MDA can be used to assess the degree of membrane lipid peroxidation, indirectly determining the extent of membrane system damage and plant stress resistance. The MDA content of maize seedlings irrigated with *P. simplex M1* bacteria is lower than that in the maize seedlings under 400 mM $NaHCO_3$ stress. *P. simplex M1* bacteria can reduce stress damage to plant membranes.

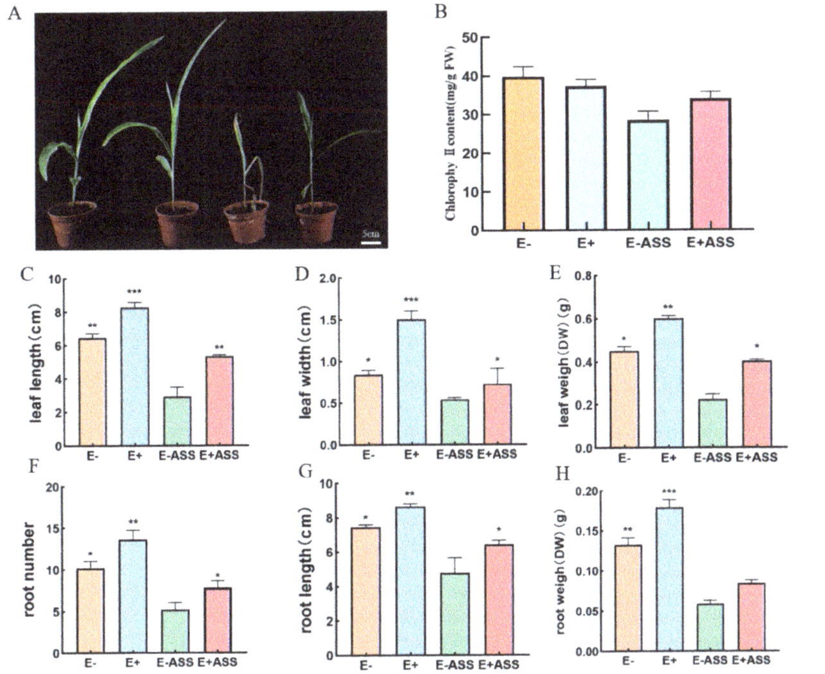

Figure 9. Effect of *P. simplex M1* strain on maize growth parameters after 7 days of cultivation under saline conditions. (**A**) The growth of maize seedling, from leaf to right: uninoculated maize seedling growth under free salt, inoculated maize seedling with *P. simplex M1*, uninoculated maize seedling growth under $NaHCO_3$ stress, and inoculated maize seedling with *P. simplex M1* under $NaHCO_3$ stress; (**B**) chlorophyll content index; (**C**) leaf length; (**D**) leaf depth; (**E**) leaf dry weight; (**F**) root number; (**G**) root length; (**H**) root dry weight. E- represents uninoculated maize seedlings, E+ represents inoculated maize seedlings, E-ASS represents uninoculated maize seedlings irrigated with 400 mM $NaHCO_3$, and E + ASS represents inoculated maize seedlings irrigated with 400 mM $NaHCO_3$. The values represent the means of replicates (n = 4) ± standard deviations. Asterisks in superscript indicate a significant difference from the control at 95% between treatments. Each data point is the average of five replicates, and error bars represent ± SE. Error bars indicate ± SD. * Significance at $p < 0.05$, ** Significance $p < 0.01$, *** Significance $p < 0.001$.

The expression of *CAT, POD,* and *SOD* genes was increased under the conditions of 400 mM $NaHCO_3$, which is nearly 2–5 times higher than free salt. The CAT activity, the POD activity, and the SOD activity of the maize seedlings inoculated with *P. simplex M1* are higher than that of maize seedlings under stress; these important antioxidant enzymes can clear free radicals in the body, thereby protecting cells from oxidative damage, indicating that *P. simplex M1* significantly increases antioxidant activity in maize seedlings and reduce plant damage.

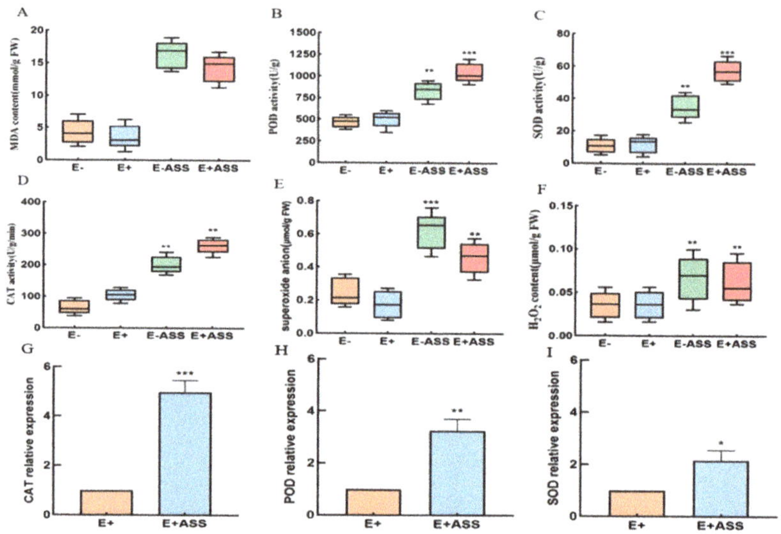

Figure 10. Antioxidant enzyme activity determination in maize seeding. (**A**) MDA content; (**B**) POD activity; (**C**) SOD activity; (**D**) CAT activity; (**E**) superoxide anion; (**F**) H_2O_2 content. (**G**) *CAT* relative expression; (**H**) *POD* relative expression; (**I**) *SOD* relative expression. * Significance at $p < 0.05$, ** Significance $p < 0.01$, *** Significance $p < 0.001$.

3. Discussion

Soil alkalinity has a notable effect on organic matter, nitrogen, carbon, and microbial biomass [34]. The utilization of beneficial microbes to enhance crop productivity under stressed environmental conditions is an eco-friendly approach [35,36]. The microbes produce advantageous effects on plants, such as phosphate solubilization and nitrogen fixation [21,37]. PGPB for not only enhancing plant growth but also mitigating abiotic and biotic stresses is essential [38,39].

A bacterium was isolated from the roots of maize grown in saline–alkali soil with a pH value of 8.5. Through the genomic research of bacterial colonies, a total of 515 nitrate process genes, 19 iron ion-binding, and 60 IAA production genes were predicted in the *P. simplex M1* genome sequence (Table 1). The presence of these genes indicates that this bacterium has the molecular basis for promoting plant growth. Maize requires the highest amount of nitrogen, so the N content in plants is a critical growth macro-element. The *P. simplex M1* bacterial isolate can convert nitrogen that plants cannot use in the environment into nitrogen that plants can absorb. There were 515 nitrogen compound metabolic process genes in the *P. simplex M1* genome (Table 1), such as the nitrogen fixation (*nifS*) genes and Nitrogen control A (*NtcA*) genes which were annotated in the *P. simplex M1* genome. *nifS* are required components of the enzymatic module encoding nitrogenase [40]. *NtcA* transcription factor operates global nitrogen regulation in these photosynthetic organisms [41]. When *P. simplex M1* colonizes the maize, these growth-related genes also enter the maize plant and may be expressed during seed germination and plant growth; these genes help nitrogen utilization and cycling in plants, thereby enhancing plant growth. *P. simplex M1* promote plant growth through processes similar to these molecular mechanisms.

Phosphorus is often one of the limiting factors for high crop yields [22]. A total of 160 genes in the *P. simplex M1* genome are involved in the phosphorus metabolic process; *P. simplex M1* has the ability to dissolve phosphorus (Figure 3), making it easier for plants to absorb and utilize the dissolved phosphorus. Growth stimulation has most commonly been attributed to direct growth promotion via IAA or siderophore secretion [24,42]. *P. simplex M1* capacity to produce IAA and iron carriers (Figure 3C,D) to directly promote plant growth. The bacteria were able to grow normally under 400 mM $NaHCO_3$ or pH value from 5 to 9 (Figure 4), indicating that the bacteria can survive and function under saline–alkali conditions.

P. simplex has demonstrated a good root colonization potential and persistence in several commercial plants [43]. *P. simplex M1* has the ability to colonize the roots of the maize (Figure 8). *P. simplex* has a positive effect on the growth of cultivated plants [20,22,25,27–30,44]. The maize seed inoculated with *P. simplex M1* showed stronger TTC staining (Figure 7) and germinated faster (Figure 6) than the controls (distilled water). *P. simplex M1* exerted a significant positive influence on early plant growth, namely germination rate. Seed germination was modulated by various factors including epigenetic modifications, hormone transport, and ROS signaling [45]. Saline irrigation negatively affected plant growth compared to the set irrigated by non-saline water (Figure 9). These results demonstrated that *P. simplex M1* significantly increased seed vigor and decreased ROS levels during seed germination.

There are a total of 16 antioxidant activity genes in the genome of bacteria, including catalase genes, superoxide dismutase genes, thiol-disulfide oxidoreductase resA, thioredoxin-dependent thiol peroxidase, and so on. CAT is an H_2O_2 scavenger that dissociates H_2O_2 into O_2 and H_2O [46]. Peroxidase can remove H_2O_2 by oxidizing the substrate. The activities of SOD and POD play an important role in scavenging ROS. The infection of *P. simplex* significantly increased the activity of the antioxidant enzymes in maize seedlings, which indicated that the addition of microbial inoculum was helpful for maize to remove excessive ROS.

A resistant maize *P. simplex M1* strain promotes growth by fixing atmospheric nitrogen, solubilizing phosphorus, producing IAA, producing siderophore, and alleviating saline and alkaline stress by increasing oxidative mechanisms (Figure 11).

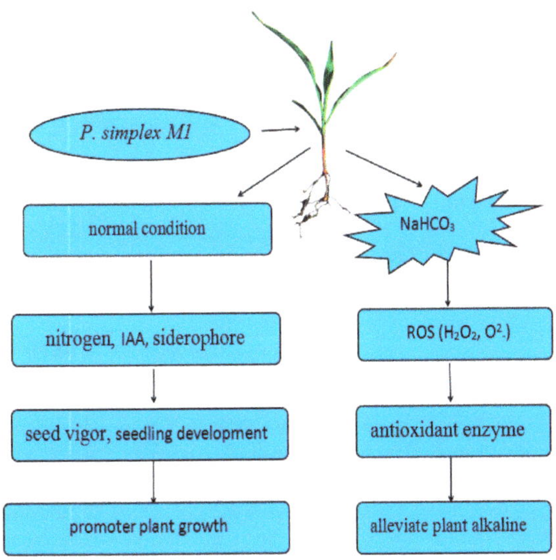

Figure 11. A model for the mechanism underlying *P. simplex M1* promoted maize development. The *P. simplex M1* has the ability to fix atmospheric nitrogen, produce IAA, produce siderophore, and enhance the antioxidant enzyme activity. When the maize is inoculated with *P. simplex M1*, *P. simplex M1* can increase seed vigor and seedling development (root length, leave length, and so on), thereby promoting maize growth. When the maize was treated with $NaHCO_3$, with the increase in ROS in plants, the activity of antioxidant enzymes caused by *P. simplex M1* also enhanced, which alleviated the saline–alkaline stress on plants.

4. Materials and Methods

4.1. Isolation and Identification of Selected Bacterium

The endophytic bacterial strains were selected from the roots of maize (B46, Salt tolerant variety) grown in saline–alkali soil with pH 8.5 on the Songnen Plain of Heilongjiang Province in Northeast China. The maize was dug out from the growing field, the roots were washed with sterile water for 5 min and cleaned with NaClO for 1 min, and then the roots were ground in a sterile centrifuge tube and resuspended in sterile water to obtain a bacterial suspension. A volume of 200 μL of the bacterial supernatant was spread onto NA (Nutrient Agar) solid medium added with 400 mM $NaHCO_3$ and inverted in an incubator at 30 °C for 3 days, and after the colonies had grown, they were purified separately on new NA solid medium. The 16S rRNA gene was amplified by PCR for these colonies. The sequences of primers used for amplification were universal primers 27F (5′-AGAGTTTGATCCTGGCTAG-3′) and 1492R (5′-CTACGGCTACCTTGTTACGA-3′), and the amplified DNA PCR products (~1540 bp for 16S rRNA sequences) were commercially sequenced (Kumei Biotechnology Co., LTD, Changchun, China). The sequenced results were compared to those in the GenBank database using the BLAST algorithm (National Center for Biotechnology Information, Bethesda, MD, USA).

4.2. The Bacterium Genome Analysis

We sent the selected *P. simplex M1* strain to Magi Gene Biotechnology Company (Guangzhou, China) for whole genome sequencing, and the genomic DNA was extracted by Guangdong Magigene Biotechnology Co., Ltd. (Guangzhou, China) using a DNA extraction kit of the corresponding sample. The ALFA-SEQ DNA Library Prep Kit (was used for qualified DNA samples for library construction. The PE150 sequencing of the constructed amplicon libraries was performed using Illumina or MGI platforms (Guangdong Magigene Biotechnology Co., Ltd., Guangzhou, China). Quality control was carried out on the raw data, and the high-quality sequence obtained was used for downstream data analysis. The tool SPAdes v3.13.0 was used to perform de novo splicing on the clean data after data filtering and quality control to obtain high-quality contig fragments. The prediction of genome components includes the prediction of coding genes, repeat sequences, non coding RNA, and prophage. Genome-wide Blast searches were performed on the following databases: NR (Non-Redundant Protein Database), Swiss-Prot, GO (Gene Ontology), KEGG (Kyoto Encyclopedia of Genes and Genomes), COG (Clusters of Orthologous Groups), etc., to predict gene function.

4.3. Evaluation of the Bacterium Traits

Referring to the classification explanation of BERGEY'S MANUAL (Second Edition Volume Three, pp. 115–116), we selected indicators such as catalase characteristics, methyl red test, Voges-Proskauer, and gelatin hydrolysis to detect the physiological and biochemical characteristics of bacteria. The experimental method is as follows: Inoculate the bacterial colonies in a tube and add 2 mL of 3% hydrogen peroxide solution; if bubbles appear within half a minute, it is considered positive. Some bacteria can decompose sulfur-containing amino acids, producing hydrogen sulfide and lead in the culture medium, forming black precipitates of lead sulfide, which are positive for the hydrogen sulfide test. The methyl red test is used to detect the ability of bacteria to resolve glucose and produce organic acids; the culture medium turns red, indicating a positive result in the methyl red test. The Voges-Proskauer test is used to determine the ability of bacteria to produce acidic products using glucose; the bacteria generate a red compound, indicating a positive VP reaction. Escherichia coli does not produce red compounds, indicating a negative reaction. Some bacteria can produce gelatinase. Gelatinase can dissolve gelatin, causing it to lose its coagulation power.

The ability of *P. simplex M1* to fix atmospheric nitrogen on an Ashby nitrogen-free culture medium was tested [47]. Inoculate 2 μL of activated bacterial solution onto a culture medium plate at 30 °C for 24 h; if there is a transparent circle around the colony, it indicates that the bacterium has nitrogen fixation ability. The ability to solubilize inorganic phosphate was tested by inoculating the bacterial isolates in NBRIP (The National Botanical Research Institute's phosphate) culture medium according to Mehta S [48]. Inoculate 2 μL of activated bacteria onto a NBRIP plate at 30 °C for 24 h. If there is a transparent circle around the colony, it indicates that the bacterium has the ability to dissolve phosphorus. The estimation of IAA was carried out by the method that has been described in Bric et al. [49]. Add 40 mL of the 2.5 mg/mL tryptophan solution into 160 mL DF liquid culture medium. Inoculate the activated bacteria into the above culture medium at 30 °C and 135 rpm for 24 h, take 2 mL of supernatant, and add 50 μL of 50% orthophosphate solution and 4mL of Salkowski reagent. If the supernatant turns red, an IAA enzyme is produced. Siderophore production was detected on CAS (Chromogenic Agar for Spore) culture medium by the method that has been described in Yeole, R et al. [50]. Inoculate 2 μL of the strain on a CAS plate. Incubate at 30 °C for 3 days. The formation of a yellow green halo indicates the production of iron carriers.

4.4. *Abiotic Stress-Tolerance Estimation for P. simplex M1*

To assess the salinity tolerance of the isolated strain, 5 μL *P. simplex M1* (OD_{600} = 0.5) were spotted on solid LB medium plates supplemented with various $NaHCO_3$ concentrations (0–500 mM) and incubated for 48 h at 30 °C.

To estimate the maximal growth and minimal inhibitory pH value, *P. simplex M1* was further tested in liquid LB. The bacterial isolates were subjected to evaluate their abiotic stress tolerance including pH (3–10). Add 200 μL *P. simplex M1* (OD_{600} = 0.5) liquid to 10 mL sterilized LB liquid medium added with different pH values (3–10) and mix well to cultivate the bacterial isolates for 24 h at 30 °C with constant shaking at 130 rpm. The OD_{600} value was measured using an enzyme-linked immunosorbent assay (ELISA) reader every hour and measured continuously for 24 h. The experiments were performed in triplicate.

4.5. *Effects of P. simplex M1 on Maize Plants Using Inoculation Experiments*

4.5.1. *P. simplex M1* Affect Maize Seed Germination Assay

The cell pellet of *P. simplex M1* suspension was obtained from a 10 mL fresh overnight culture by 5 min centrifugation at 4000 r/min 10 min. The bacterial pellet was resuspended in 10 mL of distilled sterile water (OD_{600} = 0.8) and vortexed before being used for seed treatment.

Maize seeds (*Zea mays* cv. NC236) were obtained from the Heilongjiang Academy of Agricultural Sciences. NC236 is a salt-sensitive cultivar. To observe the germination of seeds under normal and $NaHCO_3$ stress conditions, the seeds were sterilized with 75% alcohol, and 4 maize seeds were planted in sterilized 1/2 MS culture or 1/2 MS culture added with 10 mM $NaHCO_3$ medium. The control group did not add *P. simplex M1* solution, while the experimental group needed to use a pipette to aspirate 100 μL of actively growing bacterial suspensions and draw lines on the culture medium of the experimental group. The plates were incubated in a growth chamber at 25 °C under a 16 h light/8 h dark cycle. When the radicle protruded from the seed coat, the seeds were considered germinated. After 10 days, observe the germination of maize seeds and take photos to record.

Maize seeds were sterilized and divided into four groups. The first group seeds were treated with H_2O for 6 h; the second group seeds were treated with *P. simplex M1* bacterial soaking for 6 h; the third group seeds were treated with H_2O for 6 h first, followed by 100 mM $NaHCO_3$ stress for 30 min; and the fourth group seeds were treated with *P. simplex M1* bacterial soaking for 6 h first, followed by 100 mM $NaHCO_3$ stress for 30 min. The maize seeds were immersed in different solutions for testing seed vigor (TTC) or ROS content (DAB and NBT).

The TTC is a lipid-soluble photosensitive complex that can be used to detect the viability of seeds. The detection mechanism is that TTC itself can serve as a redox indicator, and dehydrogenases can reduce TTC in living cells. For seeds or plant tissues, the staining results show that living tissues are stained to varying degrees of red, while dead or lifeless tissues are not stained. The control and treated maize seeds were longitudinally split into two parts, then immersed in 0.5% TTC solution at 30 °C for 2 h in a dark chamber, and then rinsed with double distilled water, and finally photographed [51].

Triplicates of the treatment were maintained. For hydrogen peroxide (H_2O_2), DAB staining was used to detect the active site of peroxidase in cells, and was conducted as described previously [52]. The control and treated maize seeds were longitudinally split into two parts, immersed in 1% DAB solution at 30 °C for 12 h, and then photographed.

The NBT staining method is used to detect superoxide anions in plant living tissues. Superoxide anion is an oxygen-containing free radical that can reduce NBT to form insoluble blue formamide compounds. After staining, the areas where superoxide anions aggregate appear blue to dark blue. The maize seeds were immersed in 0.02% NBT solution at room temperature for 15 min in darkness. The histochemical detection of O_2^- was conducted as described previously [52].

4.5.2. *P. simplex M1* Affect Maize Seedlings Using Pot Experiment

To prove that *P. simplex M1* can colonize the maize, the maize seeds (NC236 variety) were sown in pots with soil, and then placed in a growth chamber for cultivation. Relative humidity (RH) varied around 60%, with 25 °C and photoperiod of 14 h light and 10 h dark. The GFP (green fluorescent protein) was used as a test. The construction of GFP-labeled *P. simplex M1* referred to Lifeng Guo [53]. The cell pellet of GFP-labeled *P. simplex M1* suspension (OD_{600} = 0.8) was obtained from a 50 mL fresh overnight culture by 5 min centrifugation at 4000 r/min 10 min. The bacterial pellet was resuspended in 50 mL of distilled sterile water. The maize seedlings (at the four-leaf stage) in the pot were irrigated with the 50 mL GFP-labeled *P. simplex M1* solution for 7 days, and the maize root was visualized by a confocal scanning laser microscope (CSLM) after successful transformation with GFP-labeled *P. simplex M1* solution.

To test *P. simplex M1*'s effect on maize seedlings, the cell pellet of *P. simplex M1* suspension (OD_{600} = 0.8) was obtained from a 50 mL fresh overnight culture by 5 min centrifugation at 4000 r/min 10 min. The bacterial pellet was resuspended in 50 mL of distilled sterile water. The maize seedlings (at the four-leaf stage) with similar growth and development levels were watered with four irrigation treatments. The first group of plants was irrigated into the pots only with 50 mL water (non-saline, as the control); the second group of plants was irrigated with the 50 mL resuspended *P. simplex M1* solution; the third group of plants was irrigated with 50 mL water for 5 days (maintain consistency with the fourth salt treatment time), then the pot was irrigated with the 50 mL saline solution of 400 mM $NaHCO_3$; and the fourth group of plants was inoculated with 50 mL *P. simplex M1* solution for 5 days, then irrigated with the 50 mL saline solution of 400 mM $NaHCO_3$.

A total of 20 days after four irrigation treatments, the leaf chlorophyll content index was recorded using a chlorophyll content meter (Hansatech instruments, Model CL-01). Leave and root lengths were assessed using a ruler, and the dry weights (DW) of the roots and leaves were measured after drying in an oven at 70 °C for 72 h. Student's *t*-test was conducted to test for significant differences. In order to evaluate enzymatic antioxidants, the activity of SOD [54], CAT [55], and POD [55] was determined using the methods described earlier, respectively.

In order to observe the differential antioxidant gene expression in the presence and absence of stress, RNA was extracted using the RNeasy Plant Mini Kit (Qiagen, Hilden, Germany) from the leaves of maize which were treated with four different methods. The full-length cDNA was obtained using the Takara reverse transcription kit (Takara, Tokyo, Japan), and the expression of *CAT* (the primer sequence *CAT* qPCRF: AGCGTTCTCTCCAAGTG-CAA and *CAT* qPCRR:CGCAGTAAGCTGGGTCTTCA), *SOD* (the primer sequence *SOD* qPCR-F:CGGTGAGCCTAATGGAGACG, *SOD* qPCRR:ACTCTGGGCGCCTATTTTGA), and *POD* genes (the primer sequence *POD* qPCRF:TAAAGGCAACCCGATGACCC, *POD*PCRR:AGATCCGCGCTAATGGTCAG) in the *P. simplex M1* genome was analyzed by qRT-PCR (reverse transcription-quantitative polymerase chain reaction). As an internal reference, the expression of the *Maize18S* gene (the primer sequences were Maize18SF:5′-CAACCATAAACGATGCCGA-3′ and Maize18SR: 5′-AGCCTTGCGACCATACTCC-3) was examined. RT-qPCR was carried out in IQ5 real-time PCR equipment (Bio-Rad, Hercules, CA, USA) with SYBR green (Takara, Tokyo, Japan).

4.6. Statistical Analysis

All the data in the manuscript were analyzed by *t*-tests using the SPSS 22.0 software (SPSS Inc., Chicago, IL, USA). For appropriate statistical power, all the assays were repeated three times and the values are presented as means standard errors. Differences were considered significant when the *p* value was less than or equal to 0.05.

Author Contributions: Conceptualization, G.L. and S.J. (Shumei Jin); investigation, G.L. and M.S.; validation, G.L., W.W., Z.W., S.J. (Shangwei Ji) and F.Y.; formal analysis, G.L., M.S. and W.W.; resources, G.L. and S.J. (Shumei Jin) and F.Y.; data curation, G.L., M.S., W.W., Z.W. and S.J. (Shangwei Ji); writing—original draft preparation, G.L. and S.J. (Shumei Jin); writing—review and editing, G.L., S.J. (Shumei Jin), and J.Z.; supervision, G.L., S.J. (Shumei Jin), and J.Z.; funding acquisition, S.J. (Shumei Jin) and J.Z. All authors have read and agreed to the published version of the manuscript.

Funding: This work was funded by the National Key Research and Development Program of China (2021YFD1201001); the Innovation Project of Heilongjiang Academy of Agricultural Sciences (CX23ZD05, CX23JQ04); the Heilongjiang Provence Agriculture Research System-Ecological Agriculture ([2023] 1197); Heilongjiang Province "Double First Class" Discipline Collaborative Innovation Achievement Project [LJGXCG2023-036].

Institutional Review Board Statement: Not applicable.

Informed Consent Statement: Not applicable.

Data Availability Statement: All the data supporting the findings of this article can be found in the paper.

Conflicts of Interest: The authors declare no conflict of interest.

References

1. Liu, L.; Wang, B. Protection of halophytes and their uses for cultivation of saline-alkali soil in China. *Biology* **2021**, *10*, 353. [CrossRef]
2. Srivastava, S.; Bist, V.; Srivastava, S.; Singh, P.C.; Trivedi, P.K.; Asif, M.H.; Chauhan, P.S.; Nautiyal, C.S. Unraveling aspects of *Bacillus amyloliquefaciens* mediated enhanced production of rice under biotic stress of *Rhizoctonia solani*. *Front. Plant Sci.* **2016**, *7*, 587. [CrossRef]
3. Ludwig, M.; Wilmes, P.; Schrader, S. Measuring soil sustainability via soil resilience. *Sci. Total Environ.* **2018**, *626*, 1484–1493. [CrossRef]
4. Galvez Ranilla, L. The application of metabolomics for the study of cereal corn (*Zea mays* L.). *Metabolites* **2020**, *10*, 300. [CrossRef]
5. Ma, Y.; Tashpolat, N. Current status and development trend of soil salinity monitoring research in China. *Sustainability* **2023**, *15*, 5874. [CrossRef]
6. Revilla, P.; Alves, M.L.; Andelković, V.; Balconi, C.; Dinis, I.; Mendes-Moreira, P.; Redaelli, R.; Ruiz de Galarreta, J.I.; Vaz Patto, M.C.; Žilić, S. Traditional foods from maize (*Zea mays* L.) in Europe. *Front. Nutr.* **2022**, *8*, 683399. [CrossRef]
7. Notununu, I.; Moleleki, L.; Roopnarain, A.; Adeleke, R. Effects of plant growth-promoting rhizobacteria on the molecular responses of maize under drought and heat stresses: A review. *Pedosphere* **2022**, *32*, 90–106. [CrossRef]
8. Chávez-Arias, C.C.; Ligarreto-Moreno, G.A.; Ramírez-Godoy, A.; Restrepo-Díaz, H. Maize responses challenged by drought, elevated daytime temperature and arthropod herbivory stresses: A physiological, biochemical and molecular view. *Front. Plant Sci.* **2021**, *12*, 702841. [CrossRef]
9. Mahdi, I.; Allaoui, A.; Fahsi, N.; Biskri, L. *Bacillus velezensis* QA2 potentially induced salt stress tolerance and enhanced phosphate uptake in quinoa plants. *Microorganisms* **2022**, *10*, 1836. [CrossRef]
10. Etesami, H.; Maheshwari, D.K. Use of plant growth promoting rhizobacteria (PGPRs) with multiple plant growth promoting traits in stress agriculture: Action mechanisms and future prospects. *Ecotoxicol. Environ. Saf.* **2018**, *156*, 225–246. [CrossRef]
11. Andrés-Barrao, C.; Lafi, F.F.; Alam, I.; de Zélicourt, A.; Eida, A.A.; Bokhari, A.; Alzubaidy, H.; Bajic, V.B.; Hirt, H.; Saad, M.M. Complete genome sequence analysis of *Enterobacter* sp. SA187, a plant multi-stress tolerance promoting endophytic bacterium. *Front. Microbiol.* **2017**, *8*, 2023. [CrossRef] [PubMed]
12. Joshi, H.; Bisht, N.; Mishra, S.K.; Prasad, V.; Chauhan, P.S. *Bacillus amyloliquefaciens* modulate carbohydrate metabolism in rice-PGPR cross-talk under abiotic stress and phytohormone treatments. *J. Plant Growth Regul.* **2023**, *42*, 4466–4483. [CrossRef]

13. Song, X.; Fu, C.-H.; Li, J.-H.; Sun, X.-T.; Wei, Y.-Z.; Xiao, H.-C.; Li, W.-Y.; Qin, L.-G. Research Progress on Molecular Mechanism of Endophytes Improving the Drought Resistance and Salt Tolerance of Plant. *Chinede J. Grassl.* **2024**, *32*, 13.
14. Manetsberger, J.; Caballero Gómez, N.; Benomar, N.; Christie, G.; Abriouel, H. Characterization of the culturable sporobiota of Spanish olive groves and its tolerance toward environmental challenges. *Microbiol. Spectr.* **2023**, *11*, e04013–e04022. [CrossRef] [PubMed]
15. Manetsberger, J.; Caballero Gómez, N.; Soria-Rodríguez, C.; Benomar, N.; Abriouel, H. Simply Versatile: The Use of *Peribacillus simplex* in Sustainable Agriculture. *Microorganisms* **2023**, *11*, 2540. [CrossRef]
16. Fajardo-Cavazos, P.; Maughan, H.; Nicholson, W.L. Evolution in the *Bacillaceae*. *Microbiol. Spectr.* **2016**, *2*, 21–58. [CrossRef]
17. Gottheil, O. Botanische beschreibung einiger bodenbakterien. *Zentbl. Bakteriol. Parasitenkd. Infekt. Hyg. Abt. II* **1901**, *7*, 680–691.
18. Validation of the publication of new names and new combinations previously effectively published outside the IJSB. *Int. J. Syst. Bacteriol.* **1988**, *38*, 220–222. [CrossRef]
19. Patel, S.; Gupta, R.S. A phylogenomic and comparative genomic framework for resolving the polyphyly of the genus *Bacillus*: Proposal for six new genera of *Bacillus* species, *Peribacillus* gen. nov., *Cytobacillus* gen. nov., *Mesobacillus* gen. nov., *Neobacillus* gen. nov., *Metabacillus* gen. nov. and *Alkalihalobacillus* gen. nov. *Int. J. Syst. Evol. Microbiol.* **2020**, *70*, 406–438.
20. Martínez-Hidalgo, P.; Flores-Félix, J.D.; Sánchez-Juanes, F.; Rivas, R.; Mateos, P.F.; Santa Regina, I.; Peix, Á.; Martínez-Molina, E.; Igual, J.M.; Velázquez, E. Identification of canola roots endophytic bacteria and analysis of their potential as biofertilizers for canola crops with special emphasis on sporulating bacteria. *Agronomy* **2021**, *11*, 1796. [CrossRef]
21. Santoyo, G.; Moreno-Hagelsieb, G.; del Carmen Orozco-Mosqueda, M.; Glick, B.R. Plant growth-promoting bacterial endophytes. *Microbiol. Res.* **2016**, *183*, 92–99. [CrossRef] [PubMed]
22. Yao, Z.; Chen, Y.; Luo, S.; Wang, J.; Zhang, J.; Zhang, J.; Tian, C.; Tian, L. Culturable screening of plant growth-promoting and biocontrol bacteria in the Rhizosphere and Phyllosphere of wild Rice. *Microorganisms* **2022**, *10*, 1468. [CrossRef] [PubMed]
23. Hassen, A.I.; Labuschagne, N. Root colonization and growth enhancement in wheat and tomato by rhizobacteria isolated from the rhizoplane of grasses. *World J. Microbiol. Biotechnol.* **2010**, *26*, 1837–1846. [CrossRef]
24. Erturk, Y.; Ercisli, S.; Haznedar, A.; Cakmakci, R. Effects of plant growth promoting rhizobacteria (PGPR) on rooting and root growth of kiwifruit (*Actinidia deliciosa*) stem cuttings. *Biol. Res.* **2010**, *43*, 91–98. [CrossRef]
25. Schwartz, A.R.; Ortiz, I.; Maymon, M.; Herbold, C.W.; Fujishige, N.A.; Vijanderan, J.A.; Villella, W.; Hanamoto, K.; Diener, A.; Sanders, E.R. *Bacillus simplex*—A little known PGPB with anti-fungal activity—Alters pea legume root architecture and nodule morphology when coinoculated with *Rhizobium leguminosarum* bv. viciae. *Agronomy* **2013**, *3*, 595–620. [CrossRef]
26. Al-Sman, K.M.; Abo-Elyousr, K.; Eraky, A.; El-Zawahry, A. Potential activities of *Bacillus simplex* as a biocontrol agent against root rot of *Nigella sativa* caused by *Fusarium camptoceras*. *Egypt. J. Biol. Pest Control.* **2019**, *29*, 1–6. [CrossRef]
27. Chandra, P.; Khobra, R.; Sundha, P.; Sharma, R.K.; Jasrotia, P.; Chandra, A.; Singh, D.P.; Singh, G.P. Plant growth promoting Bacillus-based bio formulations improve wheat rhizosphere biological activity, nutrient uptake and growth of the plant. *Acta Physiol. Plant.* **2021**, *43*, 1–12. [CrossRef]
28. Cochard, B.; Giroud, B.; Crovadore, J.; Chablais, R.; Arminjon, L.; Lefort, F. Endophytic PGPR from tomato roots: Isolation, in vitro characterization and in vivo evaluation of treated tomatoes (*Solanum lycopersicum* L.). *Microorganisms* **2022**, *10*, 765. [CrossRef]
29. Mushtaq, Z.; Nazir, A.; Asghar, H.N.; Zahir, Z.A. Interactive effect of siderophore-producing bacteria and l-tryptophan on physiology, tuber characteristics, yield, and iron concentration of potato. *Potato Res.* **2022**, *65*, 1015–1027. [CrossRef]
30. Akinrinlola, R.J.; Yuen, G.Y.; Drijber, R.A.; Adesemoye, A.O. Evaluation of *Bacillus* strains for plant growth promotion and predictability of efficacy by in vitro physiological traits. *Int. J. Microbiol.* **2018**, *2018*, 5686874. [CrossRef]
31. Senger, M.; Moresco, E.; Dalbosco, M.; Santin, R.; Inderbitzin, P.; Barrocas, E.N. Methods to quantify *Bacillus simplex*-based inoculant and its effect as a seed treatment on field-grown corn and soybean in Brazil. *J. Seed Sci.* **2022**, *44*, e202244041. [CrossRef]
32. Misra, S.; Chauhan, P.S. ACC deaminase-producing rhizosphere competent *Bacillus* spp. mitigate salt stress and promote *Zea mays* growth by modulating ethylene metabolism. *3 Biotech* **2020**, *10*, 119. [CrossRef] [PubMed]
33. Delcher, A.L.; Bratke, K.A.; Powers, E.C.; Salzberg, S.L. Identifying bacterial genes and endosymbiont DNA with Glimmer. *Bioinformatics* **2007**, *23*, 673–679. [CrossRef] [PubMed]
34. Ahmad, P.; Azooz, M.M.; Prasad, M.N.V. *Ecophysiology and Responses of Plants under Salt Stress*; Springer Science & Business Media: New York, NY, USA, 2012.
35. Mishra, J.; Prakash, J.; Arora, N.K. Role of beneficial soil microbes in sustainable agriculture and environmental management. *Clim. Change Environ. Sustain.* **2016**, *4*, 137–149. [CrossRef]
36. Ojuederie, O.B.; Babalola, O.O. Microbial and plant-assisted bioremediation of heavy metal polluted environments: A review. *Int. J. Environ. Res. Public Health* **2017**, *14*, 1504. [CrossRef]
37. Wang, Y.; Xu, C.; Wuriyanghan, H.; Lei, Z.; Tang, Y.; Zhang, H.; Zhao, X. *Exogenous selenium* endows salt-tolerant and salt-sensitive soybeans with salt tolerance through plant-microbial coactions. *Agronomy* **2023**, *13*, 2271. [CrossRef]
38. Hayat, R.; Ali, S.; Amara, U.; Khalid, R.; Ahmed, I. Soil beneficial bacteria and their role in plant growth promotion: A review. *Ann. Microbiol.* **2010**, *60*, 579–598. [CrossRef]
39. Yadav, A.N.; Kour, D.; Rana, K.L.; Kumar, V.; Dhaliwa, S.; Verma, P.; Singh, B.; Chauahan, V.S.; Sugitha, T.; Saxena, A.K. Plant microbiomes and its beneficial multifunctional plant growth promoting attributes. *Int. J. Environ. Sci. Nat. Resour.* **2017**, *3*, 1–8. [CrossRef]

40. Ito, Y.; Yoshidome, D.; Hidaka, M.; Araki, Y.; Ito, K.; Kosono, S.; Nishiyama, M. Improvement of the nitrogenase activity in *Escherichia coli* that expresses the nitrogen fixation-related genes from *Azotobacter vinelandii*. *Biochem. Biophys. Res. Commun.* **2024**, *728*, 150345. [CrossRef]
41. Tanigawa, R.; Shirokane, M.; Maeda, S.-I.; Omata, T.; Tanaka, K.; Takahashi, H. Transcriptional activation of NtcA-dependent promoters of *Synechococcus* sp. PCC 7942 by 2-oxoglutarate in vitro. *Proc. Natl. Acad. Sci. USA* **2002**, *99*, 4251–4255. [CrossRef]
42. Rezakhani, L.; Motesharezadeh, B.; Tehrani, M.M.; Etesami, H.; Hosseini, H.M. Phosphate–solubilizing bacteria and silicon synergistically augment phosphorus (P) uptake by wheat (*Triticum aestivum* L.) plant fertilized with soluble or insoluble P source. *Ecotoxicol. Environ. Saf.* **2019**, *173*, 504–513. [CrossRef] [PubMed]
43. Erturk, Y.; Ercisli, S.; Cakmakci, R. Yield and growth response of strawberry to plant growth-promoting rhizobacteria inoculation. *J. Plant Res.* **2012**, *35*, 817–826. [CrossRef]
44. Wang, Y.; Sun, X.; Peng, J.; Li, F.; Ali, F.; Wang, Z. Regulation of seed germination: ROS, epigenetic, and hormonal aspects. *J. Adv. Res.* **2024**, *in press*. [CrossRef] [PubMed]
45. Alori, E.T.; Glick, B.R.; Babalola, O.O. Microbial phosphorus solubilization and its potential for use in sustainable agriculture. *Front. Microbiol.* **2017**, *8*, 971. [CrossRef] [PubMed]
46. Olson, K.R.; Gao, Y.; DeLeon, E.R.; Arif, M.; Arif, F.; Arora, N.; Straub, K.D. Catalase as a sulfide-sulfur oxido-reductase: An ancient (and modern?) regulator of reactive sulfur species (RSS). *Redox Biol.* **2017**, *12*, 325–339. [CrossRef]
47. AlAli, H.A.; Khalifa, A.; Almalki, M. Plant growth-promoting rhizobacteria from *Ocimum basilicum* improve growth of *Phaseolus vulgaris* and *Abelmoschus esculentus*. *S. Afr. J. Bot.* **2021**, *139*, 200–209. [CrossRef]
48. Mehta, S.; Nautiyal, C. An efficient method for qualitative screening of phosphate-solubilizing bacteria. *Curr. Microbiol.* **2001**, *43*, 51–56. [CrossRef]
49. Bric, J.M.; Bostock, R.M.; Silverstone, S.E. Rapid in situ assay for indoleacetic acid production by bacteria immobilized on a nitrocellulose membrane. *Appl. Environ. Microbiol.* **1991**, *57*, 535–538. [CrossRef]
50. Yeole, R.D.; Dave, B.P.; Dube, H.C. Siderophore production by fluorescent pseudomonads colonizing roots of certain crop plants. *Indian J. Exp. Biol.* **2001**, *39*, 464–468.
51. Guo, Z.; Zhao, J.; Wang, M.; Song, S.; Xia, Z. Sulfur dioxide promotes seed germination by modulating reactive oxygen species production in maize. *Plant Sci.* **2021**, *312*, 111027. [CrossRef]
52. Kumar, D.; Yusuf, M.A.; Singh, P.; Sardar, M.; Sarin, N.B. Histochemical detection of superoxide and H2O2 accumulation in *Brassica juncea* seedlings. *Bio-Protocol* **2014**, *4*, e1108. [CrossRef]
53. Wang, P.; Qi, S.; Wang, X.; Dou, L.; Jia, M.-a.; Mao, T.; Guo, Y.; Wang, X. The OPEN STOMATA1–SPIRAL1 module regulates microtubule stability during abscisic acid-induced stomatal closure in *Arabidopsis*. *Plant Cell* **2023**, *35*, 260–278. [CrossRef] [PubMed]
54. Beyer, W.F., Jr.; Fridovich, I. Assaying for superoxide dismutase activity: Some large consequences of minor changes in conditions. *Anal. Biochem.* **1987**, *161*, 559–566. [CrossRef] [PubMed]
55. Nakano, Y.; Asada, K. Hydrogen peroxide is scavenged by ascorbate-specific peroxidase in spinach chloroplasts. *Plant Cell Physiol.* **1981**, *22*, 867–880.

International Journal of
Molecular Sciences

MDPI

Article

Electron-Transferring Flavoprotein and Its Dehydrogenase Required for Fungal Pathogenicity in *Arthrobotrys oligospora*

Yonglan Liu †, Zhangyu Li †, Junjie Liu, Xiqi Zhang and Xin Wang *

State Key Laboratory for Conservation and Utilization of Bio-Resources in Yunnan, Yunnan University, Kunming 650091, China
* Correspondence: xinwang2@ynu.edu.cn
† These authors contributed equally to this work.

Abstract: Electron transfer flavoprotein (ETF) plays an important function in fatty acid beta oxidation and the amino acid metabolic pathway. It can provide pathogenicity to some opportunistic fungi via modulating cellular metabolite composition. *Arthrobotrys oligospora* is a typical invasion fungus to nematodes. Its ETF characterization is still unknown. Here, we showed that the mutations of *A. oligospora* ETF (*Aoetfα* and *Aoetfβ*) and its dehydrogenase (*Aoetfdh*) led to severe defects in mitochondrial integrity and blocked fatty acid metabolism. The pathogenicity-associated trap structures were completely suppressed when exposed to nematode-derived ascarosides and nutrition signals, including ammonia and urea. Compared to the wild-type strain, the nematode predatory activity was significantly reduced and delayed. But surprisingly, the rich nutrition could restore the massive trap and robust predatory activity in the mutant *Aoetfβ* beyond all induction cues. Moreover, the deletion of *Aoetfβ* has led to the accumulation of butyrate-like smell, which has a strong attraction to *Caenorhabditis elegans* nematodes. Ultimately, ETF and its dehydrogenase play a crucial role in nematode-trapping fungi, highlighting mitochondrial metabolite fluctuations that are connected to pathogenesis and further regulating the interactions between fungi and nematodes.

Keywords: electron transfer flavoprotein; nematode–fungus interaction; trap

1. Introduction

Electron transfer flavoproteins (ETF) transfer electrons from mitochondrial dehydrogenases, such as acyl-CoA dehydrogenases, to the electron transport chain, which facilitates the production of ATP. This protein is conserved across a wide range of organisms and exists as a heterodimeric complex, consisting of ETFα and ETFβ subunits. ETF complexes are vital to fatty acid β-oxidation and amino acid metabolism. Both bacteria and eukaryotes contain ETF, that binds FAD and AMP as cofactors [1]. Flavoproteins rely on ETF to transfer electrons to the respiratory chain. Extensive research has been conducted on ETF in yeast, animals, and humans [2–4]. Mutations in the *ETFα* or *ETFβ* genes can lead to multiple acyl-CoA dehydrogenase deficiency (MADD) and other metabolic disorders [5]. Similarly, in plants, these mutations result in the accumulation of reactive oxygen species (ROS), which subsequently damage cellular components and disrupt plant growth and development [6]. It is therefore crucial to understand the function of ETF in order to gain a deeper understanding of cellular energy metabolism and diseases related to it.

Nematode-trapping fungi (NT fungi) are critical biological factors in regulating nematode populations within soil ecosystems [7–9]. These fungi capture, eliminate, and digest nematodes by forming various predatory structures, including adhesive hyphal networks, adhesive knobs, and constricting rings [10–13]. Signals, toxins, and effector proteins produced by fungi directly impact nematodes at each process of the interaction [14,15]. Approximately 380 species of nematode-trapping fungi have been identified, with most belonging to the genera *Arthrobotrys*, *Cystopage*, *Dactylellina*, *Dactylella*, *Drechslerella*, *Hohenbuehelia*, and so on [16].

After nematodes are captured by *A. oligospora*, they provide nutrients that facilitate the growth of more hyphae and spores [17]. Early studies indicate that nematode-derived nutrients are first converted into lipids by the fungus and then degraded via the β-oxidation pathway to support new hyphal growth [18]. During this process, fatty acid length determines whether degradation occurs in the mitochondria or peroxisomes: fatty acids with fewer than 20 carbons are oxidized in the mitochondria, while those with more than 20 carbons are initially oxidized in peroxisomes into shorter fatty acids, which are then fully oxidized in mitochondria [19]. The first dehydrogenation step in mitochondrial β-oxidation is catalyzed by acyl-CoA dehydrogenase, which transfers electrons to ETF. Then ETF transfers electrons to ubiquinone oxidoreductase through ETF dehydrogenase (ETFdh) in the respiratory chain, where electrons are transferred to coenzyme Q10 in the mitochondrial inner membrane for ATP synthesis [20–22]. Besides electron transport, ETF participates in aerobic metabolism and the production of superoxide and hydrogen peroxide. Mutations in *ETF* are linked to fatty acid oxidation disorders, such as multiple acyl-CoA dehydrogenase deficiency [23,24]. In summary, ETF plays a critical role in electron transfer and energy synthesis.

The research on ETF in filamentous fungi is limited. In this study, we identified the functions of *etf* and *etfdh* in *A. oligospora* through gene knockout. Our results demonstrated that both *etf* and *etfdh* are involved in the formation of traps in *A. oligospora*, particularly highlighting the strong attraction of *Caenorhabditis elegans* to the Δ*Aoetfβ* mutant. In summary, our findings suggest that ETF and ETFdh contribute to predator–prey interactions between predatory fungi and nematodes, providing a theoretical basis for understanding these interactions.

2. Results

2.1. Analysis and Deletion of etf and etfdh in A. oligospora

Based on the genome annotation data of *A. oligospora* ATCC24927, we obtained probable electron transfer flavoprotein coding genes, *Aoetfα* (AOL_s00109g97), *Aoetfβ* (AOL_s00078g521), and ETF dehydrogenase coding gene *Aoetfdh* (AOL_s00215g393). A conserved domains analysis showed us that both subunits of ETF-α and ETF-β contain typical ETF domains (Figure 1A). Further modeling revealed that ETF-α associates with ETF-β to form a stable heterodimeric complex, designated AoETF. This complex binds both the FAD cofactor and an AMP molecule (Figure 1B). In contrast, ETFdh functions as a monomer and contains several cofactor binding sites for NAD, ETFQO_UQ, and an iron-sulfur cluster (4Fe-4S) (Figure 1A). So far in NT fungi, the roles of ETF heterodimer and ETFdh have not been characterized. Phylogenetic tree analysis revealed that ETF as well as ETFdh are highly conserved in NT fungi with other fungal species, suggesting that they may have similar functions in this group of fungi (Figures S1–S3). Using the homologous recombination strategy, we generated mutants for each of these three coding genes, designated as Δ*Aoetfα*, Δ*Aoetfβ*, and Δ*Aoetfdh* (Figure 1C,D). To verify protoplast transformants using PCR, the sizes of the wild-type and mutant strains for gene *Aoetfα* are 1425 bp and 2563 bp, respectively. For gene *Aoetfβ*, the sizes of the wild-type and mutant strains are 1142 bp and 2353 bp, respectively. For gene *Aoetfdh*, the sizes of the wild-type and mutant strains are 1564 bp and 2460 bp, respectively (Figure S4A). On various nutrient-rich media, including PDA, CMY, TG, and TYGA, the vegetative growth of the three mutants was comparable to that of the wild-type strain (Figure S4B).

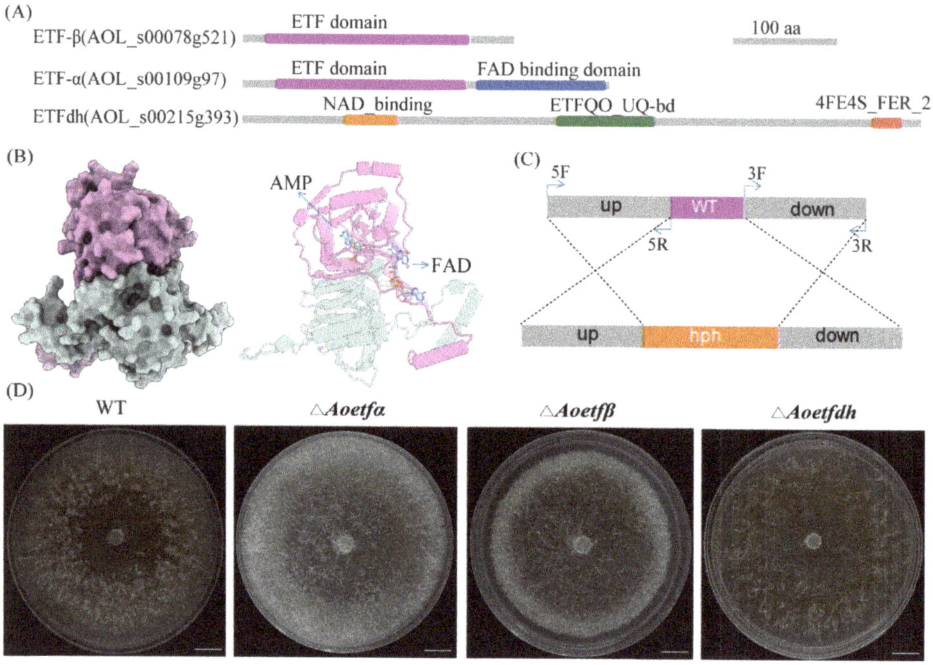

Figure 1. Bioinformatics analysis of ETF, ETFdh and gene knockouts. (**A**) The domain structure of *Arthrobotrys oligospora* ETF-α, ETF-β and ETFdh as annotated at the Broad Institute of MIT. (**B**) Structural Model of ETFα-ETFβ complex (AoETF) bound to AMP and FAD, the purple structure represents ETF-β, while the light gray structure depicts ETF-α. (**C**) Homologous recombination knockout model diagram. (**D**) Growth of wild-type strain (WT) and the mutants of *Aoetfα*, *Aoetfβ*, and *Aoetfdh* on PDA medium for 7 days; scale bar: 1 cm.

2.2. *Aoetf* and *Aoetfdh* Impair Conidiation in *A. oligospora*

We initially assessed whether *Aoetf* and *Aoetfdh* are involved in the sporulation of *A. oligospora*. Wild-type and mutant strains were cultivated on PDA medium for six days, after which conidial density and spore number were measured. The results showed a significant decrease in both conidial density and spore number in the Δ*Aoetfα*, Δ*Aoetfβ*, and Δ*Aoetfdh* compared to the wild-type (Figure 2A,B). Regarding conidial morphology, both wild-type and mutant strains produced four main types of conidia, consistent with previous findings [25]. Among these, the second type, representing mature spores in the phialides, was predominant in both wild-type and mutant strains (64%, 59%, 63.67%, and 62.33%, respectively) (Figure 2C,D). However, the deletion of Δ*Aoetfα* resulted in a notably low proportion of immature spores, constituting only 1.33% of the total. These observations suggest that the *Aoetf* and *Aoetfdh* have important involvement in conidiation in *A. oligospora*. Unexpectedly, the Δ*Aoetfβ* strain was found to spontaneously produce nematode-capturing structures, adhesive three-dimensional networks, when grown on CMY medium (Figure 2E), which contrasts with the typical formation of predatory structures only under low-nutrient conditions.

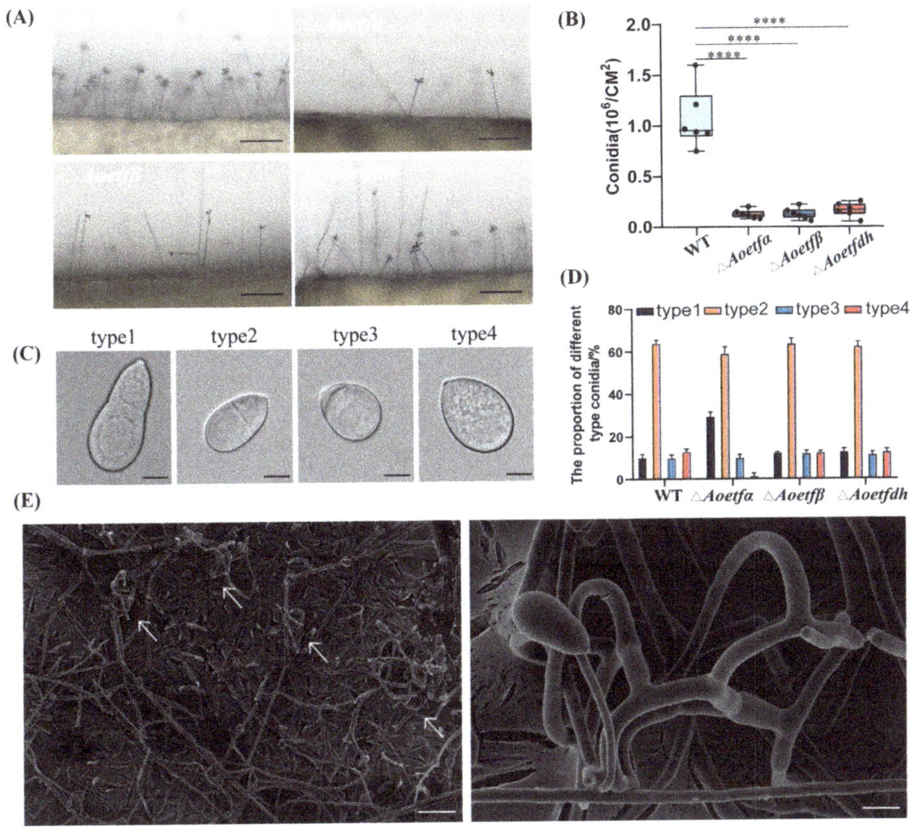

Figure 2. Mutations *Aoetf* and *Aoetfdh* affect spore production, morphology, and trap formation. (**A**) Comparison of conidia in wild-type, Δ*Aoetf*, and Δ*Aoetfdh* strains (Scale bar: 100 μm). (**B**) Comparison of conidia production in wild-type, Δ*Aoetf*, and Δ*Aoetfdh* strains. **** represents significant difference ($p < 0.0001$). (**C**,**D**) Analysis and statistics of conidia types in wild-type, Δ*Aoetf*, and Δ*Aoetfdh* strains (scale bar: 7 μm). (**E**) Scanning electron micrographs of the trapping structure in the Δ*Aoetfβ* strain at different magnifications. The left image is at 150× magnification (scale bar: 100 μm; white arrow indicates the trap), and the right image is at 1000× magnification (scale bar: 10 μm).

2.3. *Aoetf* and *Aoetfdh* Are Essential for the Formation of Traps in *A. oligospora*

To investigate whether *Aoetf* and *Aoetfdh* influence signal-induced trap formation in *A. olilgospora*, we compared trap production between wild-type and mutant strains in response to live *C. elegans*, nematode extracts, and trap-inducing signals such as ascaroside (ascr#7), urea, and ammonia (Am). Our findings revealed that the disruption of *Aoetf* or *Aoetfdh* genes delayed trap formation and reduced nematode-killing efficiency compared to the wild-type strain under same conditions. Specifically, when exposed to nematodes, the mutant strains produce traps later than the wild-type strains (Figure 3A). In the wild-type strain, only 41.67% of nematodes survived after 12 h, while the survival rates in the Δ*Aoetfα*, Δ*Aoetfβ*, and Δ*Aoetfdh* strains were 71.43%, 85.71%, and 63.49%, respectively (Figure 3B). Furthermore, when exposed to whole nematode extracts, the mutants produced fewer traps than the wild-type strain (Figure 3C). Notably, when we tested trap induction in response to nematode-derived ascaroside, urea, and ammonia, the mutants completely failed to produce any traps (Figure 3D). These results suggest that mutations in *Aoetf* or *Aoetfdh* significantly impair trap formation in *A. oligospora*.

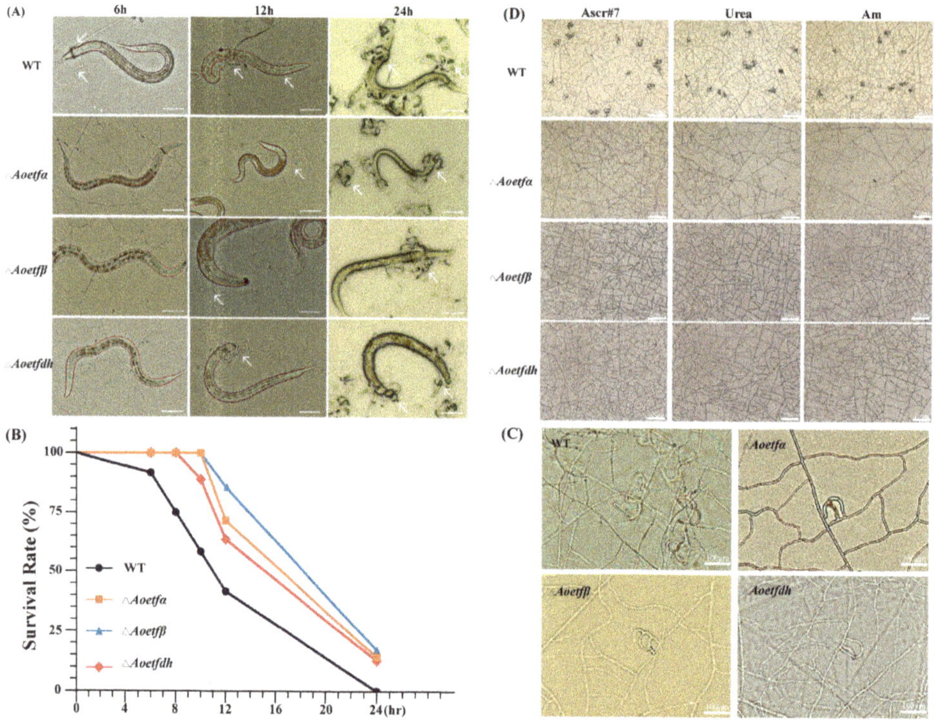

Figure 3. Trap formation in wild-type, Δ*Aoetf*, and Δ*Aoetfdh* strains. (**A**) Representative bright-field images of trap formation at different time points after nematode induction (scale bar: 200 μm; the arrows denote trap structures). (**B**) Survival rate of nematodes exposed to wild-type, Δ*Aoetfα*, Δ*Aoetfβ*, and Δ*Aoetfdh* strains. (**C**). Representative bright-field images of traps induced by nematode extracts in wild-type, Δ*Aoetfα*, Δ*Aoetfβ*, and Δ*Aoetfdh* strains (scale bar: 200 μm). (**D**) Representative bright-field images of traps induced by ascr#7, urea, and ammonia (Am) in wild-type, Δ*Aoetfα*, Δ*Aoetfβ*, and Δ*Aoetfdh* strains (scale bar: 100 μm).

2.4. The Absence of Aoetf Affects Fatty Acid Metabolism in A. oligospora

In other organisms, ETF plays crucial roles in fatty acid metabolism, amino acid metabolism, and mitochondrial function. To determine whether *Aoetf* in *A. oligospora* has similar effects, we analyzed hyphal growth in wild-type and mutant strains on MM medium supplemented with sodium carbonate and four different fatty acids (sodium butyrate, caprylic acid, lauric acid, and oleic acid) as the sole carbon source. The results indicated that the growth of the mutant strains was significantly inhibited, with lauric acid completely blocking their growth. These findings suggest that *Aoetf* and *Aoetfdh* are essential for fatty acid metabolism in *A. oligospora* (Figure 4A). Furthermore, transcriptomic sequencing revealed that the *Aoetf* mutation led to significant changes of nearly 60 genes related to lipid metabolism in expression level; the metabolism pathways involving glycerophospholipid metabolism, steroid biosynthesis, fatty acid degradation, and biosynthesis of unsaturated fatty acids were significantly enriched (Figure 4B–D). Therefore, mutations in ETFs have a significant impact on intracellular fatty acid metabolism in fungi, which can inhibit the degradation of long-chain fats such as lauric acid by fungi.

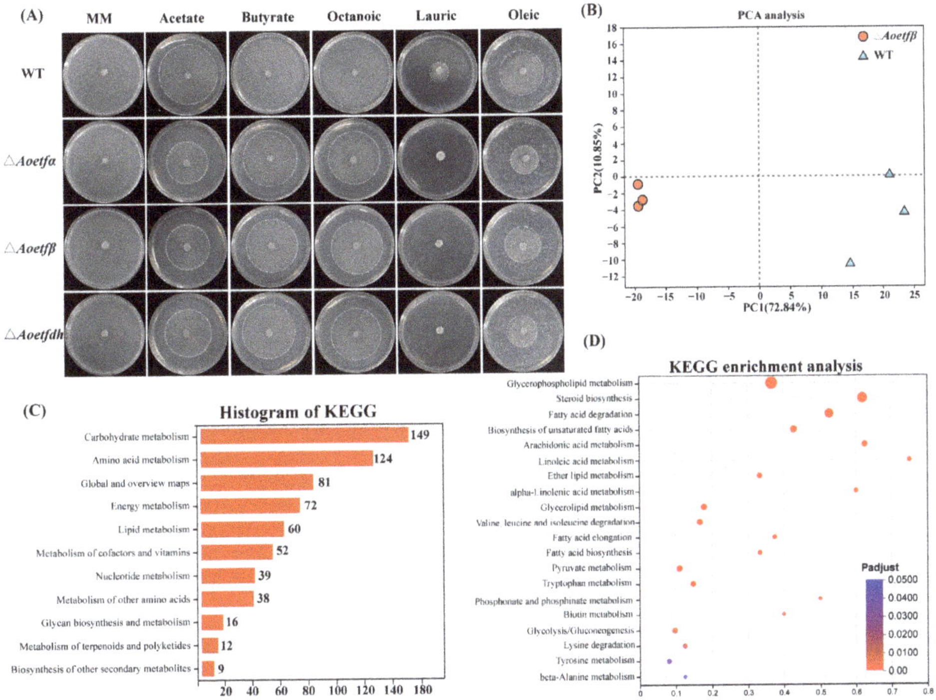

Figure 4. Δ*Aoetf* and Δ*Aoetfdh* were involved in fatty acids metabolism. (**A**) Growth conditions of WT, Δ*Aoetfα*, Δ*Aoetfβ*, and Δ*Aoetfdh* strains on MM medium with various fatty acids as the sole carbon source. (**B**) Principal component analysis (PCA) plot of *Aoetfβ* mutant and wild-type strains. (**C**) Histogram of KEGG-enriched metabolic pathways in the transcriptome, with 60 genes associated with lipid metabolism. (**D**) KEGG enrichment analysis of genes related to lipid metabolism.

2.5. Aoetf Deletion Impairs Mitochondrial Function in A. oligospra

We guessed the defects in sporulation and pathogenicity are attributed to impaired mitochondrial function, so we conducted transmission electron microscopy (TEM) observation. The TEM results revealed significant mitochondrial damage in the mutant strains compared to the wild-type strain. Specifically, the mitochondria in the Δ*Aoetfα* strain exhibited a complete loss of cristae, while those in the Δ*Aoetfβ* strain displayed fragmented cristae (Figure 5A–C). Additionally, mitochondrial staining demonstrated that the wild-type strain had well-defined, prominent mitochondria, whereas the Δ*Aoetfα* and Δ*Aoetfβ* strains showed reduced staining intensity (Figure 5D–F), indicating a decrease in mitochondrial viability due to the mutations.

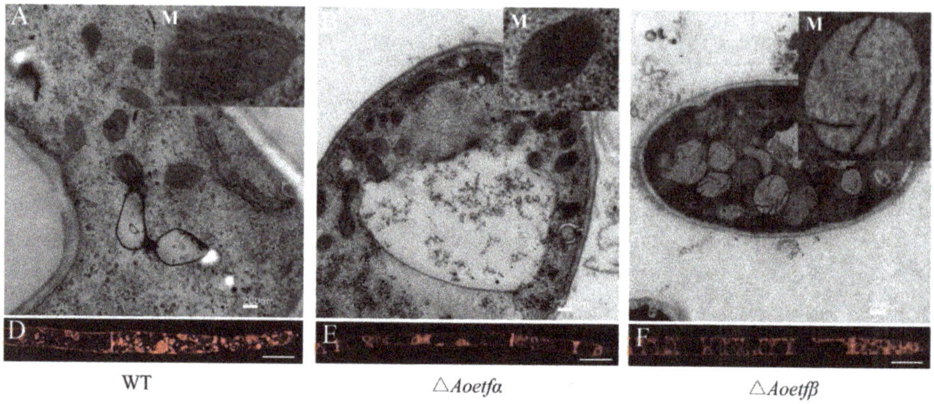

Figure 5. *Aoetf* deletion impairs mitochondria function. (**A–C**) Ultrastructural analysis of wild-type, Δ*Aoetfα*, and Δ*Aoetfβ* strains using transmission electron microscopy (scale bars: 200 nm; M denotes mitochondria). (**D–F**) Comparison of mitochondrial of wild-type and Δ*Aoetf* strains were stained with MitoTracker® Red (scale bars: 3 μm).

2.6. The Aoetf Mutant Strain Has the Ability to Attract C. elegans

ETF dysfunction has been implicated in a variety of metabolic disorders involving fatty acids and amino acids, particularly in the context of glutaric acidemia in humans [26]. During the growth of the mutant strains, a strong odor reminiscent of sweaty feet was detected. Considering the well-established attraction of NT fungi to nematodes, we conducted chemotaxis assays using the Δ*Aoetfβ* strain and the wild-type strain in the presence of *C. elegans*. The results showed that Δ*Aoetfβ* has a significant attraction to *C. elegans*, with a chemotaxis index of 0.92 (Figure 6A). Further, we evaluated the chemotactic response of *C. elegans* to glutaric acid and observed that this compound exhibited a higher attraction index of 0.8, significantly greater than that of the control (Figure 6B). In the metabolomics analysis, several significantly upregulated compounds were observed (Table 1, Figure S5A), among which the markedly upregulated compound cetraxate demonstrated an attraction to *C. elegans* with a value of 0.70 (Figure 6C). Given the role of fungal extracellular vesicles (exosomes) as mediators for shuttling chemical signals to communicate with host cells [27], we isolated exosomes from both the wild-type and Δ*Aoetfβ* strains for further attraction assays. Nanoparticle Tracking Analysis (NTA) detection and TEM imaging revealed that the exosomes from the mutant strains were larger in size and present in higher concentrations compared to those from the wild-type strain (Figure S5B,C). Chemotaxis assays with exosomes demonstrated that exosomes from both the wild-type and Δ*Aoetfβ* strains could attract *C. elegans*; however, the exosomes from the mutant strain exhibited a higher attraction index of 0.81, compared to 0.63 for the wild-type exosomes (Figure 6D,E). Furthermore, a comparison of the relative chemotactic strength revealed that the Δ*Aoetfβ* strain had a significantly higher chemotaxis index of 0.73 compared to 0.26 for the wild-type strain (Figure 6F). Based on these observations, we conclude that the *Aoetf* mutation leads to the accumulation of metabolites with nematode-attracting activity, which may be secreted via fungal extracellular vesicles.

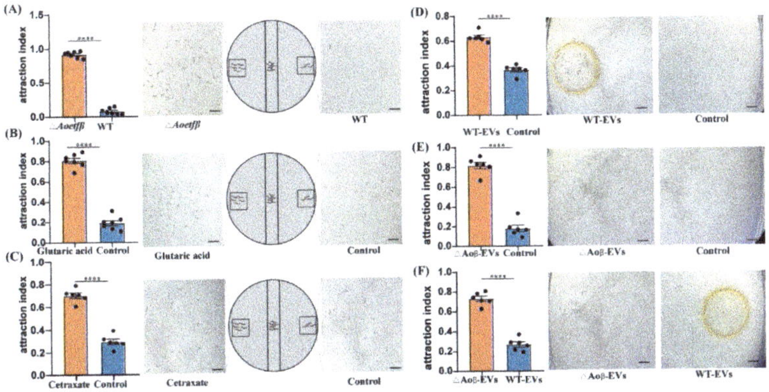

Figure 6. Chemotaxis observation of *C. elegans*. (**A**) Chemotaxis statistics of *C. elegans* towards fermentation broth of wild-type and Δ*Aoetfβ* strains. (**B**) Chemotaxis statistics of *C. elegans* towards glutaric acid and control. (**C**) Chemotaxis statistics of *C. elegans* towards cetraxate and control (**D**) Chemotaxis statistics of *C. elegans* towards exosomes from wild-type strains. (**E**) Chemotaxis statistics of *C. elegans* towards exosomes from Δ*Aoetfβ* strains. (**F**) Comparative chemotaxis statistics of *C. elegans* towards exosomes from wild-type and Δ*Aoetfβ* strains. **** represents significant difference ($p < 0.0001$; scale bars: 2 mm).

Table 1. Metabolomics analysis of differentially upregulated compounds between wild-type and Δ*Aoetfβ* strains.

Metabolism Name	FC	KEGG Pathway Description
Cetraxate	47.8841	-
S-Lactoylglutathione	1.4921	Pyruvate metabolism
Kynurenic acid	1.2806	Tryptophan metabolism
Gamma-glutamyl-L-putrescine	1.3035	Arginine and proline metabolism
2-Isopropylmalic acid	1.1205	Biosynthesis of secondary metabolites Valine, leucine and isoleucine biosynthesis 2-Oxocarboxylic acid
Theophylline	1.1008	Biosynthesis of secondary metabolites Caffeine metabolism
6-Lactoyltetrahydropterin	1.1103	Folate biosynthesis
Homogentisic acid	1.1126	Ubiquinone and other terpenoid-quinone biosynthesis Tyrosine metabolism Biosynthesis of cofactors
Imidazole lactic acid	1.1166	Histidine metabolism
D-xylonic acid	1.0826	Pentose and glucuronate interconversions Ascorbate and aldarate metabolism
N-carbamoylputrescine	1.0977	Arginine and proline metabolism
5-methoxyindoleacetate	1.0983	Tryptophan metabolism
lactose	1.0749	ABC transporters Galactose metabolism
Indole-3-acetaldehyde	1.0808	Tryptophan metabolism
Glucosamine	1.0607	Amino sugar and nucleotide sugar metabolism Biosynthesis of nucleotide sugars
LL-2,6-diaminopimelic acid	1.0692	Biosynthesis of secondary metabolites D-amino acid metabolism Biosynthesis of amino acids Lysine biosynthesis
Biotin	1.0596	Biotin metabolism;ABC transporters Biosynthesis of cofactors
Kynurenine	1.0558	Tryptophan metabolism Biosynthesis of cofactors

FC (Δ*Aoetfβ*/WT): fold change in the expression of this metabolite between the two groups. WT: expression level of this metabolite in the control group WT. Δ*Aoetfβ*: expression level of this metabolite in the experimental group Δ*Aoetfβ*. Enriched KEGG metabolic pathways.

2.7. Nematode ETF Exhibited Similar Roles in Mitochondrial Function to A. oligospora

Nematode-NT fungi are consistently present in soil; we wondered about the impact of homologous genes of *A. oligospora* ETF on the mitochondrial function of nematodes. Based on the observation of the TEM, *C. elegans* exhibited mitochondrial swelling and cristae disruption when the ETF coding was interfered with (Figure 7A). Additionally, the behavior assay showed that obvious aggregation, nearly 100%, occurred following the interference of ETFα, while the aggregation ratio of ETFβ was 60% compared to N2 *C. elegans* (Figure 7B).

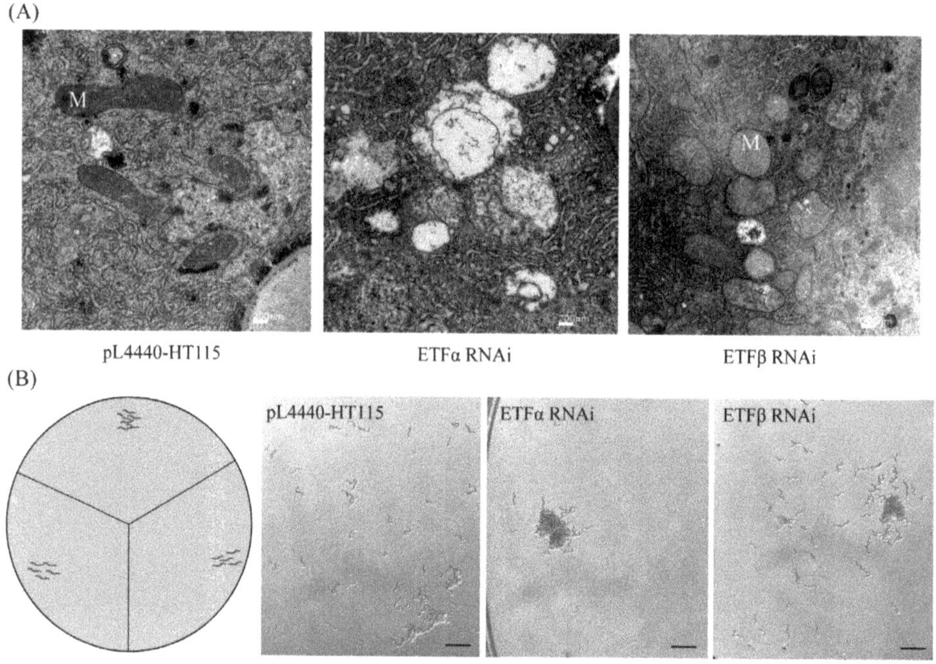

Figure 7. TEM observations and phenotypes of *ETF* Genes in *C. elegans*. (**A**) Transmission electron microscopy images of mitochondria after interfering with the *ETF* gene in *C. elegans* (scale bars: 200 nm). M denotes mitochondria. (**B**) Aggregation phenomenon observed in *C. elegans* after interfering with the *ETFα* gene (scale bars: 2 mm).

3. Discussion

ETF and ETFdh are highly conserved across bacteria, filamentous fungi, and animals. In our study, AoETF and AoETFdh also exhibited a high degree of conservation with other filamentous fungi (Figure 1 and Figure S1). We established multiple roles of *Aoetf* and *Aoetfdh* in the phenotypic regulation of conidiation and pathogenesis in *A. oligospora* (Figures 2 and 3). In the *Aoetf* mutant strains, transcriptional profiles and metabolic omics revealed distinctive differences in fatty acid metabolism and amino acid metabolism, which were further confirmed by the growth defect against long chain fatty acids (Figure 4, Table 1). Similar results have been previously demonstrated in the rice blast fungus *Magnaporthe oryzea* and in the entomopathogenic fungus *Beauveria bassiana* [2,3].

In humans, *ETF* and *ETFdh* variants are typical causes of multiple acyl-CoA dehydrogenase deficiency (MADD) and glutaric aciduria type II [28]. In cases of ETF or ETFdh deficiency, the acyl-CoA dehydrogenases cannot transfer electrons from several dehydrogenases involved in fatty acid oxidation, as well as choline and amino acid metabolism [29]. This process would lead to the deficiencies of multiple acyl-CoA dehydrogenases and disruption of ATP production, causing a large accumulation of glutaric acid, lactic, ethylmalonic, butyric, isobutyric, 2-methyl-butyric, and isovaleric acids [30,31]. As we know, these compounds are volatile and emit unpleasant odors. Coincidentally, our *A. oligospora* mutant of *Aoetf* also produced a similar foul smell, which we suspected was glutaric acid. For fungi to capture nematodes, fatal attraction is an important process. In this study, we found that glutaric acid was strongly attracted to *C. elegans*, although we did not obtain exact compound information based on metabolomics (Figure 6). Moreover, the mutant and its exosomes showed stronger attraction to nematodes than the wild-type strain, suggesting that the disruption of *Aoetf* led to the secretion of certain functional metabolites (Figure 6). Therefore, the metabolic regulatory role of *Aoetf* and *Aoetfdh* in the fungus *A. oligospora* may be evolutionarily conserved with that in higher animals.

Few studies have described the integrity of mitochondrial morphology in ETF-ETFdh deletions [3,32]. In this study, we found that metabolic disorder-produced esters and acids might impair mitochondrial fitness in *A. oligospora* (Table 1). TEM images showed defective mitochondria with fragmented, cristae-free, or swollen mitochondria exist in the mutants of *Aoetf* (Figure 5). Since mitochondria are the key organelles for energy production, mutations or deletions in ETF and ETFdh in *Magnaporthe oryzae* lead to reduced ATP levels, affecting its energy metabolism, which may further impact its growth, development, and pathogenicity [3]. Based on these observations, mitochondrial dysfunction may be the reason why fungi fail to respond to signals to produce predatory organs. In most eukaryotic cells, mitochondria are responsible for the delicate metabolic activities and stress responses that maintain cellular stability [33,34]. Our research provides the possibility that inhibition of mitochondrial function could be a potential strategy for combating fungal infections in nematodes.

We have identified the multifaceted functions of *Aoetf* and *Aoetfdh* that include maintaining mitochondrial structure, fatty acid metabolism, and further controlling pathogenicity and sporulation. The lifestyle of NT fungi includes vegetative growth and parasitic growth, but the latter only occurs in nutrient-deprived conditions [35]. Unexpectedly, we identified a mutant strain of *Aoetfβ* that grew autonomously in the form of trapping hypha under rich nutrition, which can actively kill nematodes (Figure 2). In addition, *Aoetf* mutations acquired a strong ability to attract nematodes. From the perspective of the host nematode, *Aoetf* also plays a conservative role in maintaining mitochondrial integrity and has a significant regulatory effect on nematode aggregation behavior (Figure 7). This *Aoetf* mutation-induced change in animal behavior has been observed in other animal models, where ETF-resulted metabolic dysfunction is associated with aberrant neural phenotypes and paralysis [36]. The significance of these findings should further facilitate mutant strains' potential in controlling nematodes in the field.

Faced with low or poor nutrition, the NT fungi can respond to context signals to form infection structures to capture and digest nematodes [35,37]. The involved signaling substances currently cover a wide range, including nematode or nematode-derived chemicals and nutrition signals like urea or ammonia [10,38,39]. The G protein signaling pathway plays an especially vital role for trap formation in NT fungi [39,40]. Unexpectedly, we discovered that changes in intracellular metabolic or energy levels induced by genetic manipulation might activate the regulation of trap formation in *A. oligospora*. Our future work will aim to link cellular metabolic fluctuations to signal transduction, extending the understanding of intracellular signal regulation to trap development and pathogenesis in the NT fungi.

4. Materials and Methods

4.1. Strains, Plasmids, and Culture Conditions

The wild-type (WT) strain of *A. oligospora* (ATCC 24927) and the mutant strains *Aoetfα*, *Aoetfβ*, and *Aoetfdh* were inoculated on PDA medium (200 g potato, 20 g glucose, and 18 g agar) and incubated at 28 °C. The plasmids pCSN44 and pCE-Zero (Vazyme Biotech Co., Ltd., Nanjing, China) were used for the amplification of the hygromycin fragment and the construction of knockout vectors, respectively. *Escherichia coli* strain (DH5α)

(TransGen Biotech, Beijing, China) was used for plasmid cloning and storage, grown in LB medium at 37 °C. ssPDA medium (PDA supplemented with sucrose) was used for protoplast regeneration, while CMY medium (cornmeal-molasses-yeast), TG medium (tryptone-glucose), and TYGA medium (tryptone-yeast-extract-glucose-agar) were used to analyze the phenotypic characteristics of hyphae and spores [41]. *C. elegans* was cultured on NGM medium at 20 °C for bioassays.

4.2. Targeted Gene Deletion

The sequences of *Aoetfα* (AOL_s00109g97), *Aoetfβ* (AOL_s00078g521), and *Aoetfdh* (AOL_s00215g393) were obtained from the *A. oligospora* genome in the NCBI database. Primers for the upstream fragment, hygromycin fragment, and downstream fragment were designed using the NEB England online tool (https://nebuilderv1.neb.com). Paired primers were used to amplify the upstream and downstream fragments of *Aoetfα*, *Aoetfβ*, and *Aoetfdh* from *A. oligospora* by PCR, and the hygromycin fragment was amplified using the pCSN44 plasmid as a template. The pCE-Zero vector was digested with *Eco*RV, and the three fragments were ligated to the vector and transformed into *E. coli* competent cells. Primers for gene knockout are listed in Table S1. Subsequently, the fragments were transformed into *A. oligospora* protoplasts using the PEG/$CaCl_2$-mediated method. Positive transformants were selected on TYGA medium containing 100 μg/mL hygromycin and verified by PCR.

4.3. Phenotypic Characterization of Strains

The wild-type and mutant strains were inoculated on PDA, CMY, TYGA, and TG media, and the colony diameters were measured every 24 h using a vernier caliper for a duration of 6 days. To determine conidial production, the wild-type and mutant strains were inoculated on CMY medium and cultured for 7 days. The spores were washed with 5 mL of double-distilled water, and 10 μL of the spore suspension was taken for counting using an optical microscope, with each experiment repeated three times [42]. Furthermore, different fatty acids were used as the sole carbon source to observe differences in fatty acid metabolism in the mutant strains. To observe conidia, the wild-type and mutant strains were inoculated on PDA medium and cultured for three days, then transferred to water agar plates for 24 h and observed under an optical microscope (Olympus, Tokyo, Japan).

4.4. Microscopic Observation

To observe the mitochondria, the hyphal cells were stained with 10 μg/mL mitochondrial staining (Mito Tracker™ Red. Invitrogen, Thermo Fisher Scientific, Waltham, MA, USA), and images were captured using a confocal laser scanning microscope (Zeiss, Oberkochen, Germany). Traps produced on nutrient-rich medium were scanned using a cryo-scanning electron microscope. Mycelia cultured on CMY were collected and fixed with glutaraldehyde electron microscopy fixation solution, then sent to the Kunming Institute of Zoology, Chinese Academy of Sciences, for transmission electron microscopy imaging. The interfered nematodes were fixed with glutaraldehyde and imaged using Transmission Electron Microscope (TEM) (Zeiss, Oberkochen, Germany).

4.5. Difference in Trap Formation

Spore suspensions containing 2000–3000 spores of the wild-type and mutant strains, cultured for 7 days, were spread onto WA plates and grown for 30 h. Different compounds (ascr#7, urea, and ammonia) were added to induce trap formation. The number of traps was counted, and photographs were taken 48 h later. Similarly, 1000 *C. elegans* nematodes were added to induce trap formation in the wild-type and mutant strains, and the traps were photographed and counted at 6, 12, and 24 h.

4.6. Chemotaxis Assay

The mutant and WT strains were inoculated on CMY solid and liquid media and cultured for 7 days. *C. elegans* nematodes were synchronized and allowed to mature to adulthood beforehand. Blocks of mutant and wild-type cultures, along with their fermentation broth, were placed on opposite sides of 6 cm NGM agar plates (without peptone), with 10 μL drops of nematode suspension (approximately 200 nematodes) added in the center. The direction of nematode movement was observed, and the number of nematodes moving toward the experimental group and the control group was counted, respectively. Perform a behavioral assay on the interfered nematodes, as described previously. Primers for gene interference are listed in Table S2 [43].

4.7. Exosome Assay

The mutant and WT strains were inoculated into TG medium and cultured for 7 days. Subsequently, the culture supernatant was collected and filtered sequentially through
0.45 μm and 0.22 μm filters. The filtered supernatant was subjected to centrifugation at 4 °C for 30 min at 100,000× *g*; the supernatant was discarded, and the pellet was collected. After resuspension in PBS, exosomes derived from mutant and wild-type strains were separately applied to opposite sides of 6 cm NGM agar plates (without peptone), with 10 μL drops containing approximately 200 adult-stage nematodes each added to the center. Nematode movement and direction were observed and quantified. Furthermore, the isolated exosomes were sent to Yanzai Biotechnology (Shanghai, China) Co., Ltd. for imaging and particle size measurement.

4.8. Transcriptome and Metabolome

The mutant and wild-type strains were cultured in CMY liquid medium for 7 days. Mycelia and fermentation broth were collected separately, flash-frozen in liquid nitrogen, and sent to Shanghai Majorbio Bio-Pharm Technology Co., Ltd. (Majorbio, Shanghai, China). Gene Ontology (GO) function and Kyoto Encyclopedia of Genes and Genomes (KEGG) pathway enrichment were analyzed using the Majorbio cloud platform (https://www.majorbio.com. accessed on 19 February 2025).

5. Conclusions

This study highlights the complications of ETF and its dehydrogenase ETFdh in the pathogenesis and sporulation of NT fungus *A. oligospora*. Further, the disruptions of *Aoetf* and *Aoetfdh* can alter metabolic and energy levels, possibly leading to mitochondrial dysfunction and modifying the interaction between NT fungi and nematodes. Moving forward, future research should focus on the regulatory mechanisms by which metabolic alterations influence signal transduction and phenotypic regulation in the NTF group.

Supplementary Materials: The supporting information can be downloaded at: https://www.mdpi.com/article/10.3390/ijms252010934/s1.

Author Contributions: Conceptualization, X.W.; writing—original draft preparation, Y.L. and Z.L.; writing—review and editing, J.L. and X.Z.; supervision, X.W. All authors have read and agreed to the published version of the manuscript.

Funding: This research was funded by the Science and Technology Major Project from Yunnan Province (202201BC070004), the National Natural Science Foundation of China (32370138 and 32260029), the Applied Basic Research Foundation of Yunnan Province (202201AT070089), the special fund from the Yunnan University "double first-class" construction, and the Scientific Research Fund Project of Yunnan Provincial Department of Education (2024Y019).

Institutional Review Board Statement: Not applicable.

Informed Consent Statement: Not applicable.

Data Availability Statement: The original contributions presented in the study are included in the article and Supplementary Material, further inquiries can be directed to the corresponding author, Xin Wang.

Acknowledgments: We thank Ying-qi Guo (Kunming Institute of Zoology, Chinese Academy of Sciences) for their help with taking and analyzing the TEM images, and we would like to thank Zhi-jia Gu (Kunming Institute of Botany, Chinese Academy of Scienses) for making sample and taking images.

Conflicts of Interest: The authors declare no conflicts of interest.

References

1. Henriques, B.J.; Katrine Jentoft Olsen, R.; Gomes, C.M.; Bross, P. Electron transfer flavoprotein and its role in mitochondrial energy metabolism in health and disease. *Gene* **2021**, *776*, 145407. [CrossRef] [PubMed]
2. Lin, D.; Wan, M.; Fan, Y. Electron-transferring flavoprotein and its dehydrogenase contributed to growth development and virulence in *Beauveria bassiana*. *J. Invertebr. Pathol.* **2024**, *205*, 108141. [CrossRef] [PubMed]
3. Li, Y.; Zhu, J.; Hu, J.; Meng, X.; Zhang, Q.; Zhu, K.; Chen, X.; Chen, X.; Li, G.; Wang, Z.; et al. Functional characterization of electron-transferring flavoprotein and its dehydrogenase required for fungal development and plant infection by the rice blast fungus. *Sci. Rep.* **2016**, *6*, 24911. [CrossRef]
4. Toplak, M.; Brunner, J.; Tabib, C.R.; Macheroux, P. Closing the gap: Yeast electron-transferring flavoprotein links the oxidation of d-lactate and d-α-hydroxyglutarate to energy production via the respiratory chain. *FEBS J.* **2019**, *286*, 3611–3628. [CrossRef] [PubMed]
5. Wakitani, S.; Torisu, S.; Yoshino, T.; Hattanda, K.; Yamato, O.; Tasaki, R.; Fujita, H.; Nishino, K. Multiple Acyl-CoA Dehydrogenation Deficiency (Glutaric Aciduria Type II) with a Novel Mutation of Electron Transfer Flavoprotein-Dehydrogenase in a Cat. *JIMD Rep.* **2014**, *13*, 43–51.
6. Hasanuzzaman, M.; Bhuyan, M.; Parvin, K.; Bhuiyan, T.F.; Anee, T.I.; Nahar, K.; Hossen, M.S.; Zulfiqar, F.; Alam, M.M.; Fujita, M. Regulation of ROS Metabolism in Plants under Environmental Stress: A Review of Recent Experimental Evidence. *Int. J. Mol. Sci.* **2020**, *21*, 8695. [CrossRef]
7. Su, H.; Zhao, Y.; Zhou, J.; Feng, H.; Jiang, D.; Zhang, K.Q.; Yang, J. Trapping devices of nematode-trapping fungi: Formation, evolution, and genomic perspectives. *Biol. Rev. Camb. Philos. Soc.* **2017**, *92*, 357–368. [CrossRef]
8. Soliman, M.S.; El-Deriny, M.M.; Ibrahim, D.S.S.; Zakaria, H.; Ahmed, Y. Suppression of root-knot nematode Meloidogyne incognita on tomato plants using the nematode trapping fungus *Arthrobotrys oligospora* Fresenius. *J. Appl. Microbiol.* **2021**, *131*, 2402–2415. [CrossRef]
9. Pires, D.; Vicente, C.S.L.; Menéndez, E.; Faria, J.M.S.; Rusinque, L.; Camacho, M.J.; Inácio, M.L. The Fight against Plant-Parasitic Nematodes: Current Status of Bacterial and Fungal Biocontrol Agents. *Pathogens* **2022**, *11*, 1178. [CrossRef]
10. Wang, X.; Li, G.H.; Zou, C.G.; Ji, X.L.; Liu, T.; Zhao, P.J.; Liang, L.M.; Xu, J.P.; An, Z.Q.; Zheng, X.; et al. Bacteria can mobilize nematode-trapping fungi to kill nematodes. *Nat. Commun.* **2014**, *5*, 5776. [CrossRef]
11. Zhang, F.; Yang, Y.Q.; Zhou, F.P.; Xiao, W.; Boonmee, S.; Yang, X.Y. Morphological and Phylogenetic Characterization of Five Novel Nematode-Trapping Fungi (Orbiliomycetes) from Yunnan, China. *J. Fungi* **2023**, *9*, 735. [CrossRef] [PubMed]
12. Zhang, D.; Zhu, X.; Sun, F.; Zhang, K.; Niu, S.; Huang, X. The roles of actin cytoskeleton and actin-associated protein Crn1p in trap formation of *Arthrobotrys oligospora*. *Res. Microbiol.* **2017**, *168*, 655–663. [CrossRef] [PubMed]
13. Wang, B.L.; Chen, Y.H.; He, J.N.; Xue, H.X.; Yan, N.; Zeng, Z.J.; Bennett, J.W.; Zhang, K.Q.; Niu, X.M. Integrated Metabolomics and Morphogenesis Reveal Volatile Signaling of the Nematode-Trapping Fungus *Arthrobotrys oligospora*. *Appl. Environ. Microbiol.* **2018**, *84*, e02749. [CrossRef] [PubMed]
14. Yu, X.; Hu, X.; Pop, M.; Wernet, N.; Kirschhöfer, F.; Brenner-Weiß, G.; Keller, J.; Bunzel, M.; Fischer, R. Fatal attraction of *Caenorhabditis elegans* to predatory fungi through 6-methyl-salicylic acid. *Nat. Commun.* **2021**, *12*, 5462. [CrossRef] [PubMed]
15. Youssar, L.; Wernet, V.; Hensel, N.; Yu, X.; Hildebrand, H.G.; Schreckenberger, B.; Kriegler, M.; Hetzer, B.; Frankino, P.; Dillin, A.; et al. Intercellular communication is required for trap formation in the nematode-trapping fungus *Duddingtonia flagrans*. *PLoS Genet.* **2019**, *15*, e1008029. [CrossRef]
16. Jiang, X.; Xiang, M.; Liu, X. Nematode-Trapping Fungi. *Microbiol. Spectr.* **2017**, *5*, 963–974. [CrossRef]
17. Heintz, C.E.; Pramer, D. Ultrastructure of nematode-trapping fungi. *J. Bacteriol.* **1972**, *110*, 1163–1170. [CrossRef]
18. Veenhuis, M.; Harder, W.; Nordbring-Hertz, B. Occurrence and metabolic significance of microbodies in trophic hyphae of the nematophagous fungus *Arthrobotrys oligospora*. *Antonie Leeuwenhoek.* **1989**, *56*, 241–249. [CrossRef]
19. Talbot, N.J.; Foster, A.J. Genetics and genomics of the rice blast fungus *Magnaporthe grisea*: Developing an experimental model for understanding fungal diseases of cereals. *Adv. Bot. Res.* **2001**, *34*, 263–287.
20. Watmough, N.J.; Frerman, F.E. The electron transfer flavoprotein: Ubiquinone oxidoreductases. *Biochim. Biophys. Acta* **2010**, *1797*, 1910–1916. [CrossRef]
21. Wang, Y.; Palmfeldt, J.; Gregersen, N.; Makhov, A.M.; Conway, J.F.; Wang, M.; McCalley, S.P.; Basu, S.; Alharbi, H.; St Croix, C.; et al. Mitochondrial fatty acid oxidation and the electron transport chain comprise a multifunctional mitochondrial protein complex. *J. Biol. Chem.* **2019**, *294*, 12380–12391. [CrossRef] [PubMed]

22. Chokchaiwong, S.; Kuo, Y.T.; Hsu, S.P.; Hsu, Y.C.; Lin, S.H.; Zhong, W.B.; Lin, Y.F.; Kao, S.H. ETF-QO Mutants Uncoupled Fatty Acid β-Oxidation and Mitochondrial Bioenergetics Leading to Lipid Pathology. *Cells* **2019**, *8*, 106. [CrossRef] [PubMed]
23. Salerno, K.M.; Domenico, J.; Le, N.Q.; Stiles, C.D.; Solov'yov, I.A.; Martino, C.F. Long-Time Oxygen Localization in Electron Transfer Flavoprotein. *J. Chem. Inf. Model* **2022**, *62*, 4191–4199. [CrossRef] [PubMed]
24. Herrero Martín, J.C.; Salegi Ansa, B.; Álvarez-Rivera, G.; Domínguez-Zorita, S.; Rodríguez-Pombo, P.; Pérez, B.; Calvo, E.; Paradela, A.; Miguez, D.G.; Cifuentes, A.; et al. An ETFDH-driven metabolon supports OXPHOS efficiency in skeletal muscle by regulating coenzyme Q homeostasis. *Nat. Metab.* **2024**, *6*, 209–225. [CrossRef]
25. Liu, Q.; Bai, N.; Duan, S.; Shen, Y.; Zhu, L.; Yang, J. Characterizing the Role of AosfgA and AofluG in Mycelial and Conidial Development in *Arthrobotrys oligospora* and Their Role in Secondary Metabolism. *Microorganisms* **2024**, *12*, 615. [CrossRef]
26. Cheng, Y.Y.; Tang, Y.; Liu, A.J.; Wei, L.; Lin, L.; Zhang, J.; Zhi, L. Clinical features and ETFDH mutations of children with late-onset glutaric aciduria type II: A report of two cases. *Zhongguo Dang Dai Er Ke Za Zhi = Chin. J. Contemp. Pediatr.* **2017**, *19*, 975–978.
27. Stotz, H.U.; Brotherton, D.; Inal, J. Communication is key: Extracellular vesicles as mediators of infection and defence during host-microbe interactions in animals and plants. *FEMS Microbiol. Rev.* **2022**, *46*, fuab044. [CrossRef]
28. Macchione, F.; Salviati, L.; Bordugo, A.; Vincenzi, M.; Camilot, M.; Teofoli, F.; Pancheri, E.; Zordan, R.; Bertolin, C.; Rossi, S.; et al. Multiple acyl-COA dehydrogenase deficiency in elderly carriers. *J. Neurol.* **2020**, *267*, 1414–1419. [CrossRef]
29. Brandao, S.R.; Ferreira, R.; Rocha, H. Exploring the contribution of mitochondrial dynamics to multiple acyl-CoA dehydrogenase deficiency-related phenotype. *Arch. Physiol. Biochem.* **2021**, *127*, 210–216. [CrossRef]
30. Angelini, C.; Burlina, A.; Blau, N.; Ferreira, C.R. Clinical and biochemical footprints of inherited metabolic disorders: X. Metabolic myopathies. *Mol. Genet. Metab.* **2022**, *137*, 213–222. [CrossRef]
31. Szrok-Jurga, S.; Czumaj, A.; Turyn, J.; Hebanowska, A.; Swierczynski, J.; Sledzinski, T.; Stelmanska, E. The Physiological and Pathological Role of Acyl-CoA Oxidation. *Int. J. Mol. Sci.* **2023**, *24*, 14857. [CrossRef] [PubMed]
32. Xiao, C.; Astiazaran-Symonds, E.; Basu, S.; Kisling, M.; Scaglia, F.; Chapman, K.A.; Wang, Y.; Vockley, J.; Ferreira, C.R. Mitochondrial energetic impairment in a patient with late-onset glutaric acidemia Type 2. *Am. J. Med. Genet. A* **2020**, *182*, 2426–2431. [CrossRef] [PubMed]
33. Yuan, Q.; Zeng, Z.L.; Yang, S.; Li, A.; Zu, X.; Liu, J. Mitochondrial Stress in Metabolic Inflammation: Modest Benefits and Full Losses. *Oxid. Med. Cell. Longev.* **2022**, *2022*, 8803404. [CrossRef]
34. Picard, M.; McEwen, B.S.; Epel, E.S.; Sandi, C. An energetic view of stress: Focus on mitochondria. *Front. Neuroendocrinol.* **2018**, *49*, 72–85. [CrossRef]
35. Zhu, M.C.; Li, X.M.; Zhao, N.; Yang, L.; Zhang, K.Q.; Yang, J.K. Regulatory Mechanism of Trap Formation in the Nematode-Trapping Fungi. *J. Fungi* **2022**, *8*, 406. [CrossRef]
36. Angelini, C.; Tavian, D.; Missaglia, S. Heterogeneous Phenotypes in Lipid Storage Myopathy Due to ETFDH Gene Mutations. *JIMD Rep.* **2018**, *38*, 33–40. [PubMed]
37. Chen, S.A.; Lin, H.C.; Schroeder, F.C.; Hsueh, Y.P. Prey sensing and response in a nematode-trapping fungus is governed by the MAPK pheromone response pathway. *Genetics* **2021**, *217*, iyaa008. [CrossRef] [PubMed]
38. Huang, J.; Zheng, X.; Tian, M.; Zhang, K. Ammonia and Nematode Ascaroside Are Synergistic in Trap Formation in *Arthrobotrys oligospora*. *Pathogens* **2023**, *12*, 1114. [CrossRef]
39. Hu, X.; Hoffmann, D.S.; Wang, M.; Schuhmacher, L.; Stroe, M.C.; Schreckenberger, B.; Elstner, M.; Fischer, R. GprC of the nematode-trapping fungus *Arthrobotrys flagrans* activates mitochondria and reprograms fungal cells for nematode hunting. *Nat. Microbiol.* **2024**, *9*, 1752–1763. [CrossRef]
40. Kuo, C.Y.; Tay, R.J.; Lin, H.C.; Juan, S.C.; Vidal-Diez de Ulzurrun, G.; Chang, Y.C.; Hoki, J.; Schroeder, F.C.; Hsueh, Y.P. The nematode-trapping fungus *Arthrobotrys oligospora* detects prey pheromones via G protein-coupled receptors. *Nat. Microbiol.* **2024**, *9*, 1738–1751. [CrossRef]
41. Xie, M.; Yang, J.; Jiang, K.; Bai, N.; Zhu, M.; Zhu, Y.; Zhang, K.Q.; Yang, J. AoBck1 and AoMkk1 Are Necessary to Maintain Cell Wall Integrity, Vegetative Growth, Conidiation, Stress Resistance, and Pathogenicity in the Nematode-Trapping Fungus *Arthrobotrys oligospora*. *Front. Microbiol.* **2021**, *12*, 649582. [CrossRef] [PubMed]
42. Zhou, L.; Li, M.; Cui, P.; Tian, M.; Xu, Y.; Zheng, X.; Zhang, K.; Li, G.; Wang, X. Arrestin-Coding Genes Regulate Endocytosis, Sporulation, Pathogenicity, and Stress Resistance in *Arthrobotrys oligospora*. *Front. Cell Infect. Microbiol.* **2022**, *12*, 754333. [CrossRef] [PubMed]
43. Conte, D., Jr.; MacNeil, L.T.; Walhout, A.J.M.; Mello, C.C. RNA Interference in *Caenorhabditis elegans*. *Curr. Protoc. Mol. Biol.* **2015**, *109*, 26.23.21–26.23.30. [CrossRef] [PubMed]

International Journal of
Molecular Sciences

MDPI

Article

Enhancing the Growth of Chili Plants and Soil Health: Synergistic Effects of Coconut Shell Biochar and *Bacillus* sp. Strain Ya-1 on Rhizosphere Microecology and Plant Metabolism

Shimeng Tan [1], Bao Wang [1,2], Qian Yun [1], Wanrong Yan [1], Tongbin Xiao [1] and Zhixiang Zhao [1,*]

[1] Key Laboratory of Plant Disease and Pest Control of Hainan Province, Institute of Plant Protection (Research Center of Quality Safety and Standards for Agricultural Products of Hainan Academy of Agricultural Sciences), Hainan Academy of Agricultural Sciences, Haikou 571100, China; tshmeng@126.com (S.T.); wangbao@hnaas.org.cn (B.W.); yanwanrong818@163.com (W.Y.)

[2] Key Laboratory of Green Prevention and Control of Tropical Plant Diseases and Pests Ministry of Education, Hainan University, Haikou 571100, China

* Correspondence: zhaozhixiang@hnaas.org.cn

Abstract: To mitigate soil degradation and decrease dependency on chemical inputs in agriculture, this study examined the joint effects of coconut shell biochar and *Bacillus* strain Ya-1 on soil fertility, rhizosphere bacterial communities, and the growth of chili (*Capsicum annuum* L.). A controlled pot experiment with four treatments was conducted: control (CK), biochar only (C), *Bacillus* strain Ya-1 only (B), and a combination of both (BC). The BC treatment significantly enhanced the soil carbon and available phosphorus contents by approximately 20% and the soil nitrogen content and pH by 18% and 0.3 units, respectively, compared to the control. It also increased microbial biomass carbon and nitrogen by 25% and 30%, respectively, indicating improved soil microbial diversity as shown by the highest Pielou evenness index and Shannon index values. The combined application of biochar and the Ya-1 strain resulted in a 15% increase in chili plant height and a 40% improvement in root dehydrogenase activity, suggesting enhanced nutrient uptake and metabolism. Metabolic profiling showed shifts in stress response and nutrient assimilation under different treatments. Collectively, these results indicate the potential of biochar and microbial inoculants to significantly promote soil and plant health, providing a sustainable strategy to improve agricultural productivity and reduce reliance on chemical inputs.

Keywords: biochar application; *Bacillus* inoculation; rhizosphere microbiology; soil enhancement; *Capsicum annuum*

1. Introduction

Chili (*Capsicum annuum* L.) is a globally important vegetable crop in the agriculture sector, both economically and as a cooking staple, particularly in Hainan, China, which is a major supplier of off–season vegetables. The warm and humid climate dominating this region, however, increases the prevalence of some soil–borne pathogens, such as *Meloidogyne enterolobii* [1], *Fusarium oxysporum* [2], and *Pseudomonas solanacearum* [3], by extending their lifecycle and accelerating their activity, which consequently pose a severe threat to the production of peppers, significantly reducing their yield and quality.

The pressing environmental and health issues associated with the extensive use of chemical pesticides call for the urgent need to adopt sustainable agricultural practices. In this context, there is growing popularity in the application of biological control agents (BCAs), such as *Pseudomonas fluorescens* and *Bacillus subtilis* [4], as well as soil amendments like biochar [5]. BCAs protect plants against pathogens through mechanisms such as triggering antibacterial activity, inducing systemic resistance [6], and hyperparasitism [7], offering an eco-friendly alternative to synthetic chemicals. Meanwhile, the mediation of the interactions between microbes in the rhizosphere environment and plants by a variety of secondary metabolites and signaling molecules has been suggested by some studies that have investigated the effects of rhizosphere microecology on plant immunity. Song et al. [8] found that the generation of ROS mediated by FER can regulate the levels of plant-beneficial *Pseudomonas* spp. in the Arabidopsis rhizosphere microbiome. The *Pseudomonas*-specific induction of the expression of PSKR1 in Arabidopsis roots further demonstrated the role of salicylic acid in regulating the colonization ability and growth of *Pseudomonas*, indicating the ways through which the plant's immunity is boosted in coordination with positive functions of the microbiome.

With unique properties such as high porosity, sufficiently large surface area, and nutrient retention capacity, biochar is renowned for its ability to improve soil structure, which supports the growth of beneficial microbes and enhances plant health and productivity [9]. Beyond its ample carbon and elemental contributions, biochar fortifies aggregation in finer-textured soils, ameliorating the capillary structure within compacted soils, and, consequently, bestows significant functions related to this structure, including water retention and the enhancement of microbial communities [10]. In this context, the synergistic application of biochar and BCAs has been observed to amplify the enhancement effects on farmland soil health and plant disease resistance, surpassing the benefits of biochar alone [11]. A readily observable rationale is the substantial retention of active functional groups, such as hydroxyl, carboxyl, sulfonic, and amino groups, on the biochar surface, which are highly effective for microbial cell adhesion and proliferation, facilitating biofilm formation and providing an excellent habitat and refuge for microbes [12]. On the other hand, research into rhizosphere microbial communities has revealed that the addition of biochar can enrich the basic BCA community, encompassing Pseudomonas, *Bacillus*, and Streptomyces, while also supporting the active manifestation of rhizosphere metabolic enzymes [13]. By stabilizing and supporting the functionality of microbial communities, biochar, when used as a carrier for microbial inoculants, has been demonstrated to enhance the persistence, survival, and colonization of inoculated microbes in soil and on plant roots, which is clearly advantageous in overcoming the disadvantages of low functional efficiency and short utility periods of microbial inoculants. The combined use of biochar and microbial inoculants has proven to be an effective strategy for controlling a variety of plant pathogens, highlighting its potential contribution to integrated disease management in agriculture [14].

In this study, we conducted a comparative analysis between the sole application and the combined application of coconut shell biochar and *Bacillus* strain Ya-1 in pot experiments, integrating the analysis of rhizosphere microbial diversity and plant metabolism to explore potential synergistic interactions and, more importantly, to analyze the impact of these measures on plant metabolism. The pot experiment design, conducted under controlled conditions, allows for the precise manipulation and monitoring of both environmental variables and treatment effects. Through this design, we aim to elucidate the precise effects of these combined application measures on plant immunity before further pursuing the microbial agents carried by biochar. This experimental setup provides profound insights into the interactions established between biochar, *Bacillus* strain Ya-1, and the plant, laying the groundwork for the potential future co-application of biochar and microorganisms under field conditions.

2. Results

2.1. Combined Effects of Biochar and Strain Ya-1 on the Soil and Rhizosphere Bacterial Communities

2.1.1. Soil Physicochemical Properties

Figure 1 shows the results of a comparative analysis of the soil's physicochemical properties in different treatments. The control group (CK) was used to establish a baseline for all evaluated parameters. Significant increases in the carbon (C) and available phosphorus (AP) contents were recorded with biochar (C) addition, while other parameters, such as the nitrogen (N) content and microbial biomass carbon (C), moderately changed. The sole application of strain Ya-1 (B), however, notably increased the microbial biomass N and P contents, indicating the enhancement of microbial activity, and also significantly improved the water content. The biochar treatment in combination with Ya-1 (BC) resulted in the maximum contents of total P and C as well as available P and potassium (K). Moreover, compared to the treatment of plants with only Ya-1 (B), a noticeable increase was shown in microbial C and P and water content with BC, which indicates the synergistic effect of the combined use of biochar and Ya-1.

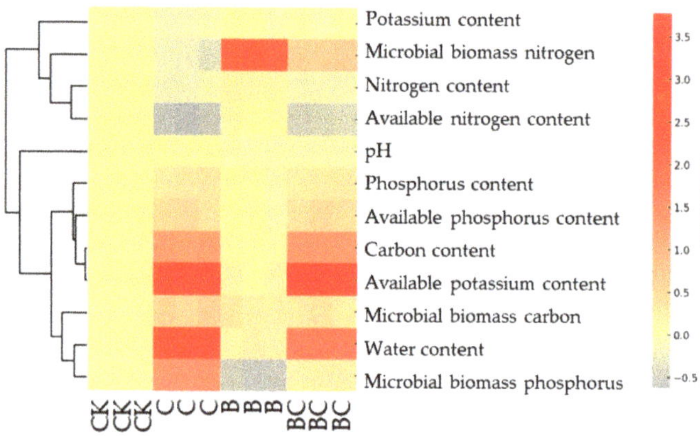

Figure 1. Heatmap analysis of the relationships between soil physicochemical properties and microbial biomass in different treatments.

The heatmap shows the effects of different treatments, including the control (CK), biochar addition alone (C), Ya-1 addition alone (B), and biochar combined with Ya-1 (BC), on soil physicochemical properties. The parameters evaluated included the potassium (K), nitrogen (N), phosphorus (P), and carbon (C) contents; microbial biomass N; available K, N, and P contents; microbial biomass C and P; pH; and water content. The data were normalized and color coded, with a yellow color indicating higher contents, and purple indicating lower contents.

2.1.2. The Rhizosphere Bacterial Diversity of Chili Plants

The boxplots in Figure 2 show the effects of different treatments on bacterial diversity indices. In the BC treatment, significantly higher values of Pielou's evenness index compared to those of B and CK were obtained, which demonstrated a more even distribution of the bacterial community. The intermediate evenness of the bacterial community was observed for the C treatment, which was on par with that for BC. In terms of species richness, the plants exposed to both BC and C treatments exhibited significantly higher values than those under B and CK, and plants grown under the BC treatment achieved the highest richness, suggesting that biochar could promote bacterial diversity. Similarly, the BC group, in tandem with the C group, exhibited the highest Shannon diversity index, indicating that biochar induced a comprehensive enhancement in bacterial diversity. It is important to highlight that the integration of Ya-1 with biochar did not create a significant divergence in diversity indices compared to the biochar-only treatment. Furthermore, the introduction of Ya-1 did not upset the community equilibrium, even in large quantities, suggesting that biochar plays a role in the stabilization of microbial communities.

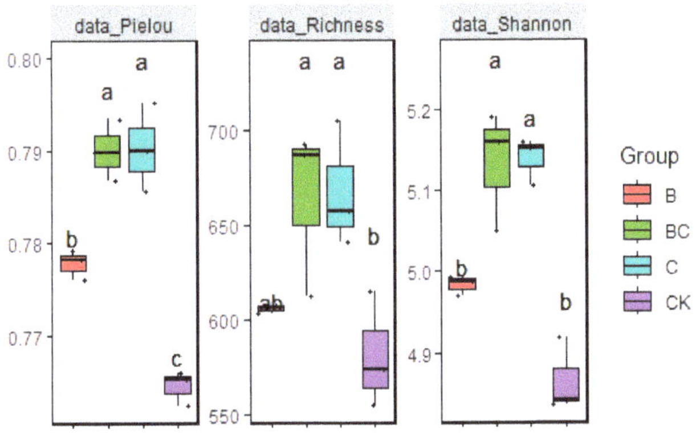

Figure 2. Boxplots of bacterial diversity indices in the rhizosphere soil. The boxplots represent the effects of different treatments, including the control (CK), biochar alone (C), Ya-1 alone (B), and biochar combined with Ya-1 (BC), on bacterial diversity indices (Pielou's evenness, species richness, and Shannon diversity index) in the rhizosphere soil. Different letters above the boxes denote significant differences ($p < 0.05$) between treatment groups, $^{+}$ $p \leq 0.05$.

2.2. Combined Effects of Biochar and Strain Ya-1 on the Growth of Chili Plants

The effects of biochar and the strain Ya-1 on the growth of chili plants were evaluated by measuring the total height of the plant, including the root and stem lengths, in different treatments, as shown in Figures 3 and 4. Figure 3E demonstrates that the control group (CK) plants exhibited a mean total height of 375 mm, with average values of 150 mm and 225 mm for the roots and stems, respectively. When applied alone, biochar (C) caused a significant increase in the total plant height, averaging 475 mm, with the roots and stems having lengths of 175 mm and 300 mm, respectively. In contrast, a total plant height similar to that of the control group was shown when Ya-1 (B) was applied alone, suggesting that plant growth was negligibly influenced by the bacterial strain. However, the co-application of biochar and Ya-1 (BC) exerted the most pronounced effect, with average values of over 230 mm and 215 mm for the stem length and root length of the longest plant, respectively, which points out their synergistic influence on plant growth.

The leaf surface area, an important determinant of photosynthetic capacity in plants, displayed variations among different treatment groups. A median leaf surface area of approximately 1200 mm^2 was achieved for the control plants (CK), within the range of 1000 mm^2 to 1600 mm^2 (Figure 3F). The biochar treatment (C) caused a reduction of about 1000 mm^2 in the median leaf surface area, with a narrower range of 800 mm^2 to 1200 mm^2. Conversely, an increase in the median leaf area to approximately 1400 mm^2, within a broader range of 800 mm^2–1600 mm^2, was seen with the sole application of Ya-1 (B). The leaf surface area underwent the highest variation, with a median of 1200 mm^2 and a range of 800 mm^2–1600 mm^2, as a result of the co-treatment of biochar and Ya-1 (BC).

The dehydrogenase activity of plant roots, which is a biological indicator of root health, was assessed in all treatment groups, as presented in Figure 3G. The plant roots in the control group (CK) displayed an average dehydrogenase activity of 0.045 mg/(g·h). A slight enhancement of this activity (an average of 0.048 mg/(g·h)) was recorded for the biochar treatment (C), while the addition of Ya-1 (B) slightly decreased the dehydrogenase activity, with a mean value of 0.042 mg/(g·h). Most importantly, when biochar and Ya-1 (BC) were applied in combination, significant increases in the dehydrogenase activity of the roots, with an average of 0.065 mg/(g·h), resulted, indicating that together they could promote the metabolic activity in plant roots.

Figure 3. Plant growth pot illustrations and parameters in different treatments. (**A**) Full plant of control (CK); (**B**) full plant of treatment biochar application alone (**C**); (**C**) full plant of Ya-1 application alone (**B**); (**D**) full plant of biochar co-applied with Ya-1 (BC); (**E**) bar graph showing the total height of chili plants, including the root and stem lengths, in different treatments, including the control (CK), biochar application alone (**C**), Ya-1 application alone (**B**), and biochar co-applied with Ya-1 (BC); error bars represent the standard error of the mean. (**F**) The boxplot of the leaf surface area of chili plants subjected to different treatments; in the boxplot, the box shows the interquartile range (IQR) of the data, while the line inside the box represents the median values, and whiskers indicate the range; dots denote the outliers. (**G**) Bar graph for the dehydrogenase activity of the roots of chili plants grown under different treatments; error bars represent the standard error of the mean ($p < 0.05$). (**a**) Underground part of control (CK); (**b**) underground part of biochar application alone (**C**); (**c**) underground part of Ya-1 application alone (**B**); (**d**) underground part of biochar co-applied with Ya-1 (BC).

The results revealed a significant improvement in the growth parameters of chili plants with the combined application of biochar and Ya-1. In the BC treatment, not only was the maximum plant height achieved but also the metabolic activity of roots was enhanced. This indicated that it was a particularly effective treatment and was manifested by the elevated dehydrogenase activity of roots. These findings suggest that the combined use of biochar and Ya-1 could synergistically promote the physical and metabolic development of chili plants, potentially leading to a boost in plant health and productivity.

2.3. Effects of the Co-Application of Biochar and Ya-1 on Metabolic Changes in Chili Plants

The comparative metabolomic analysis of chili plants in different treatments, including biochar (C), *Bacillus* strain (B), and biochar–*Bacillus* combination (BC), revealed significant metabolic shifts. Figure 4 represents the key findings from the comparative metabolomic analysis.

Figure 4A presents the volcano plot of differential metabolites between the C and BC treatments, which exhibited a stark difference, and 7282 metabolites were upregulated, while 2816 metabolites were downregulated in the BC treatment compared to those in C. Metabolites with a p-value less than 0.05 were marked as statistically significant, and red and blue represent the upregulated and downregulated metabolites, respectively. Furthermore, metabolites with high variable importance in the projection (VIP) values were prominently highlighted. Similarly, the volcano plot illustrating the differential metabolites between the B and BC treatments, with 4424 upregulated metabolites but 6384 downregulated metabolites in the treatment BC compared to those in B, is shown in Figure 4B.

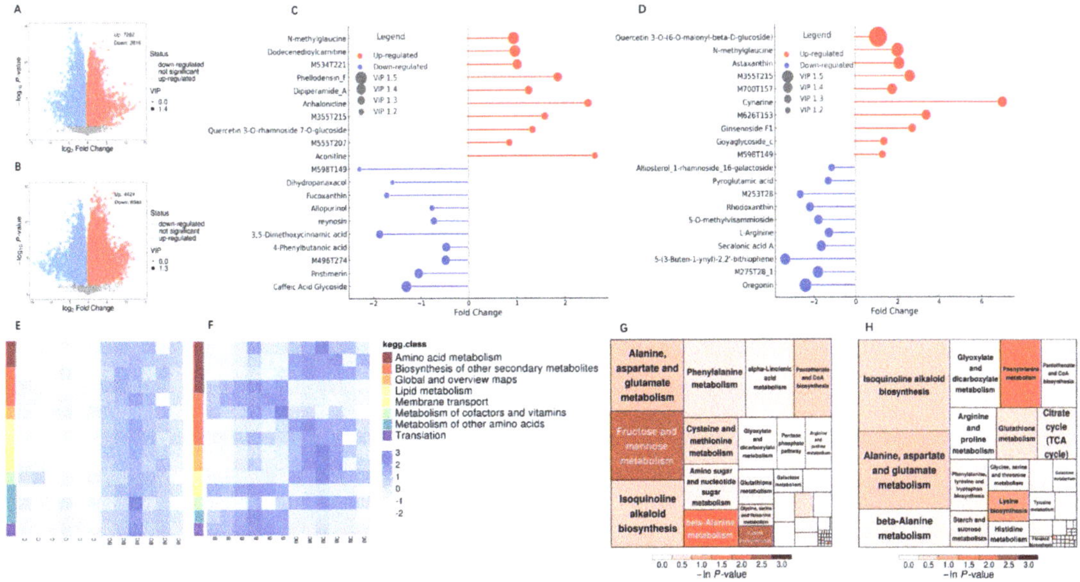

Figure 4. Metabolic profiling and KEGG pathway enrichment analysis of chili plants grown under different treatments. (**A**) The volcano plot of differential metabolites between the treatment C and the treatment BC; (**B**) the volcano plot of differential metabolites between the B and BC treatments; (**C**) the matchstick diagram showing the top 10 differential metabolites between the C and BC treatments; (**D**) matchstick plot of the top 10 differential metabolites between the B and BC treatments; (**E**) heatmap of the KEGG pathway enrichment analysis of differential metabolites between the C and BC treatments; (**F**) heatmap of differential metabolites between the B and BC treatments annotated by KEGG; (**G**) treemap showing the KEGG pathway enrichment analysis of differential metabolites between the treatment C and the treatment BC; (**H**) treemap of differential metabolites between the B and BC treatments as detected by the KEGG pathway enrichment analysis.

The matchstick diagrams represent the top differential metabolites. The top 10 differential metabolites between the C and BC treatments are shown in Figure 4C, with the significant upregulation of N-methylglycine and dodecanoylcarnitine, whereas pristimerin and caffeic acid glycoside exhibited a notable downregulation. The length of the lines in the plot corresponds to the magnitude of fold change, and the red color indicates upregulation, while blue denotes downregulation. Likewise, Figure 4D demonstrates the key upregulated differential metabolites identified, such as quercetin 3-O-(6-O-malonyl-β-D-glucoside) and N-methylglycine, and downregulated metabolites, including ginsenoside F1 and allopurinol, between the B and BC treatments.

The pathway enrichment analysis further revealed the induction of metabolic changes by different treatments. As seen in Figure 4E, a heatmap of the KEGG pathway enrichment analysis of differential metabolites between the C and BC treatment groups represents significantly enriched pathways associated with amino acid metabolism and the KEGG pathway of biosynthesis of various other secondary metabolites. In the heatmap, the intensity of the color reflects the degree of enrichment, with a darker shade indicating a higher level of significance. Similarly, Figure 4F illustrates the enrichment analysis of KEGG pathways related to differential metabolites identified between the *Bacillus* treatment (B) and the Biochar–*Bacillus* treatment (BC). This heatmap depicts several significantly enriched metabolic pathways, particularly "amino sugar and nucleotide sugar metabolism" and "isoquinoline alkaloid biosynthesis".

Treemaps provide a visual representation of statistically significant pathways, highlighting the relative contribution of these pathways to the overall metabolic change in plants. In the treemap, the size of each rectangle corresponds to the significant enrichment of the KEGG pathway, with larger rectangles indicating greater impacts, and larger and darker squares denoting more significantly enriched pathways. Figure 4G shows the tree map, which offers the representation of the KEGG pathway enrichment analysis of differential metabolites between the treatments C and BC, emphasizing the relative importance of the two pathways of alanine, aspartate, and glutamate metabolism and isoquinoline alkaloid biosynthesis. Similarly, a tree map of the KEGG pathway enrichment analysis of differential metabolites between the B and BC treatments is presented in Figure 4H. Notably, the largest and darkest rectangles in the treemap represent pathways such as "amino sugar and nucleotide sugar metabolism" and "isoquinoline alkaloid biosynthesis", underscoring their importance to the induction of metabolic shifts by the BC treatment.

Substantial metabolic changes were observed in chili plants when biochar was integrated with *Bacillus* (BC) compared to those in the individual treatments of biochar (C) and *Bacillus* (B). The BC treatment triggered significant alterations in key metabolites and metabolic pathways, suggesting the synergistic effect of the combined application of biochar and *Bacillus* on the metabolism of chili plants, which could be a promising strategy to improve metabolic processes, potentially promoting plant growth and stress resilience.

2.4. Integrative Analysis of the Effects of the Co-Application of Biochar and Bacillus

The integrative analysis investigates the biochemical and microbial processes taking place in chili plants treated with biochar combined with *Bacillus* (BC) compared to those under the sole applications of C and B. Pearson correlation heatmaps (Figure 5) revealed how plant metabolites and soil microbial diversity were influenced by the BC treatment, highlighting the synergistic effects of biochar and *Bacillus* on plant growth and soil health. Some plant growth-promoting metabolites, such as "3,5-Dimethoxycinnamic acid" in Group B, exhibited significant positive correlations with key growth parameters like the root length and leaf area, which the biochar–*Bacillus* combination treatment effectively mitigated. Similarly, in Group C, the treatment notably potentiated the beneficial effects of "4-Phenylbutanoic Acid" on these traits. This suggests that the combined application of biochar and bacterial strains not only enhances the production of specific beneficial metabolites but also has the potential to improve overall plant growth vigor and mitigate some of the negative impacts of certain metabolites on plant growth. Additionally, the correlations between the Shannon index and metabolites such as "pyroglutamic acid" and "5-(3-Buten-1-ynyl)-2,2′-bithiophene" indicated an increase in the soil bacterial diversity by the BC treatment, implying that biochar and *Bacillus* could together improve the soil environment, which supports better plant growth.

The heatmap (Figure 6) shows the coefficients of the Pearson correlation between differential metabolites and growth parameters for the treatment groups C versus BC (left) and B versus BC (right). The intensity of the color of each cell reflects the magnitude of the correlation coefficient, with blue indicating a negative correlation, red denoting a positive correlation, and white representing the non-significant correlation. The absolute value of the coefficient close to 1 indicates a strong linear relationship between the variables. Correlations between the top 20 metabolites and the leaf area, root vitality (indicated by dehydrogenase activity), and morphological traits such as root and stem lengths, were observed.

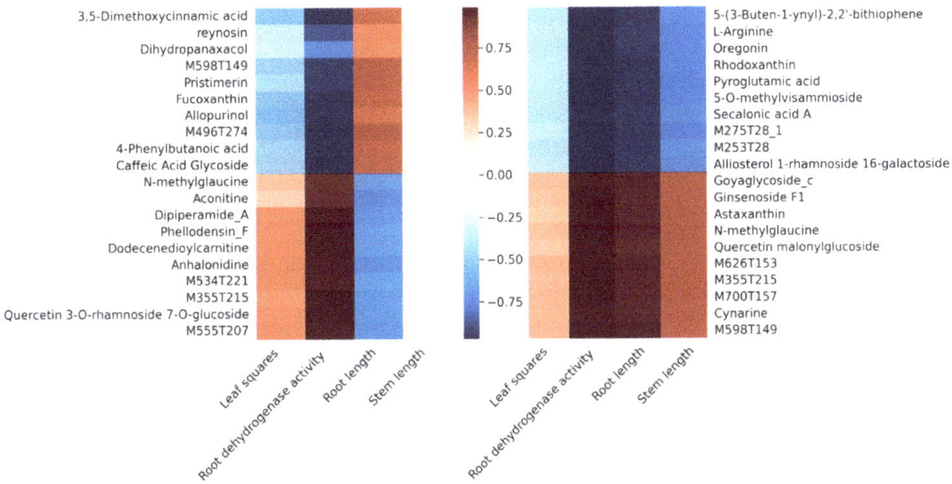

Figure 5. The heatmap of the Pearson correlation between differential metabolites and phenotypic parameters for growth.

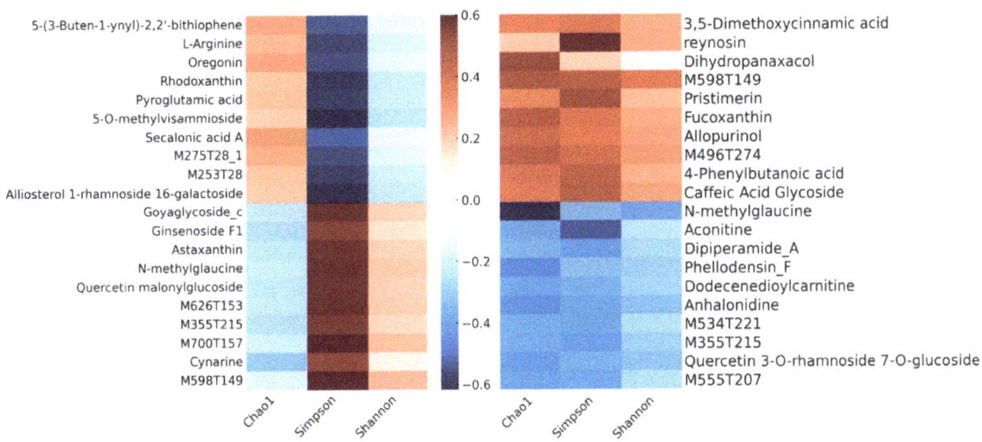

Figure 6. The heatmap of the Pearson correlation between differential metabolites and statistical parameters for the biodiversity indices of soil bacterial community.

The heatmap illustrates the coefficients of the Pearson correlation between differential metabolites in chili plants and soil bacterial diversity indices for two sets of treatment comparisons: C versus BC (left) and B versus BC (right). The saturation of the color of each cell represents the magnitude of the correlation coefficient, with the blue color indicating negative correlations, the red color denoting positive correlations, and white signifying non-significant correlations. A high absolute value of the Pearson correlation coefficient suggests a strong linear relationship between the variables. The heatmap shows the correlations between the top 20 most variable metabolites and soil bacterial diversity indices.

3. Discussion

3.1. Impact of Biochar on Soil Properties

The targeted soil analysis revealed that the application of biochar alone resulted in a significant increase in soil carbon and available phosphorus, a finding of paramount importance for the enhancement of nutrient availability for plant growth. A longitudinal study has underscored the role of soil carbon accumulation in elevating the rhizosphere microbial community, as elucidated in the work of Tian et al. [15], which highlights the microbially mediated mechanisms underlying the soil carbon accrual fostered by conservation agriculture under prolonged warming. The significance of phosphorus in plant development is manifold, with the dynamics of available phosphorus in the soil being intricately linked to factors such as the soil pH, aggregate structure, and elemental composition, as reviewed by Jindo et al. [16].

Our detection results indicate that the direct replenishment of the soil's elemental components through the addition of coconut shell biochar notably enhanced the availability of plant nutrients. There was a slight decrease in the pH value, which, while not statistically significant, hinted at the positive impact of biochar on the soil's physical structure. This partly explains the observed increase in available phosphorus in the soil treated with biochar. The correlation between the rise in the soil carbon content and available phosphorus and the level of plant growth in this study aligns with previous research, such as [17], which demonstrates the positive influence of biochar on soil carbon and phosphorus levels. Furthermore, studies on soil amendment for crops [18] also acknowledge the fertilizing effects of biochar when applied independently. Although these studies do not directly link soil nutrient replenishment to microbial communities, they support our observations and lay the groundwork for understanding the subsequent changes in microbial communities and plant metabolic differences within the soil.

Research also indicates that the impact of biochar application on soil carbon sequestration and plant nutrition is subject to structural and functional changes over time, an issue known as biochar aging, as analyzed by Meng et al. [19]. However, for short-cycle vegetable cultivation, such as chili pepper cultivation, the issue of biochar aging is largely circumvented.

3.2. *Impact of Bacillus Strain Ya-1 on Microbial Activity*

The standalone application of *Bacillus* strain Ya-1 was found to increase soil microbial biomass nitrogen (N) and phosphorus (P), highlighting the strain's role in enhancing microbial activity. This activity is essential for nutrient cycling and for suppressing pathogenic microorganisms through competitive exclusion and the production of antimicrobial compounds. Our findings echo those in [20,21], which also emphasized the strain's role in activating plant roots and promoting the recruitment of beneficial microbial populations. Furthermore, [22] discussed the mechanism by which antagonistic bacteria can enhance plants' resistance to soil-borne pathogens by activating plant immune pathways through root colonization.

3.3. *Synergistic Effects of Biochar and Bacillus Strain Ya-1*

When biochar and *Bacillus* strain Ya-1 were applied together, a synergistic enhancement of soil nitrogen content, pH, and microbial biomass carbon was observed. This suggests an interaction between the physical properties of biochar and the biological functions of *Bacillus* Ya-1. Study [23] has shown that the microstructure of biochar can positively influence microbial habitation, particularly in the formation of biofilms by plant growth-promoting rhizobacteria (PGPR). Infrared spectroscopy and community diversity data from [24] further demonstrated the recruitment of beneficial microorganisms by different surface functional groups on biochar. The provision of a conducive habitat by biochar likely enhances the effectiveness of rhizosphere-dwelling strains like *Bacillus* Ya-1 [25]. The integration of biochar and *Bacillus* strain Ya-1 also significantly impacted rhizosphere bacterial diversity, as evidenced by positive effects on Pielou's evenness and Shannon diversity indices, which are critical determinants of a healthy and resilient microbial community.

3.4. Enhancement of Microbial Diversity and Soil Health

The integration of biochar with microbial inoculants, such as *Bacillus* strain Ya-1, has been demonstrated to significantly enhance microbial diversity and soil health, contributing to improved plant nutrient uptake and reduced susceptibility to soil-borne diseases [26]. Biochar's role in amending soil and facilitating microbial activity is particularly notable, as it can alter microbial habitats and interactions, thereby enhancing the efficacy of biological control agents [27]. Recent studies have highlighted the potential synergistic effect of biochar in combination with microbial agents for improving soil health and plant growth. For instance, the addition of biochar to wheat–maize systems have been shown to stimulate microbial richness and enzyme activities, leading to increased soil organic carbon and total nitrogen [28]. Moreover, modified biochar has been effective in enhancing soil fertility and rice yield in mercury-contaminated soils, underscoring its role in mitigating the impacts of soil pollution [29]. The combined application of biochar with bacteria and plants has also shown promise in the remediation of oil-contaminated soils, indicating the broad application potential of this approach [30].

Furthermore, the interaction between biochar, soil, and plants is crucial for sustainable agriculture under changing climate conditions [31]. The synergistic effect of biochar and microorganisms has been shown to greatly improve the vegetation and microbial structure in degraded alpine grasslands, contributing to the restoration of these fragile ecosystems [32]. These findings underscore the importance of biochar and microbial interactions in enhancing soil health and microbial diversity, which is essential for maintaining soil fertility and supporting plant growth.

3.5. Metabolite Regulation of Plant Growth

In this study, we observed that key metabolites, particularly cinnamic acid and its derivatives, significantly promoted the growth of chili plants under the combined application of biochar and *Bacillus* strain Ya-1. As a natural plant growth-promoting compound, cinnamic acid has been demonstrated to enhance plant growth by modulating phytohormone signaling and strengthening the antioxidant defense system [33]. In this research, the notable increase in the cinnamic acid levels under the BC treatment was closely correlated with the improvement in plant growth parameters. Furthermore, the antioxidant and pro-oxidant activities of cinnamic acid derivatives within the plant body are considered key factors in regulating plant cell growth [34], affecting cell division and elongation by influencing the levels of reactive oxygen species (ROS) and the activity of antioxidant enzymes.

Moreover, cinnamic acid and its derivatives also play a significant role in the plant's response to environmental stress. These metabolites are involved in intracellular signaling pathways, aiding plants in adapting to variable environmental conditions [35]. Under the conditions of combined biochar and microbial application, these metabolites may enhance the antioxidant system of chili plants, improve the plants' tolerance to harmful substances in the soil, and, thus, promote healthy plant growth.

4. Materials and Methods

4.1. Pot Experiment

4.1.1. Experimental Groups

The pot experiment was carried out in a room under controlled temperature and light conditions and lasted 52 days. "Boyou" (*Capsicum annuum* Var. *longum*) was the variety of pepper used in the present study. The seeds were pre-germinated on moist filter paper in a dark environment at 25 °C, and then seedlings with uniform germination were transplanted into plastic pots (10 cm in diameter and 12.5 cm in height) containing 0.95 L of soil. The soil used in the experiment was collected from the experimental field of the Hainan Academy of Agricultural Sciences (110.37° E, 20.01° N), followed by its crushing and the removal of stones and debris. The biochar used was produced via the pyrolysis of coconut coir at 650 °C for 30 min and had a pH value of 8.16. The pot experiment consisted of four treatment groups, including the control group (CK), bacterial inoculant treatment group (B), biochar treatment group (C), and biochar–bacteria treatment group (BC), with three biological replicates per treatment group. For the C and BC groups, biochar and soil were uniformly mixed at a mass ratio of 2%. During the experiment, the temperature was set to 25 °C, with a 16 h light/8 h dark cycle. The watering of the plants was performed every 2–3 days to ensure that the soil moisture was maintained above 30%, and it ceased three days before the end of the experiment.

4.1.2. Preparation of the Inoculum of the *Biochar*-Immobilized Bacillus Strain Ya-1

The previously isolated *Bacillus* strain Ya-1 used in this study was kept in our laboratory [15,36,37]. A single colony of bacterium grown on the plate was inoculated into 100 mL of the liquid LB medium containing 1 g NaCl, 1 g tryptone, and 0.5 g yeast extract dissolved in 100 mL of water and incubated at 28 °C for 24 h with constant shaking at 200 rpm to obtain the pure culture of this bacterial strain. The concentration was confirmed by measuring the OD600 of the bacterial culture solution with a spectrophotometer, and fresh LB medium was used to adjust the bacterial solution to a concentration of 10^8 cfu/mL. Thereafter, 20 mL of this culture inoculum was added to a flask containing 200 mL of fresh LB liquid medium and incubated at 28 °C with a shaking speed of 150 rpm. The flask received 20 g of biochar every 8 h, with the pH value being maintained below 7.5 by using 0.1 mol/L of the NaOH solution and dilute hydrochloric acid. The cultures were incubated for 48 h, and to remove the extra moisture, the inoculum was air-dried.

4.2. *The Measurement of Soil Edaphic Factors*

After carrying out the pot experiment, the careful removal of soil surrounding the plant roots using a brush was performed, followed by the discarding of plant tissues. The methods used for the determination of the physicochemical properties of the soil are listed in Table 1.

Table 1. Methods used for the analysis of soil physicochemical properties.

Detection Index	Detection Method
pH	Electrode pH meter
Water content	Oven
Carbon	Elemental analyzer
Nitrogen	Elemental analyzer
Quick-acting nitrogen	Alkaline diffusion
Quick-acting phosphorus	Molybdenum antimony colorimetry
Quick-acting potassium	Flame photometric
Microbial carbon	Chloroform fumigation and leaching
Microbial nitrogen	Chloroform fumigation and extraction
Microbial phosphorus	Chloroform fumigation and extraction
Phosphorus, potassium, sodium, calcium, magnesium, sulfur, iron, manganese, copper, zinc, molybdenum, and boron	ICP-MS/ICP-OES

Note: ICP-MS stands for "Inductively Coupled Plasma-Mass Spectrometry", while ICP-OES stands for "Inductively Coupled Plasma-Optical Emission Spectrometry".

4.3. *The Investigation of Bacterial Diversity via 16S Amplicon Sequencing*

The TGuide S96 Magnetic Soil/Stool DNA Kit (Tiangen Biotech (Beijing) Co., Ltd., Beijing, China) was used to extract the soil's total DNA. The extracted DNA served as a template for the amplification of 16S rDNA with the primers, including 338F (5′-ACTCCTACGGGAGGCAGCA-3′) and 806R (5′-GGACTACHVGGGTWTCTAAT-3′) [38] using a PCR machine. The PCR products resulting from amplification were collected and sent to Beijing Tsingke Biotech Co., Ltd., Guangzhou Sequencing Centre (Guangzhou, China), and then they were purified, quantified, normalized, and used for library construction for next-generation sequencing. Qualified libraries were then subjected to the Illumina NovaSeq 6000 platform (Illumina, Inc., San Diego, CA, USA) for sequencing.

Trimmomatic v0.33 was used to filter raw sequencing reads, and thereafter, the identification and removal of primer sequences were performed using cutadapt v1.9.1, which resulted in the generation of clean reads that did not contain primer sequences. The DADA2 method (divisive amplicon denoising algorithm) [39] implemented in QIIME2 2020.6 [40] was employed to perform denoising. To obtain the final valid data (non-chimeric reads), the first paired-end sequencing reads were merged, and then chimeric sequences were identified and removed. The analysis of diversity was performed on the sequencing data.

4.4. The Measurement of the Agronomic Traits of Chili Plants

After carefully removing the plants from the soil using a brush, they were thoroughly cleaned. The entire plant, as well as the separated individual roots, stems, and leaves, was photographed. The determination of the plant height and leaf surface area was performed using MATO V 4.2 software [41]. The roots were stained using the triphenyl tetrazolium chloride (TTC) reduction method, and their vitality was assessed.

4.5. The Extraction of Metabolites and LC-MS/MS Analysis

The fresh aerial parts of plants from each treatment group were harvested, transferred into centrifuge tubes, and quenched in liquid nitrogen. The plant samples were freeze-dried and ground, and 25 mg of each was then weighed into EP tubes at a low temperature. Homogenizer beads and 1000 μL of the extraction solution, a mixture of methanol/acetonitrile/water at a ratio of 2:2:1 v/v, which contains an isotope-labeled internal standard, were together added to the tubes. The samples were vortexed for 30 s, followed by homogenization using a homogenizer operated at 35 Hz for 4 min and the ultrasonic extraction in an ice-water bath for 5 min three times. The samples were then allowed to stand for 1 h at −40 °C, and next, they were centrifuged at 4 °C and 12,000 rpm with a relative centrifugal force of 13,800× g and rotational radius of 8.6 cm for 15 min. The resulting supernatant was collected and centrifuged again at 4 °C and 12,000 rpm for 15 min with the same conditions. Finally, the supernatant was transferred into new centrifuge vials for LC-MS/MS analysis, and an equal volume of it was obtained from all samples and was mixed and subjected to quality control (QC) testing using the LC-MS/MS equipment (Thermo Fisher Scientific, Waltham, MA, USA).

To analyze the non-polar metabolites, a Thermo Fisher Scientific Vanquish ultra-high-performance liquid chromatography (UHPLC) system (Thermo Fisher Scientific, Waltham, MA, USA) was used, which allows for the chromatographic separation of target compounds with a liquid chromatography (LC) column (Phenomenex Kinetex 2.6 μm C18 (2.1 mm × 50 mm). To acquire primary-ion and secondary-ion mass spectrometry data, we employed an Orbitrap Exploris 120 mass spectrometer (Thermo Fisher Scientific, Waltham, MA, USA). The instrumental parameters of UHPLC included a sheath gas flow rate of 50 Arb, an auxiliary gas flow rate of 15 Arb, a capillary temperature of 320 °C, and spray voltages of 3.8 kV (positive) and −3.4 kV (negative). The collected raw data were subjected to filtration and standardization, followed by annotation, screening, and analysis by comparison with the database.

4.6. Statistical Analysis Methods

The measurement results of the soil pH, moisture content, elemental content, and other parameters were organized, and the Log2FC of each group's parameter changes were calculated based on the CK group, followed by the creation of a heatmap. The diversity index was calculated based on the results of microbial diversity sequencing, and the Kruskal–Wallis test was used to compare the diversity indices between groups. Tukey's Honest Significant Difference was used for multiple comparisons and the ranking of means ($p < 0.05$). The collected plant growth parameters were organized, and ANOVA was used to compare the mean differences of parameters between groups ($p < 0.05$). All tests in this study were calculated using Rstudio (Posit Software, Boston, MA, USA) based on R version 4.4.1, using the packages pheatmap, vegan, agricolae, and ggplot2.

5. Conclusions

This study provides compelling evidence that the integration of coconut shell biochar and *Bacillus* strain Ya-1 significantly enhances soil quality and microbial diversity, concurrently promoting the growth of chili plants. The combination leverages the unique properties of biochar and the biological control potential of Ya-1, offering synergistic benefits for soil health and plant performance. This strategy could substantially reduce our reliance on chemical inputs by improving soil properties and bolstering microbial activity, thereby contributing to sustainable agricultural practices.

The notable increases in soil nutrient content, along with improvements in microbial diversity and plant growth parameters, position the integrated application of biochar and Ya-1 as a promising approach for sustainable agriculture. These benefits are especially significant in regions susceptible to soil degradation and intensive chemical use, where sustainable management practices are essential for long-term productivity. The treatment's profound impact on plant physiology is further indicated by metabolic changes that may enhance crop resilience and yield. While this research offers a comprehensive understanding of the effects of biochar and bacterial strains on chili plants under controlled conditions, the translation of these findings to field applications necessitates further investigation. This includes addressing the challenges of scalability and economic viability.

Future research should focus on the long-term impacts of biochar and *Bacillus* treatments across various soil types and climates, ensuring the universality of their benefits. A deeper exploration of plant–microbe interactions in the rhizosphere will elucidate the mechanisms behind the enhancement of soil health and plant growth through the integrated use of biochar and *Bacillus* strains. Such studies will inform the scaling up of this approach and its integration into broader sustainable agricultural practices.

In summary, the combined application of biochar and microbial agents emerges as a viable strategy to improve soil and plant health, laying the groundwork for developing more diversified and resilient agricultural systems. This approach holds potential for addressing global challenges such as climate change and food security.

Author Contributions: Conceptualization, S.T. and Z.Z.; Formal analysis, W.Y.; Investigation, Q.Y.; Methodology, B.W.; Project administration, Z.Z.; Software, B.W.; Supervision, S.T.; Validation, Q.Y.; Writing—original draft, S.T.; Writing—review and editing, T.X. and Z.Z. All authors have read and agreed to the published version of the manuscript.

Funding: This research was funded by the Technical System of Winter Melon and Vegetable Industries in Hainan Province (HNARS-05-ZJ04) and Innovative Team Project of Hainan Academy of Agricultural Sciences (HAAS2023TDYD12).

Institutional Review Board Statement: Not applicable.

Informed Consent Statement: Not applicable.

Data Availability Statement: Microbial diversity sequencing data (SAMN44111606, SAMN44111607, SAMN44111608, SAMN44111609) can be found at the following links: https://www.ncbi.nlm.nih.gov/biosample/44111606 (accessed on 9 October 2024); https://www.ncbi.nlm.nih.gov/biosample/44111607 (accessed on 9 October 2024); https://www.ncbi.nlm.nih.gov/biosample/44111608 (accessed on 9 October 2024); https://www.ncbi.nlm.nih.gov/biosample/44111609 (accessed on 9 October 2024). The metabolite data can be found on MetaboLights as MTBLS11042, at https://www.ebi.ac.uk/metabolights/editor/study/MTBLS11042 (accessed on 6 September 2024).

Acknowledgments: We would like to express our gratitude to Tsingke Guangzhou Sequencing Centre for providing technical support for soil microbial diversity testing and to Biotree Mass Spectrometry Platform for providing technical support for the non-targeted LC-MS testing of plant metabolites.

Conflicts of Interest: The authors declare no conflicts of interest. The funders had no role in the design of the study; in the collection, analyses, or interpretation of the data; in the writing of the manuscript; or in the decision to publish the results.

References

1. Sikandar, A.; Jia, L.; Wu, H.; Yang, S. *Meloidogyne enterolobii* risk to agriculture, its present status and future prospective for management. *Front. Plant Sci.* **2023**, *13*, 1093657. [CrossRef] [PubMed]
2. El-Kazzaz, M.K.; Ghoneim, K.E.; Agha, M.K.M.; Helmy, A.; Behiry, S.I.; Abdelkhalek, A.; Saleem, M.H.; Al-Askar, A.A.; Arishi, A.A.; Elsharkawy, M.M. Suppression of pepper root rot and wilt diseases caused by *Rhizoctonia solani* and *Fusarium oxysporum*. *Life* **2022**, *12*, 587. [CrossRef] [PubMed]
3. Jiang, G.; Wei, Z.; Xu, J.; Chen, H.; Zhang, Y.; She, X.; Macho, A.P.; Ding, W.; Liao, B. Bacterial wilt in China: History, current status, and future perspectives. *Front. Plant Sci.* **2017**, *8*, 1549. [CrossRef]
4. Singh, S.; Balodi, R.; Meena, P.N.; Singhal, S. Biocontrol activity of *Trichoderma harzianum*, *Bacillus subtilis* and *Pseudomonas fluorescens* against *Meloidogyne incognita*, *Fusarium oxysporum* and *Rhizoctonia solani*. *Indian Phytopathol.* **2021**, *74*, 703–714. [CrossRef]
5. Nepal, J.; Ahmad, W.; Munsif, F.; Khan, A.; Zou, Z. Advances and prospects of biochar in improving soil fertility, biochemical quality, and environmental applications. *Front. Environ. Sci.* **2023**, *11*, 1114752. [CrossRef]
6. Lee, J.; Kim, S.; Jung, H.; Koo, B.-K.; Han, J.A.; Lee, H.-S. Exploiting bacterial genera as biocontrol agents: Mechanisms, interactions and applications in sustainable agriculture. *J. Plant Biol.* **2023**, *66*, 485–498. [CrossRef]
7. Boro, M.; Sannyasi, S.; Chettri, D.; Verma, A.K. Microorganisms in biological control strategies to manage microbial plant pathogens: A review. *Arch. Microbiol.* **2022**, *204*, 666. [CrossRef]
8. Song, Y.; Wilson, A.J.; Zhang, X.; Thoms, D.; Sohrabi, R.; Song, S.; Geissmann, Q.; Liu, Y.; Walgren, L.; He, S. FERONIA restricts Pseudomonas in the rhizosphere microbiome via regulation of reactive oxygen species. *Nat. Plants* **2021**, *7*, 644–654. [CrossRef]
9. Singh, H.; Northup, B.K.; Rice, C.W.; Prasad, P.V.V. Biochar applications influence soil physical and chemical properties, microbial diversity, and crop productivity: A meta-analysis. *Biochar* **2022**, *4*, 8. [CrossRef]
10. Wang, D.; Fonte, S.J.; Parikh, S.J.; Six, J.; Scow, K.M. Biochar additions can enhance soil structure and the physical stabilization of C in aggregates. *Geoderma* **2017**, *303*, 110–117. [CrossRef]
11. Bolan, S.; Hou, D.; Wang, L.; Hale, L.; Egamberdieva, D.; Tammeorg, P.; Li, R.; Wang, B.; Xu, J.; Wang, T.; et al. The potential of biochar as a microbial carrier for agricultural and environmental applications. *Sci. Total Environ.* **2023**, *886*, 163968. [CrossRef] [PubMed]
12. Nyumah, F.; Pang, Z.; Zhang, C.; Tayyab, M.; Yang, Z.; Lin, Z.; Lin, W.; Ishimwe, C.; Ntambo, M.S.; Zhang, H. Complementary effects of biochar, secondary metabolites, and bacteria biocontrol agents rejuvenate ratoon sugarcane traits and stimulate soil fertility. *Ind. Crops Prod.* **2023**, *202*, 117081.
13. Iacomino, G.; Idbella, M.; Laudonia, S.; Vinale, F.; Bonanomi, G. The suppressive effects of biochar on aboveand belowground plant pathogens and pests: A review. *Plants* **2022**, *11*, 3144. [CrossRef] [PubMed]
14. Arshad, U.; Azeem, F.; Mustafa, G.; Bakhsh, A.; Toktay, H.; McGiffen, M.; Nawaz, M.A.; Naveed, M.; Ali, M.A. Combined application of biochar and biocontrol agents enhances plant growth and activates resistance against *Meloidogyne incognita* in tomato. *Gesunde Pflanz.* **2021**, *73*, 591–601. [CrossRef]
15. Tian, J.; Dungait, J.A.J.; Hou, R.; Deng, Y.; Hartley, I.P.; Yang, Y.; Kuzyakov, Y.; Zhang, F.; Cotrufo, M.F.; Zhou, J. Microbially mediated mechanisms underlie soil carbon accrual by conservation agriculture under decade-long warming. *Nat. Commun.* **2024**, *15*, 377. [CrossRef]
16. Jindo, K.; Audette, Y.; Olivares, F.L.; Canellas, L.P.; Smith, D.S.; Voroney, R.P. Biotic and abiotic effects of soil organic matter on the phytoavailable phosphorus in soils: A review. *Chem. Biol. Technol. Agric.* **2023**, *10*, 29. [CrossRef]
17. Wang, S.; Wang, L.; Li, S.; Zhang, T.; Cai, K. The Win–Win Effects of an invasive plant biochar on a soil–crop system: Controlling a bacterial soilborne disease and stabilizing the soil microbial community network. *Microorganisms* **2024**, *12*, 447. [CrossRef]
18. Shahzadi, A.; Noreen, Z.; Alamery, S.; Zafar, F.; Haroon, A.; Rashid, M.; Aslam, M.; Younas, A.; Attia, K.A.; Mohammed, A.A. Effects of biochar on growth and yield of Wheat (*Triticum aestivum* L.) under salt stress. *Sci. Rep.* **2024**, *14*, 20024. [CrossRef]
19. Meng, Y.; Shen, Y.; Meng, W.; Wang, X.; Li, Z.; Liu, K.; Dai, H. Effect of Biochar on Agricultural Soil Aggregates and Organic Carbon: A Meta-analysis. *Environ. Sci.* **2023**, *44*, 6847–6856.
20. Guan, T.; Wang, Q.; Li, J.; Yan, H.; Chen, Q.; Sun, J.; Liu, C.; Han, Y.; Zou, Y.; Zhang, G. Biochar immobilized plant growth-promoting rhizobacteria enhanced the physicochemical properties, agronomic characters and microbial communities during lettuce seedling. *Front. Microbiol.* **2023**, *14*, 1218205. [CrossRef]
21. Ramírez-Pool, J.A.; Calderón-Pérez, B.; Ruiz-Medrano, R.; Ortiz-Castro, R.; Xoconostle-Cazares, B. *Bacillus* strains as effective biocontrol agents against phytopathogenic bacteria and promoters of plant growth. *Microb. Ecol.* **2024**, *87*, 76. [CrossRef] [PubMed]
22. Wang, D.; Wei, L.; Ma, J.; Wan, Y.; Huang, K.; Sun, Y.; Wen, H.; Chen, Z.; Li, Z.; Yu, D. *Bacillus cereus* NJ01 induces SA-and ABA-mediated immunity against bacterial pathogens through the EDS1-WRKY18 module. *Cell Rep.* **2024**, *43*, 113985. [CrossRef] [PubMed]
23. Wang, S.; Shi, F.; Li, P.; Yang, F.; Pei, Z.; Yu, Q.; Zuo, X.; Liu, J. Effects of rice straw biochar on methanogenic bacteria and metabolic function in anaerobic digestion. *Sci. Rep.* **2022**, *12*, 6971.

24. Fan, Q.; Sun, J.; Chu, L.; Cui, L.; Quan, G.; Yan, J.; Hussain, Q.; Iqbal, M. Effects of chemical oxidation on surface oxygen-containing functional groups and adsorption behavior of biochar. *Chemosphere* **2018**, *207*, 33–40. [CrossRef] [PubMed]
25. Hei, J.; Yin, X.; Wang, W.; Sardans, J.; Wang, C.; Chen, X.; Tariq, A.; Zeng, F.; Alrefaei, A.F.; Peñuelas, J. N-enriched biochar increases carbon, nitrogen, and phosphorus accumulation associated with changes in plant ecological stoichiometry in subtropical rice paddy fields. *Exp. Agric.* **2023**, *59*, e11. [CrossRef]
26. Glaser, B.; Lehr, V.-I. Biochar effects on phosphorus availability in agricultural soils: A meta-analysis. *Sci. Rep.* **2019**, *9*, 9338. [CrossRef]
27. Ajeng, A.A.; Abdullah, R.; Ling, T.C. Biochar-*Bacillus* consortium for a sustainable agriculture: Physicochemical and soil stability analyses. *Biochar* **2023**, *5*, 17. [CrossRef]
28. Bai, J.; Huang, Y.; Bai, Y.; Chen, D.; Haider, S.; Song, J.; Moreira, B.R.D.A.; Ren, G.; Yang, G.; Feng, Y. Impact of straw-biochar amendments on microbial activity and soil carbon dynamics in wheat-maize system. *Soil Tillage Res.* **2024**, *244*, 106284. [CrossRef]
29. Li, H.; Yang, L.; Mao, Q.; Zhou, H.; Guo, P.; Agathokleous, E.; Wang, S. Modified biochar enhances soil fertility and nutrient uptake and yield of rice in mercury-contaminated soil. *Environ. Technol. Innov.* **2023**, *32*, 103435. [CrossRef]
30. Fang, X.; Zhang, M.; Zheng, P.; Wang, H.; Wang, K.; Lv, J.; Shi, F. Biochar-bacteria-plant combined potential for remediation of oil-contaminated soil. *Front. Microbiol.* **2024**, *15*, 1343366. [CrossRef]
31. Murtaza, G.; Ahmed, Z.; Eldin, S.M.; Ali, B.; Bawazeer, S.; Usman, M.; Iqbal, R.; Neupane, D.; Ullah, A.; Khan, A. Biochar-Soil-Plant interactions: A cross talk for sustainable agriculture under changing climate. *Front. Environ. Sci.* **2023**, *11*, 1059449. [CrossRef]
32. Li, J.; Li, H.; Shang, J.; Liu, K.; He, Y.; Shao, X. The synergistic effect of biochar and microorganisms greatly improves vegetation and microbial structure of degraded alpine grassland on qinghai–tibet plateau. *Agronomy* **2023**, *13*, 2203. [CrossRef]
33. Steenackers, W.; El Houari, I.; Baekelandt, A.; Witvrouw, K.; Dhondt, S.; Leroux, O.; Gonzalez, N.; Corneillie, S.; Cesarino, I.; Inzé, D. cis-Cinnamic acid is a natural plant growth-promoting compound. *J. Exp. Bot.* **2019**, *70*, 6293–6304. [CrossRef] [PubMed]
34. Lee, B.-H.; Choi, H.-S.; Hong, J. Roles of anti- and pro-oxidant potential of cinnamic acid and phenylpropanoid derivatives in modulating growth of cultured cells. *Food Sci. Biotechnol.* **2022**, *31*, 463–473. [CrossRef] [PubMed]
35. Shuab, R.; Lone, R.; Koul, K.K. Cinnamate and cinnamate derivatives in plants. *Acta Physiol. Plant.* **2016**, *38*, 64. [CrossRef]
36. Zhao, Z.; Yan, W.; Chen, Y.; Xiao, T.; Xiao, M. Screening, identification and fermentation condition optimization of biocontrol bacteria YA-1 against ginger bacterial wilt. *J. South. Agric.* **2015**, *46*, 421–427.
37. Zhao, Z.; Yan, W.; Wang, B.; Chen, Y.; Wang, H.; Xiao, T. Increase of defense enzyme activity and expression of resistance-related genes in chilli induced by *Bacillus* sp. Ya-1. *Mol. Plant Breed.* **2022**, *20*, 2699–2706.
38. Huse, S.M.; Dethlefsen, L.; Huber, J.A.; Welch, D.M.; Relman, D.A.; Sogin, M.L. Exploring microbial diversity and taxonomy using SSU rRNA hypervariable tag sequencing. *PLoS Genet.* **2008**, *21*, e1000255. [CrossRef]
39. Callahan, B.; McMurdie, P.; Rosen, M.; Han, A.W.; Johnson, A.J.A.; Holmes, S.P. DADA2: High-resolution sample inference from Illumina amplicon data. *Nat. Methods* **2016**, *13*, 581–583. [CrossRef]
40. Bolyen, E.; Rideout, J.R.; Dillon, M.R.; Bokulich, N.A.; Abnet, C.C.; Al-Ghalith, G.A.; Alexander, H.; Alm, E.J.; Arumugam, M.; Asnicar, F.; et al. Reproducible, interactive, scalable and extensible microbiome data science using QIIME 2. *Nat. Biotechnol.* **2019**, *37*, 852–857. [CrossRef]
41. Liu, L.; Wang, Q.; Zhang, Z.; He, X.; Yu, Y. MATO: An updated tool for capturing and analyzing cytotaxonomic and morphological data. *Innov. Life* **2023**, *1*, 100010. [CrossRef]

International Journal of *Molecular Sciences*

MDPI

Article

Impact of Agroforestry Practices on Soil Microbial Diversity and Nutrient Cycling in Atlantic Rainforest Cocoa Systems

Sayure Mariana Raad Nahon [1,2], Felipe Costa Trindade [2], Caio Augusto Yoshiura [2], Gabriel Caixeta Martins [2], Isa Rebecca Chagas da Costa [2], Paulo Henrique de Oliveira Costa [2], Héctor Herrera [3,4], Diego Balestrin [5], Tiago de Oliveira Godinho [5], Bia Makiyama Marchiori [5] and Rafael Borges da Silva Valadares [2,*]

1 Programa de Pós-Graduação em Biotecnologia Aplicada à Agropecuária, Universidade Federal Rural da Amazônia (UFRA), Belém 66077-830, PA, Brazil
2 Instituto Tecnológico Vale (ITV), Rua Boaventura da Silva, 955, Belém 66050-090, PA, Brazil
3 Departamento de Ciencias Forestales, Facultad de Ciencias Agropecuarias y Medioambiente, Universidad de La Frontera, Temuco 4811230, Chile
4 Center for Biodiversity and Ecological Sustainability, Facultad de Ciencias Agropecuarias y Medioambiente, Universidad de La Frontera, Temuco 4811230, Chile
5 Reserva Natural Vale, Rodovia BR 101, km 122 s/n Zona Rural, Linhares 29900-111, ES, Brazil
* Correspondence: rafael.borges.valadares@itv.org

Abstract: Microorganisms are critical indicators of soil quality due to their essential role in maintaining ecosystem services. However, anthropogenic activities can disrupt the vital metabolic functions of these microorganisms. Considering that soil biology is often underestimated and traditional assessment methods do not capture its complexity, molecular methods can be used to assess soil health more effectively. This study aimed to identify the changes in soil microbial diversity and activity under different cocoa agroforestry systems, specially focusing on taxa and functions associated to carbon and nitrogen cycling. Soils from three different cocoa agroforestry systems, including a newly established agroforestry with green fertilization (GF), rubber (*Hevea brasiliensis*)–cocoa intercropping (RC), and cocoa plantations under Cabruca (cultivated under the shave of native forest) (CAB) were analyzed and compared using metagenomic and metaproteomic approaches. Samples from surrounding native forest and pasture were used in the comparison, representing natural and anthropomorphic ecosystems. Metagenomic analysis revealed a significant increase in Proteobacteria and Basidiomycota and the genes associated with dissimilatory nitrate reduction in the RC and CAB areas. The green fertilization area showed increased nitrogen cycling activity, demonstrating the success of the practice. In addition, metaproteomic analyses detected enzymes such as dehydrogenases in RC and native forest soils, indicating higher metabolic activity in these soils. These findings underscore the importance of soil management strategies to enhance soil productivity, diversity, and overall soil health. Molecular tools are useful to demonstrate how changes in agricultural practices directly influence the microbial community, affecting soil health.

Keywords: metabarcoding; metagenomics; proteomics; soil microorganisms

1. Introduction

Brazil is widely recognized as one of the world's most important biodiversity hotspots, with the Amazon, Cerrado, Atlantic Forest, and Pantanal biomes being particularly rich in biodiversity and playing a critical role in soil carbon sequestration [1–3]. However, these ecosystems are under significant threat from a variety of disturbances, including deforestation, mining, cattle ranching, and conversion to agricultural or forestry systems, which negatively impact natural forest regeneration [4,5]. Given the current rate of biodiversity loss, it is critical to develop strategies to conserve native biodiversity at multiple scales.

Agroforestry systems offer a promising approach by integrating agricultural and forestry crops to promote sustainable land management while supporting biodiversity conservation. A notable example is the cultivation of *Theobroma cacao* L. (cocoa), which is of great economic importance in Brazil. The state of Espírito Santo, particularly the municipality of Linhares, is a major cocoa producer, accounting for 70.7% of the state's total production [6]. Cocoa-based agroforestry systems have shown numerous benefits, including increased yields, improved soil micronutrient levels, and enhanced carbon storage compared to monoculture systems [7,8]. In addition to supporting economic development, these systems provide important ecological services, such as increasing soil carbon and nitrogen storage and enhancing microbial functional diversity [9–13]. Such benefits reduce the need for chemical inputs and contribute to sustainable land management. However, further research is needed to optimize productivity, particularly by exploring plant–soil interactions that can mitigate biodiversity loss, reduce production costs, and improve soil quality [14].

Soil quality refers to the ability of soil to support biological productivity, maintain environmental quality, and promote plant and animal health, with its chemical, physical, and biological components serving as key indicators [15]. Microbial diversity, which is influenced by factors such as plant species and land use, is a critical indicator of soil biological quality because of its role in biogeochemical cycles such as carbon and nitrogen [16–19]. While traditional biochemical assays or direct measurements of biological activity provide only a limited snapshot of the complex and dynamic interactions within soils, molecular tools offer deeper insights into processes such as nutrient cycling and symbiosis [20]. By integrating molecular techniques, we can achieve a more comprehensive and nuanced understanding of soil ecosystems, ultimately leading to more effective soil health management strategies [21]. Although studies using metabarcoding have examined microbial composition, approaches such as metaproteomics and metagenomics are essential for understanding the metabolic processes and microbial diversity in response to land-use changes and agroecosystems [22–24].

Few studies have analyzed the impact of cocoa agroforestry systems on the community structure of soil microorganisms in the Brazilian Atlantic Rainforest, and the identification of active microorganisms remains a largely unexplored area. To better understand the role of these microbial communities in nutrient cycling, it is crucial to identify shifts in their composition. A practical approach is to analyze the genes and proteins associated with soil processes, which reveals the functional capacity of microbial communities [25]. In this research, we hypothesized that cocoa agroforestry systems promote greater diversity and improved carbon and nitrogen cycling, thereby enhancing soil quality. Therefore, the aim of this study was to reveal the impact of cocoa agroforestry systems on the soil microbial structure and functional profiles of taxa involved in carbon and nitrogen cycling in the Southeast Atlantic Rainforests of Brazil. To the best of our knowledge, this is the first study to investigate the impact of different cocoa agroforestry systems on the structure of active microbial communities using a combination of high-throughput sequencing methods.

2. Results

2.1. Chemical and Physical Soil Characterization

The pasture (PAS) and agroforestry system with green fertilization (GF) had lower soil organic matter (SOM) and nitrogen (N) accumulations compared to the Cabruca agroforestry system (CAB), rubber tree (*Hevea brasiliensis*)–cocoa intercropping (RC), and native forest (NF) areas (Figure S1). The median $\pm$ MAD values for SOM were 1.06 $\pm$ 0.37, 1.44 $\pm$ 0.62, 2.47 $\pm$ 1.33, 2.53 $\pm$ 0.89, and 2.15 $\pm$ 0.99%, and for N were 0.08 $\pm$ 0.02, 0.08 $\pm$ 0.05, 0.23 $\pm$ 0.08, 0.21 $\pm$ 0.08, and 0.16 $\pm$ 0.08% for PAS, GF, CAB, RC, and NF, respectively. Soil pH values ranged from 4.0 to 7.2, with the lowest median values found in the NF (4.3 $\pm$ 0.15) and PAS (4.8 $\pm$ 0.15) plots. These two areas also had the highest values for aluminum saturation (NF = 61 $\pm$ 26% and PAS = 66 $\pm$ 25%) and the lowest values for base saturation (NF = 13 $\pm$ 7% and PAS = 18 $\pm$ 11%). The CAB area showed the highest values of phosphorus (P = 5.3 $\pm$ 0.9 mg/dm^3), manganese (Mn = 151 $\pm$ 37 mg/dm^3), and copper (Cu = 1.6 $\pm$ 0.7 mg/dm^3) compared to the other areas (Figure S1). For example, the PAS area presented 1.9 $\pm$ 0.6 mg/dm^3, 0.7 $\pm$ 0.6 mg/dm^3, and

0.1 ± 0.0 mg/dm^3 of P, Mn, and Cu, respectively. In general, the cation exchange capacity (CEC) values followed a similar trend as the SOM values. The soils in the Cabruca agroforestry system and rubber–cocoa intercropping areas have a clayey texture, while the other areas had a medium texture.

2.2. Total Bacterial and Fungal Diversity Identified by Metabarcoding

To detect changes in the community structure of soil microorganisms in response to different cocoa agroforestry systems, fungal and bacterial metabarcoding analyses were conducted. The metabarcoding analysis showed a total of 2,664,675 reads in the 25 16S libraries, which after quality filtering resulted in 1,578,053 sequences for analysis. Overall, there were a total of 114,068 bacterial sequences distributed across 192 distinct operational taxonomic units (OTUs). The Shannon–Wiener diversity index and Pielou Evenness were lower in the native forest, with a Shannon–Wiener index of 4.89 for bacteria and 3.26 for fungi, as well as Pielou Evenness values of 0.76 for bacteria and 0.67 for fungi. In the agroforestry system with green fertilization, the Shannon–Wiener index was 5.14 for bacteria and 3.38 for fungi, with Pielou Evenness values of 0.80 for bacteria and 0.70 for fungi (Figure S2). Proteobacteria and Acidobacteria were the most abundant phyla, accounting for 51% and 22% of the relative abundance, respectively (Figure 1A). At the genus level, uncultured, Candidatus Solibacter and *Acidibacter* were the most abundant, accounting for 52%, 5%, and 4% of the assigned sequences, respectively (Table S1). A higher number of bacterial genera were detected in the agroforestry system with green fertilization (33 genera), rubber–cocoa intercropping (37 genera), and the Cabruca agroforestry system (38 genera) areas (Table S1).

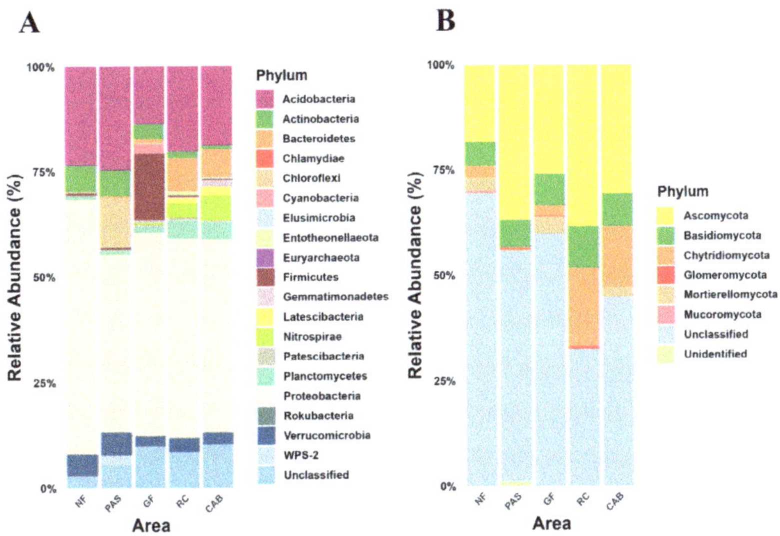

Figure 1. Relative abundance of operational taxonomic units (OTUs) at the phylum level for bacterial (**A**) and fungal (**B**) communities, identified through metabarcoding, in soil samples from native forest (NF), grassland (PAS), green fertilization (GF), rubber tree–cocoa (RC), and Cabruca (CAB) systems.

For the 25 ITS libraries, a total of 148,977 reads were obtained, which, after quality filtering, yielded 89,749 sequences for analysis. In total, there were 74,234 fungal sequences distributed among 1008 distinct OTUs. No significant differences were found between the areas regarding fungal diversity as measured by the Shannon–Wiener index (native forest: 0.53; pasture: 0.60; agroforestry system with green fertilization: 1.13; rubber–cocoa intercropping: 1.45; and Cabruca agroforestry system: 1.12). Similarly, for the Pielou Evenness index, the values were as follows: native forest: 0.36; pasture: 0.39;

agroforestry system green fertilization: 0.57; rubber–cocoa intercropping: 0.66; and Cabruca agroforestry system: 0.59 (Figure S2). At the phylum level, unclassified, Ascomycota and Chytridiomycota were the most abundant, accounting for 41%, 32%, and 9% of relative abundance, respectively (Figure 1B). Similarly, from the 34 detected genera, unidentified, *Saccharomyces* and *Mortierella* were the most abundant, accounting for 69%, 18%, and 6%, respectively (Table S2). *Mortierella* was significantly abundant in the rubber–cocoa intercropping and Cabruca agroforestry systems.

2.3. Active Microbial Diversity Identified by Shotgun Metagenomics

The active microbiota involved in carbon and nitrogen cycling across the different soils in cocoa agroforestry systems were identified using shotgun metagenomics. A total of 25 metagenomes were obtained from the different cocoa agroforestry systems and the land uses analyzed in this study. In total, there were 212,455,526 raw reads, which were reduced to 116,564,438 clean reads and 24,653,303 contigs after the quality filtering. A total of 5684 KEGG Orthology identifiers (KOs) were obtained from all samples. The Shannon–Wiener diversity indices and Pielou Evenness in the active taxa showed higher values for bacteria and fungi in the rubber–cocoa intercropping areas. For the rubber–cocoa intercropping, the Shannon–Wiener values were 5.39 for bacteria and 3.51 for fungi, while the Pielou Evenness indices were 0.84 for bacteria and 0.76 for fungi. In the Cabruca agroforestry system, the Shannon–Wiener values were 5.41 for bacteria and 3.52 for fungi, with Pielou Evenness indices of 0.84 for bacteria and 0.78 for fungi (Figure S3). PERMANOVA results for bacteria (0.46) and fungi (0.83) showed that all microbial diversity was affected by sample origin, explaining 41% and 42% of the variance for bacteria and fungi, respectively ($p < 0.001$).The RDA analysis also showed that the sampling site affected the diversity of active microbial communities (Axis 1—43.5% and Axis 2—31.8% of explained variation) (Figure 2). In addition, the samples from the Cabruca agroforestry system, rubber–cocoa intercropping, and green fertilization areas clearly differentiated from those of native forest and pasture, where the variables sand, Fe, P, Mn, and Zn were the factors that significantly shaped the microbial diversity in all the samples (Figure 2).

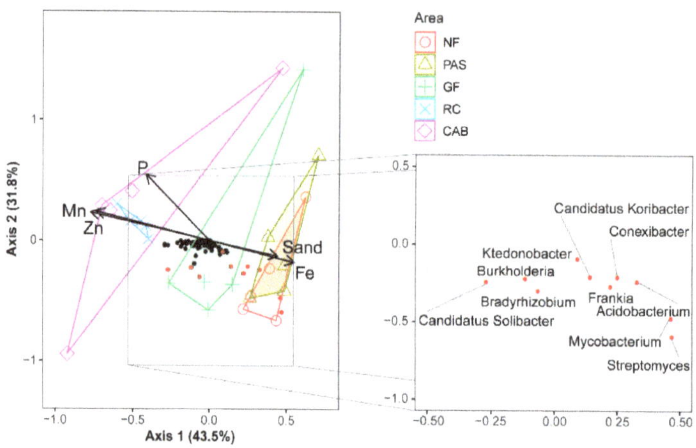

Figure 2. Redundancy analysis (RDA) illustrating the relationships between microbial communities and soil physicochemical parameters, including phosphorus (P), manganese (Mn), zinc (Zn), and iron (Fe) in the native forest (NF), pasture (PAS), green fertilization (GF), rubber tree–cocoa (RC), and Cabruca (CAB) systems. The red points represent the genera that show significant differences between the samples.

The analysis of the taxa involved in carbon and nitrogen cycling revealed differences in the relative abundance of active taxa in the different soils. Proteobacteria was one of the most abundant taxa with values of 44.87% in the native forest, 35.85% in the pasture, 40.71% in the agroforestry system with green fertilization, 50.93% in the rubber–cocoa intercropping area, and 52.98% in the Cabruca agroforestry system. Similarly, Actinobacteria showed high relative abundances with values of 30.09% in the native forest, 33.90% in the pasture, 35.36% in the agroforestry system with green fertilization, 17.74% in the rubber–cocoa intercropping, and 11.73% in the Cabruca agroforestry system. Ascomycota and Basidiomycota were the most active fungal taxa, with Ascomycota showing a relative abundance of 94.27% in the native forest, 93.95% in the pasture, 88.89% in the agroforestry system with green fertilization, 85.46% in the rubber–cocoa intercropping, and 83.78% in the Cabruca agroforestry system; Basidiomycota accounted for 5.43% in the native forest, 5.73% in the pasture, 10.23% in the agroforestry system with green fertilization, 13.60% in the rubber–cocoa intercropping area, and 15.21% in the Cabruca agroforestry system (Figures 1 and 3). Differential analysis of bacteria revealed a greater abundance of Alphaproteobacteria and Acidobacteria in the native forest (Figure S4). The pasture was significantly enriched with Actinobacteria and Ktedonobacteria, whereas green fertilization showed enrichment of Actinobacteria (Figure S4). In the rubber–cocoa intercropping area, the dominant classes were Deltaproteobacteria, Betaproteobacteria, Gammaproteobacteria, and Planctomycetacia (Figure S4). Finally, in the Cabruca agroforestry system, the most predominant classes were Deltaproteobacteria, Betaproteobacteria, Gammaproteobacteria, and Verrucomicrobiae, showing similarity with the rubber–cocoa intercropping (Figure S4). The differential analysis of fungal taxa (Figure S5) showed similarity between the rubber–cocoa intercropping and Cabruca agroforestry systems, as they presented the fungal classes Saccharomycetes, Agaricomycetes, and Schizosaccharomycetes as preferential (Figure S5). However, in addition to these, the Cabruca agroforestry system exhibits Tremellomycetes and Ustilaginomycetes (Figure S5). The native forest had the highest average difference in the proportion of the fungal classes Eurotiomycetes, Dothideomycetes, and Leotiomycetes, belonging to the phylum Ascomycota. These same classes were also abundant in the pasture area (Figure S5). The class Sordarlomycetes, also belonging to the phylum Ascomycota, showed higher abundance in the agroforestry system with green fertilization (Figure S5).

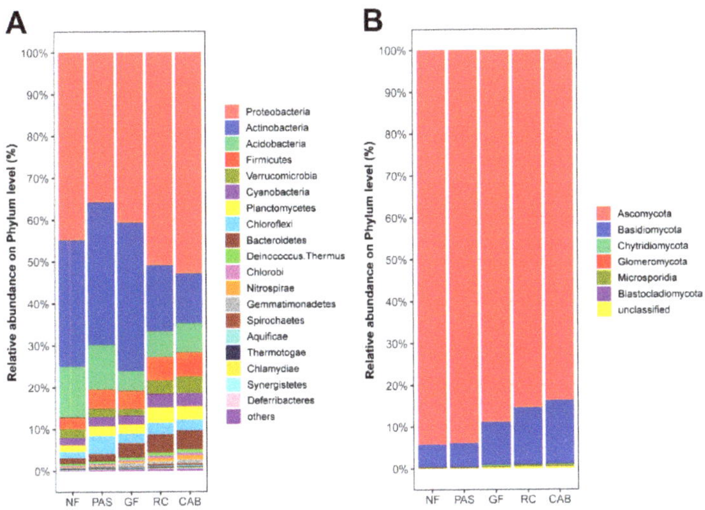

Figure 3. Relative abundance of active taxa involved in carbon and nitrogen cycling at the phylum level for bacterial (**A**) and fungal (**B**) communities, identified through shotgun metagenomics, in soil samples from native forest (NF), grassland (PAS), green fertilization (GF), rubber tree–cocoa (RC), and Cabruca (CAB) systems.

2.4. Genes Involved in C and N Cycling

A total of 33 and 13 genes involved in C and N cycling were retrieved from the 25 metagenomes analyzed, respectively. Regarding the functional genes involved in C cycling, they were mainly classified in methane metabolism, carbon fixation, and carbon degradation (Figure 4). Within the functional groups, genes associated with CO oxidation exhibited greater relative abundance: NF: 21.94%, PAS: 16.87%, GF: 15.61%, RC: 19.34%, and CAB: 19.86%. Similarly, genes related to hemicellulose were found in the following abundances: NF: 17.31%, PAS: 21.42%, GF: 20.87%, RC: 17.74%, and CAB: 18.50%. For starch, the relative abundances were as follows: NF: 14.44%, PAS: 14.04%, GF: 14.41%, RC: 16.23%, and CAB: 14.48%. Regarding lignin, the values were as follows: NF: 14.45%, PAS: 14.12%, GF: 14.99%, RC: 14.14%, and CAB: 13.63%. Lastly, carbon fixation genes showed the following abundances: NF: 38.95%, PAS: 34.75%, GF: 33.49%, RC: 38.82%, and CAB: 40.54%. Overall, there were minimal variations among the observed treatments (Figure 4). Also identified were other genes involved in carbon fixation, especially tktA in the native forest, rubber–cocoa intercropping, and Cabruca agroforestry systems. In the pasture area, there was a higher abundance of genes related to carbon degradation compared to the other areas (Figure S6). The genes involved in the N cycle were mainly related to nitrogen degradation and denitrification (Figure 5), with higher relative abundances of genes related to denitrification in the native forest, rubber–cocoa intercropping, and Cabruca agroforestry systems. The metagenome revealed a high abundance of key genes in the N cycle, such as UreC (ammonification) in the native forest, pasture, Cabruca agroforestry system, and rubber–cocoa intercropping areas, being expressed at values of 0.21 for the native forest and pasture and 0.24 for the Cabruca agroforestry system and rubber–cocoa intercropping areas, and the abundance of glutamate synthase in the agroforestry system with green fertilization, with a value of 0.55 (Figure S6).

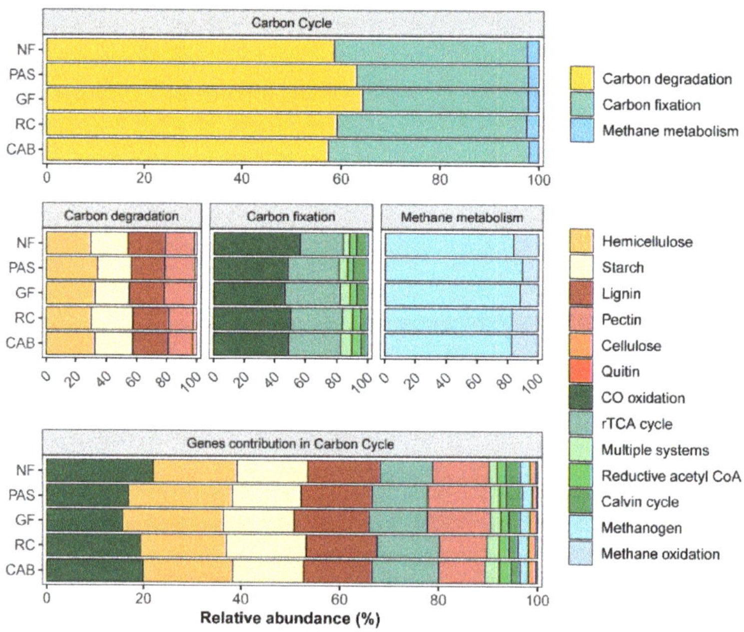

Figure 4. Relative abundance of carbon cycle genes across different land uses, including native forest (NF), pasture (PAS), green fertilization (GF), rubber tree–cocoa (RC), and Cabruca (CAB) systems.

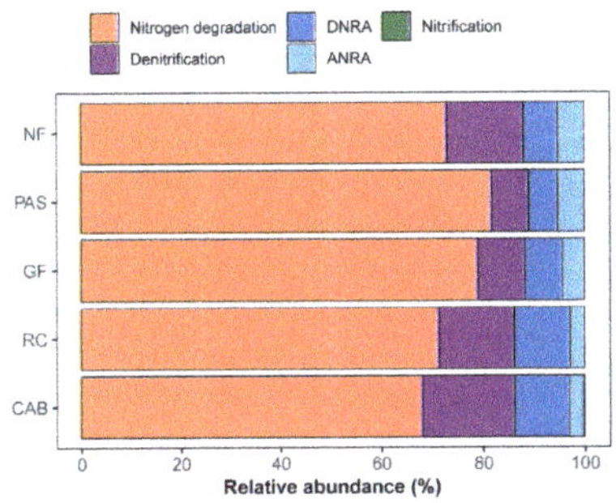

Figure 5. Relative abundance of nitrogen cycle genes across different land uses, including native forest (NF), pasture (PAS), green fertilization (GF), rubber tree–cocoa (RC), and Cabruca (CAB) systems. Nitrogen cycling pathways include DNRA (dissimilatory nitrate reduction to ammonium) and ANRA (assimilatory nitrate reduction to ammonium).

2.5. Active Soil Enzymes

Metaproteomic analysis was conducted to identify active soil enzymes involved in carbon and nutrient cycling in the different treatments. Metaproteomic analysis identified 24 enzymes associated with carbon fixation, predominantly found in native forest areas, and 6 enzymes with dehydrogenase activity. In addition, a low abundance of carbon-fixing enzymes was observed in the agroforestry system with green fertilization and rubber–cocoa intercropping areas (Figure 6A). Of the 12 enzymes related to organic matter mineralization, 10 were found in the rubber–cocoa intercropping area. Notably, β-glucosidase was identified exclusively in the green fertilization, rubber tree, and native forest areas, with the highest peptide abundances in the latter two areas (Figure 6C). A greater abundance and similarity of ABC transporters was observed in the rubber–cocoa intercropping and native forest areas compared to the other areas (Figure 6B). These transporters are capable of transporting various substrates across cellular membranes by hydrolysis of adenosine triphosphate (ATP).

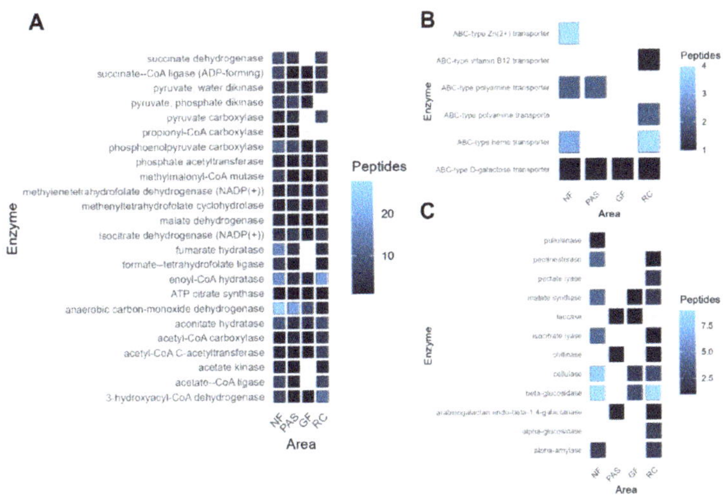

Figure 6. Abundance of soil enzymes associated with key biological processes in different land uses, including rubber–cocoa intercropping (RC), native forest (NF), green fertilization (GF), and pasture (PAS). (**A**) Enzymes responsible for carbon fixation, (**B**) enzymes responsible for nitrogen fixation, and (**C**) enzymes involved in the mineralization of soil organic matter.

3. Discussion

3.1. Soil Chemical Composition

Soil chemical analyses showed different results in the different study areas. The acidity (average pH) of the soils in the Cabruca agroforestry system and the rubber–cocoa intercropping system can be classified as weakly acidic. In contrast, the native forest and pasture have high acidity (pH < 5.0), while the agroforestry system with green fertilization falls in the range of medium acidity (pH 5.0–5.9), according to the classification of Prezotti, et al. [26]. These pH values also influenced the aluminum saturation (m), with the Cabruca agroforestry system, the rubber–cocoa intercropping area, and the agroforestry system with green fertilization showing low saturation values (<20%), whereas the native forest and the pasture showed high values (>40%). The pH values observed in the pasture are similar to those reported by da Rocha Junior, et al. [27]. In the native forest, soil acidification may be related to organic matter, which releases H+ ions during mineralization, increasing acidity. This process contributes to the typical acidity of tropical soils, which tend to be acidic due to intense leaching of base cations such as calcium (Ca), magnesium (Mg), and potassium (K) [28–30].

The highest levels of organic matter and nitrogen were observed in the native forest, rubber–cocoa intercropping, and Cabruca agroforestry systems. Forest-based cocoa agroecosystems typically produce a continuous flow of organic matter into the soil [31]. Monroe, et al. [32] found that leaf deposition in different cocoa agroforestry systems resulted in a thick litter layer, which significantly increased soil organic matter. In addition to the natural senescence of cocoa leaves, pruning practices and the deposition of prunings in the field also contribute to soil organic matter. Thomazini, et al. [33] quantified the effects of different management practices in agroforestry systems and reported similarly high nitrogen levels in forested areas. Furthermore, Schneidewind, et al. [31] demonstrated that cocoa agroforestry systems produce nitrogen-rich litter, highlighting their capacity to store this nutrient in the soil.

3.2. Changes in Community Structure of Soil Microorganisms

The results of the effect of agroforestry systems on microbial diversity, obtained through metabarcoding, showed changes in the relative abundance of microorganisms, with Proteobacteria and Ascomycota being the most abundant phyla. This finding is consistent with previous studies highlighting the important role of these groups in soil ecosystems [34,35]. Changes in the structure of soil microbial communities under different cocoa agroforestry systems have been reported previously, supporting the results of this study [36,37]. The conversion of native forest to pasture in the Atlantic rainforest has shown a particular enrichment of microbial taxa such as Firmicutes [38]. Consistent with the present study, Alphaproteobacteria were found to be highly abundant in forest soils [39]. This class includes several important genera, such as *Bradyrhizobium* and *Sphingobium*, which play key roles in nitrogen cycling and the degradation of lignin by-products, respectively [40,41]. In addition, Zhou, et al. [42] observed an increased abundance of Alphaproteobacteria with high nitrogen additions to the soil. Studies have shown that members of the Acidobacteria phylum prefer acidic conditions for growth [43], which probably explains their presence in native forest soils. They are essential for lignin degradation, a crucial process in regulating the soil carbon cycle [44,45]. Other members of Proteobacteria are involved in nitrogen fixation and often form symbiotic relationships with plants [42,46]. This phylum is strongly associated with carbon availability and cycling, showing a positive correlation with the carbon to nitrogen (C/N) ratio [47–50]. Similarly, Actinobacteria play an important role in nutrient cycling, especially carbon and nitrogen cycling [51,52]. They are essential for the decomposition of plant residues, which enhances carbon sequestration in soils [53], which explains their presence in the agroforestry system with green fertilization.

A similar trend in the relative abundance of fungal taxa was observed by metabarcoding, with many taxa remaining unclassified. However, there was a notable enrichment of Chytridiomycota in the rubber–cocoa intercropping and Cabruca agroforestry systems. Chytridiomycota have been identified as a fungal phylum with strong potential for lignin and cellulose degradation [54], which may be related to the incorporation of plant debris in both systems, and to a lesser extent in the green fertilization agroforestry system and native forest areas (Figure 2). Genera such as *Rhizophlyctis* and *Spizellomyces* are widely recognized for their ability to degrade complex organic matter through various enzymes [55,56]. Similar to the present study, Cerqueira, et al. [57] reported the presence of Ascomycota in forest and pasture areas, probably due to the ability of fungi from this phylum to inhabit diverse environments [58]. The fungal class Tremellomycetes, preferentially associated with Cabruca soils, was also reported by Yu, et al. [59], who found a positive correlation between this class and increased soil organic carbon mineralization. In pasture soils, the abundance of Dothideomycetes was observed, which can attributed to their ability to thrive in challenging substrates [60]. The fungal class Agaricomycetes, found in both Cabruca and rubber–cocoa intercropping soils, contributes to cellulose degradation, thereby increasing nutrient availability and playing a critical role in the carbon cycle [61,62]. Wang and Zhou [63] recently investigated the effect of green fertilization on the fungal composition of the rhizosphere and observed an increase in the abundance of Sordariomycetes with the application of green fertilization. Zhou, et al. [64] and Koechli, et al. [65] also found that cellulose stimulates the growth of this fungal class, which could explain the higher abundance of Sordariomycetes in the green fertilization area.

3.3. Microbial Functional Profiles Involved in Carbon and Nitrogen Cycling

The metagenomic results revealed changes in the relative abundance of microbial functional profiles involved in carbon and nitrogen cycling, with a decrease in Actinobacteria and an increase in Proteobacteria and Basidiomycota in the agroforestry systems. Members of Proteobacteria have been identified as key players in soil systems, playing crucial roles in regulating carbon and nutrient cycling, responding to fertilizer inputs, and possessing plant growth promoting traits that can directly enhance plant growth [66,67]. Similarly, Basidiomycota are abundant in forest soils and play a positive role in wood decay and litter decomposition, processes that are enhanced in agroforestry systems [68–70].

Nitrogen cycling pathways were particularly enhanced in the agroforestry system with green fertilization (Figure S6), possibly indicating a link between green fertilization and legumes. These results are in line with Hu, et al. [71], who confirmed that legumes stimulate microorganisms involved in nitrogen cycling. Similarly, Lori, et al. [72] showed that agroforestry systems improve soil biological quality. Among the genes identified in the study plots, those related to glutamate synthase showed higher abundance in the agroforestry system with green fertilization (Figure S6). Studies have shown that glutamate synthase is a key enzyme in the GS/GOGAT cycle, which is critical for carbon and nitrogen metabolism [73–75]. In addition, UreC, found in the native forest, pasture, Cabruca agroforestry system and rubber–cocoa intercropping areas, encodes urease, an important enzyme derived mainly from soil microorganisms that makes nitrogen available to plants [76,77]. This suggests that the study areas harbor microorganisms that carry this gene and contribute to the nitrogen cycle.

The results also showed that despite changes in overall microbial functional profiles, particularly for bacteria (Figure 3), these changes did not result in the loss of the essential genes involved in carbon and nitrogen cycling (Figures 4 and 5). This may be directly related to the functional redundancy of soil microbial communities, allowing the selective stimulation of microbial taxa capable of maintaining these functions, thus maintaining ecological balance [78,79].

3.4. Soil Enzymes Involved in Carbon and Nitrogen Cycling

Metaproteomics facilitated the analysis of soil enzymes, revealing their association with biological processes and positioning them as key indicators of soil quality. Our results showed that enzyme activity varied depending on the management practices applied, highlighting the importance of proper soil management. Enzymes with dehydrogenase activity were detected in rubber–cocoa intercropping and native forest soils, indicating high rates of organic matter decomposition [80]. In addition, recent studies suggest that these enzymes are sensitive indicators of microbial activity, making them valuable for monitoring biological quality and effectively detecting soil changes [81–84].

A higher abundance of ABC transporters was observed in rubber–cocoa intercropping and native forest soils compared to other areas. ABC transporters are involved in several biological processes, transporting a wide range of substrates across cell membranes by ATP hydrolysis [85]. In soil, certain bacteria possess a significant number of these transporters, which benefit bacterial communities by enhancing nutrient uptake and providing protection against environmental stress [86,87]. Similarly, the cellulase enzyme was more abundant in the native forest, rubber–cocoa intercropping and green fertilization soils, indicating active cellulose decomposition and an increase in organic matter. This finding is in agreement with Florez, et al. [88], who highlighted the influence of vegetation type on cellulose content. Another enzyme closely related to cellulose degradation, beta-glucosidase, was also identified in these areas. This enzyme plays a role in the final stages of cellulose degradation and serves as a reliable indicator of soil quality, providing early indications of changes in organic matter status and turnover [89,90].

Overall, our results show that different agroforestry practices induce changes in the community structure of soil microorganisms, while highlighting the functional redundancy of these communities. This redundancy ensures the conservation of essential genes and enzymes involved in carbon and nutrient cycling. However, this study had significant limitations, including a lack of recent research on soil biology and the effects of agroforestry practices in cocoa systems in the Atlantic rainforest. In addition, the lack of proteomic approaches to assess soil biological quality hindered theoretical development. The soil protein extraction protocol was particularly challenging in the Cabruca agroforestry system, where extraction was unsuccessful due to incompatibilities between the methods used and the soil samples.

4. Materials and Methods

4.1. Study Area

The study was conducted in areas of the municipality of Linhares, in the state of Espírito Santo, southeastern Brazil. Five different areas (land uses) were evaluated: (i) primary native forest (NF), which is a conserved area with tree growth that is part of the original vegetation; (ii) pasture (PAS), characterized by extensive livestock production with no soil management and numerous areas of exposed soil; (iii) agroforestry system under the implementation of green fertilization (GF), which began in 2020 with the planting of *Gliricidia sepium* using cuttings and mechanical and chemical weeding practices; (iv) rubber tree (*Hevea brasiliensis*)–cocoa intercropping (RC), a well-established and highly productive system combining rubber and cocoa trees, in which the soil is naturally enriched by the leaf litter from the falling rubber tree leaves; and (v) the Cabruca agroforestry system (CAB), consisting of a cocoa plantation established under thinned native forest dominated by "Boleira" (*Joannesia princeps*).

Soil samples were collected from the three different cocoa agroforestry systems: (i) the rubber–cocoa intercropping system (19°03′33″ S, 39°58′55″ W); (ii) the Cabruca agroforestry system (19°24′33″ S, 40°06′16″ W); and (iii) the agroforestry system with green fertilization (19°09′08″ S, 40°04′57″ W). Soil samples from the native forest were collected in the protected area of the Vale Natural Reserve, which is part of a mosaic of protected areas known for its high biodiversity (19°01′36″ S, 40°01′05″ W). The annual rainfall in the area is 1214 mm, with a maximum temperature of 29.9 °C.

In each area, 5 plots of 30 × 30 m were delineated and a total of 3 composite samples were collected within each plot at depths of 0–5 cm, 5–10 cm and 10–20 cm. Soil samples for chemical and physical analyses were placed in plastic bags and sent to the Laboratório Brasileiro de Análises Ambientais e Agrícolas (LABRAS) in Monte Carmelo, MG, Brazil.

Samples for biological analysis were collected at a depth of 0–5 cm, placed in a cooler and transported to the Vale Technological Institute—Sustainable Development in Belém, PA, Brazil.

4.2. Soil Chemical and Physical Analyses

Soil samples were air dried and sieved through a 2 mm mesh. Analyses were performed according to previous methods [91,92]. Briefly, pH was determined in a water/soil suspension at a ratio of 1:2.5 (v/v); phosphorus (P) and potassium (K) were extracted with a Mehlich I solution (0.05 mol·L^{-1} HCl and 0.0125 mol·L^{-1} H_2SO_4); calcium (Ca), magnesium (Mg), and aluminum (Al) were extracted with a 1 mol·L^{-1} KCl solution; potential acidity (H+Al) was extracted using a SMP buffer solution at pH 7.5; sulfur (S) was extracted with a 0.01 mol·L^{-1} calcium monobasic phosphate solution; boron (B) was extracted with hot water; and copper (Cu), iron (Fe), manganese (Mn), and zinc (Zn) were extracted with a DTPA-TEA solution. The quantifications were performed using an inductively coupled plasma optical emission spectrometer (ICP-OES). In addition, organic matter (OM) was determined by the colorimetric method, and total nitrogen (N) was determined by the Kjeldahl method. The cation exchange capacity (CEC; sum of bases [Ca^{+2} + Mg^{+2} + K^+ + Na^+] + H+Al), base saturation (V%; sum of bases/CEC), and aluminum saturation (m%; Al^{+3}/sum of bases + Al^{+3}) were also calculated. The sand fraction was determined by sieving after chemical dispersion with 0.1 mol·L^{-1} sodium hydroxide and 0.016 mol·L^{-1} sodium hexametaphosphate after stirring for 4 h. The clay fraction was determined after sedimentation using a Bouyoucos hydrometer, and the silt fraction was calculated as the difference between the total mass and the sum of the clay and sand contents.

4.3. Fungal and Bacterial Metabarcoding

Total DNA from soil samples was extracted from the top layer (0–5 cm) using the DNeasy PowerSoil Kit (Qiagen, Hilden, Germany) according to the manufacturer's recommendations. DNA samples were quantified using the Qubit® 3.0 fluorometer (Thermo Fisher Scientific, Waltham, MA, USA) and their quality was verified by 1% agarose gel electrophoresis.

The paired-end metabarcoding amplicon libraries were constructed by PCR amplification of the V3–V4 region of the bacterial 16S rRNA gene and the internal transcribed spacer (ITS) of the fungal 18S rRNA gene. The bacterial primers used were S-D-Bact-0341-b-S-17-N (5′-TCGTCGGCAGCGTCAGATGTGTATAAGAGACAGCCTACGGGNGGCWGCAG-3′) and S-D-Bact-0785-a-A-21-N (5′-GTCTCGTGGGCTCGGAGATGTGTATAAGAGACAGG ACTACHVGGGTATCTAATCC-3′), while the fungal primers used were fITS7i (5′-TCGTCGG CAGCGTCAGATGTGTATAAGAGACAGGTGARTCATCGAATCTTTG-3′) and ITS4i (5′-GTCTCGTGGGCTCGGAGATGTGTATAAGAGACAGTCCTCCGCTTATTGATATGC-3′). After amplification, the libraries were verified using the Qubit dsDNA HS Assay (Thermo Fisher Scientific) and the Agilent 2100 Bioanalyzer (Agilent Technologies, Santa Clara, CA, USA) with a DNA 1000 chip. Libraries were purified using the AMPure XP Purification Kit (Beckman Coulter, Brea, CA, USA) and processed using the Nextera XT Kit (Illumina, San Diego, CA, USA). They were then sequenced on the MiSeq-Illumina platform using the MiSeq V3 reagent kit (600 cycles; Illumina) at the Instituto Tecnológico Vale (Belém, PA, Brazil) [93].

4.4. Shotgun Metagenomics

Metagenomic libraries were constructed from 50 ng of total DNA and then subjected to a random enzymatic fragmentation reaction using the Illumina DNA Prep Kit (Illumina) according to the manufacturer's instructions. The fragmented DNA was purified using AMPure XP magnetic beads (Beckman Coulter, Brea, CA, USA) and then subjected to an amplification reaction using primers complementary to the Illumina flow cell adapters. The amplified libraries were re-purified using AM-Pure XP magnetic beads (Beckman Coulter) and quantified using the Qubit® 3.0 fluorometer (Thermo Fisher Scientific). Fragment sizes were verified using a Bioanalyzer 2100 fluorometer (Agilent Technologies, Santa Clara, CA, USA) with a DNA 1000 chip. The libraries were then adjusted to a DNA concentration of 4 nM, pooled, denatured, and diluted to a running concentration of 1.8 pM. Sequencing was performed on the Illumina NextSeq 500/550 platform using the NextSeq 500 v2.5 kit at 300 cycles.

4.5. Metaproteomics

The protein extraction from the upper soil layer (0–5 cm) was conducted using an adapted protocol from Qian and Hettich [94]. In 5 g of soil, 10 mL of lysis buffer was added. The samples were placed in a boiling water bath (95 °C) for 10 min in a 1:1 mixture of water and extraction buffer, as described by Ogunseitan [95]. Subsequently, the suspension was sonicated (5 times for 30 s each, at a power of 70 W) and centrifuged at 3200× *g* for 10 min. Ammonium acetate in methanol (0.1 M) was added to the recovered supernatant to precipitate proteins. To clean the samples, the process of adding 1.5 mL of cold acetone (80%) to the microtubes was repeated twice. The tubes were vortexed and centrifuged at 14,000 rpm for 7 min at 4 °C. The supernatant was discarded and 200 µL of 0.1% Rapigest (Waters, Milford, MA, USA) for protein solubilization was added.

The preparation for digestion involved the reduction of proteins with 2 µL of DTT (5 mM) and incubation at 56 °C for 25 min, followed by alkylation with 5.71 µL of iodoacetamide 14 mM for 30 min at room temperature. Residual iodoacetamide was removed by adding 2.09 µL of 5 mM DTT and incubating at room temperature for 15 min. Next, 2.11 µL of 1 mM calcium chloride ($CaCl_2$) and 50 µL of 20 ng·µL^{-1} trypsin were added and incubated at 37 °C for 20 h. Then, 1.05 µL of trifluoroacetic acid (TFA; 0.4%) was added and left to incubate at 37 °C for 90 min to stop the enzymatic reaction. Subsequently, the samples were centrifuged at 14,000 rpm at 6 °C for 10 min. The recovered supernatants were transferred to 10 kDa filters and centrifuged for 50 min at 14,000× *g* at room temperature. Finally, the filtered solutions were transferred to vials, dried, and suspended in ammonium formate (20 mM) for effective capture on the first-dimensional column of Ultra-Performance Liquid Chromatography (UPLC). The chromatograph was directly coupled to an ESI-Q-ToF Synapt G2S mass spectrometer (Waters) operating in positive mode and continuous fragmentation (MS E), with collision energy ranging from 5 to 40 eV [96].

4.6. Bioinformatics and Data Analysis

For the analysis of the metabarcoding raw data, the paired-end reads were trimmed and filtered using PRINSEQ v0.20.4 and merged using the Paired-End reAd mergeR (PEAR) v0.9.10. The obtained sequences were analyzed using the PIMBA pipeline [97] with the Ribosomal Database Project RDP v11.5 [98] for 16S sequences and the UNITE v8.2 [99] for 18S sequences as reference databases.

For the shotgun metagenomics analysis, the paired-end reads were assembled into contigs using MEGAHIT v1.2.9. Taxonomic classification was performed using Kaiju v1.4.4. The NCBI BLAST non-redundant protein sequences were used as the reference database for taxonomic classification to obtain relative abundance information of taxonomic groups (bacteria, archaea, and fungi). For functional analysis, the paired-end reads were merged using Paired-End reAd mergeR (PEAR) v0.9.10 and submitted to the online tool EcofunMAP http://iegst1.rccc.ou.edu:8080/ecofunmap/ (accessed on 15 May 2022) with the following parameters: 10 threads; and workflow: ultra-sensitive.

The proteomic data were analyzed in the Proteinlynx software v.3.0.2 (Waters) for protein identification and quantification using the public database of bacteria, fungi, and plants from Uniprot https://www.uniprot.org/ (accessed on 20 October 2021), as well as data based on the soil metagenomes from the Vale National Reserve. Phylogenetic and functional analyses based on peptides were performed using the Unipept 4.3 software https://unipept.ugent.be/ (accessed on 10 January 2022). In this peptide-centered functional annotation context, biological processes (BPs) related to organic matter mineralization, carbon fixation, and ABC transporters were extracted.

Quantitative data were compared by ANOVA with Tukey's post-hoc analysis using stats v.4.2.3 and agricolae v.1.3.5 in the R software [100]. The Kruskal–Wallis test followed by the Dunn test ($p < 0.05$) were used to assess the differences of the soil chemical and physical attributes between the land uses. The diversity index calculations were performed using vegan v.2.5.7 and chemodiv v.0.2.0, and then plotted using ggplot2 v3.3.5 in the R environment. Differential analyses were obtained using the software STAMP 2.1.3 [101]. The clustering analysis based on metaproteomics was performed using the Manhattan distance. The sequences obtained in this study have been deposited in the NCBI Sequence Read Archive (http://www.ncbi.nlm.nih.gov/sra/PRJNA1114758) under accession number PRJNA1114758.

5. Conclusions

Our approach revealed the effects of different cocoa agroforestry systems on the community structure and functional profiles of soil microorganisms. While several changes in community structure were observed, the functional profiles of microorganisms involved in carbon and nitrogen cycling remained active, highlighting the functional redundancy of soil microorganisms in cocoa agroforestry systems under different management strategies. In particular, the analysis of enzymes related to ABC transporters and organic matter mineralization yielded significant results, especially in rubber–cocoa intercropping, suggesting a positive impact on soil microbial diversity and overall soil health. These results indicate that different cocoa agroforestry systems can improve soil quality by promoting the abundance of active microbial taxa involved in carbon and nutrient cycling. This provides valuable insights into the beneficial role of agroforestry systems in improving soil biological quality, particularly through their effects on functional profiles and active soil enzymes. Future studies should aim to establish links between these functional microbial communities and key ecosystem services such as biomass production, crop yield, and carbon sequestration.

Supplementary Materials: The following supporting information can be downloaded at https://www.mdpi.com/article/10.3390/ijms252111345/s1.

Author Contributions: Conceptualization, S.M.R.N., R.B.d.S.V. and H.H.; methodology, S.M.R.N., P.H.d.O.C., F.C.T., I.R.C.d.C., D.B., T.d.O.G., B.M.M. and R.B.d.S.V.; validation, S.M.R.N., R.B.d.S.V. and H.H.; formal analysis, S.M.R.N., P.H.d.O.C., C.A.Y., H.H., G.C.M., R.B.d.S.V., F.C.T. and I.R.C.d.C.; resources, R.B.d.S.V.; data curation, S.M.R.N., C.A.Y., G.C.M., P.H.d.O.C. and H.H.; writing—preparation of original draft, S.M.R.N. and H.H.; writing—review and editing, S.M.R.N., H.H., G.C.M. and R.B.d.S.V.; supervision, R.B.d.S.V.; acquisition of financing, R.B.d.S.V. All authors have read and agreed to the published version of the manuscript.

Funding: This research was funded by Vale Institute of Technology, project "Reserva Natural Vale". S.M.R.N is a CNPq fellow and received funding from CAPES (88887.646418/2021-00). G.O. is a CNPq fellow and received funding from CAPES.

Institutional Review Board Statement: Not applicable.

Informed Consent Statement: Not applicable.

Data Availability Statement: The sequences obtained in this study have been deposited in the NCBI Sequence Read Archive (http://www.ncbi.nlm.nih.gov/sra/PRJNA1114758) under accession number PRJNA1114758.

Acknowledgments: We would like to thank the Vale Institute of Technology and the Vale Fund for funding and supporting this study, the Coordination for the Improvement of Higher Education Personnel (CAPES) for granting scholarships, and the Postgraduate Program in Biotechnology Applied to Agriculture at the Federal Rural University of the Amazon (UFRA).

Conflicts of Interest: The authors declare no conflicts of interest.

References

1. Jézéquel, C.; Tedesco, P.A.; Bigorne, R.; Maldonado-Ocampo, J.A.; Ortega, H.; Hidalgo, M.; Martens, K.; Torrente-Vilara, G.; Zuanon, J.; Acosta, A. A database of freshwater fish species of the Amazon Basin. *Sci. Data* **2020**, *7*, 96. [CrossRef]

2. Feeley, K.J.; Silman, M.R. Extinction risks of Amazonian plant species. *Proc. Natl. Acad. Sci. USA* **2009**, *106*, 12382–12387. [CrossRef]
3. Damasco, G.; Fontes, C.; Françoso, R.; Haidar, R. The Cerrado biome: A forgotten biodiversity hotspot. *Front. Young Minds* **2018**, *6*, 22. [CrossRef]
4. Sonter, L.J.; Herrera, D.; Barrett, D.J.; Galford, G.L.; Moran, C.J.; Soares-Filho, B.S. Mining drives extensive deforestation in the Brazilian Amazon. *Nat. Commun.* **2017**, *8*, 1013. [CrossRef]
5. Alho, C.; Sabino, J. A conservation agenda for the Pantanal's biodiversity. *Braz. J. Biol.* **2011**, *71*, 327–335. [CrossRef]
6. SEAC. Secretaria de Estado de Agricultura, Abastecimento, Aquicultura e Pesca. 2023. Espírito Santo Está Entre os Três Maiores Produtores de Cacau do Brasil. 2023. Available online: https://seag.es.gov.br/Not%C3%ADcia/espirito-santo-esta-entre-os-tres-maiores-produtores-de-cacau-do-brasil#:~:text=O%20Esp%C3%ADrito%20Santo%20%C3%A9%20o,,7%25%20da%20produ%C3%A7%C3%A3o%20estadual (accessed on 21 August 2023).
7. Tinoco-Jaramillo, L.; Vargas-Tierras, Y.; Habibi, N.; Caicedo, C.; Chanaluisa, A.; Paredes-Arcos, F.; Viera, W.; Almeida, M.; Vásquez-Castillo, W. Agroforestry systems of cocoa (*Theobroma cacao* L.) in the Ecuadorian Amazon. *Forests* **2024**, *15*, 195. [CrossRef]
8. Caicedo-Vargas, C.; Pérez-Neira, D.; Abad-González, J.; Gallar, D. Assessment of the environmental impact and economic performance of cacao agroforestry systems in the Ecuadorian Amazon region: An LCA approach. *Sci. Total Environ.* **2022**, *849*, 157795. [CrossRef]
9. Fahad, S.; Chavan, S.B.; Chichaghare, A.R.; Uthappa, A.R.; Kumar, M.; Kakade, V.; Pradhan, A.; Jinger, D.; Rawale, G.; Yadav, D.K. Agroforestry systems for soil health improvement and maintenance. *Sustainability* **2022**, *14*, 14877. [CrossRef]
10. Mosquera-Losada, M.R.; Santos, M.G.S.; Gonçalves, B.; Ferreiro-Domínguez, N.; Castro, M.; Rigueiro-Rodríguez, A.; González-Hernández, M.P.; Fernández-Lorenzo, J.L.; Romero-Franco, R.; Aldrey-Vázquez, J.A. Policy challenges for agroforestry implementation in Europe. *Front. For. Glob. Change* **2023**, *6*, 1127601. [CrossRef]
11. Muchane, M.N.; Sileshi, G.W.; Gripenberg, S.; Jonsson, M.; Pumariño, L.; Barrios, E. Agroforestry boosts soil health in the humid and sub-humid tropics: A meta-analysis. *Agric. Ecosyst. Environ.* **2020**, *295*, 106899. [CrossRef]
12. Rodríguez, L.; Suárez, J.C.; Rodriguez, W.; Artunduaga, K.J.; Lavelle, P. Agroforestry systems impact soil macroaggregation and enhance carbon storage in Colombian deforested Amazonia. *Geoderma* **2021**, *384*, 114810. [CrossRef]
13. Cardinael, R.; Chevallier, T.; Barthès, B.G.; Saby, N.P.; Parent, T.; Dupraz, C.; Bernoux, M.; Chenu, C. Impact of alley cropping agroforestry on stocks, forms and spatial distribution of soil organic carbon—A case study in a Mediterranean context. *Geoderma* **2015**, *259*, 288–299. [CrossRef]
14. Gama-Rodrigues, A.C.; Müller, M.W.; Gama-Rodrigues, E.F.; Mendes, F.A.T. Cacao-based agroforestry systems in the Atlantic Forest and Amazon Biomes: An ecoregional analysis of land use. *Agric. Syst.* **2021**, *194*, 103270. [CrossRef]
15. Doran, J.; Parkin, T. Defining and assessing soil quality. In *Defining Soil Quality for A Sustainable Environment*; Doran, J.W., Coleman, D.C., Bezdicek, D.F., Stewart, B.A., Eds.; Soil Science Society of America, Inc.: Madison, WI, USA, 1994; SSSA Special Publication Number 35.
16. Wang, X.; McConkey, B.G.; VandenBygaart, A.; Fan, J.; Iwaasa, A.; Schellenberg, M. Grazing improves C and N cycling in the Northern Great Plains: A meta-analysis. *Sci. Rep.* **2016**, *6*, 33190. [CrossRef]
17. Bhaduri, D.; Sihi, D.; Bhowmik, A.; Verma, B.C.; Munda, S.; Dari, B. A review on effective soil health bio-indicators for ecosystem restoration and sustainability. *Front. Microbiol.* **2022**, *13*, 938481. [CrossRef] [PubMed]
18. Ortiz, A.; Sansinenea, E. The role of beneficial microorganisms in soil quality and plant health. *Sustainability* **2022**, *14*, 5358. [CrossRef]
19. Almonacid, L.; Herrera, H.; Fuentes, A.; Vargas, R.; Toy, O.; Valadares, R.; De Oliveira, P. What fire didn't take away: Plant growth-promoting microorganisms await for plant species' arrival in burned old-growth forests of Los Andes Cordillera. *Plant Soil* **2024**. [CrossRef]
20. Trindade, F.C.; Gastauer, M.; Ramos, S.J.; Caldeira, C.F.; Araújo, J.F.d.; Oliveira, G.; Valadares, R.B.d.S. Soil Metaproteomics as a Tool for Environmental Monitoring of Minelands. *Forests* **2021**, *12*, 1158. [CrossRef]
21. Nascimento, S.V.d.; Costa, P.H.d.O.; Herrera, H.; Caldeira, C.F.; Gastauer, M.; Ramos, S.J.; Oliveira, G.; Valadares, R.B.d.S. Proteomic profiling and rhizosphere-associated microbial communities reveal adaptive mechanisms of dioclea apurensis kunth in eastern amazon's rehabilitating minelands. *Plants* **2022**, *11*, 712. [CrossRef]
22. Niu, X.; Song, L.; Xiao, Y.; Ge, W. Drought-tolerant plant growth-promoting rhizobacteria associated with foxtail millet in a semi-arid agroecosystem and their potential in alleviating drought stress. *Front. Microbiol.* **2018**, *8*, 2580. [CrossRef]
23. Batuecas, I.; Alomar, O.; Castañe, C.; Piñol, J.; Boyer, S.; Gallardo-Montoya, L.; Agustí, N. Development of a multiprimer metabarcoding approach to understanding trophic interactions in agroecosystems. *Insect Sci.* **2022**, *29*, 1195–1210. [CrossRef]
24. Trindade, F.C.; Ramos, S.J.; Gastauer, M.; Saraiva, A.M.M.; Caldeira, C.F.; Oliveira, G.; da Silva Valadares, R.B. Metaproteomes reveal increased capacity for stress tolerance of soil microbes in ferruginous tropical rocky outcrops. *Pedobiologia* **2020**, *81*, 150664. [CrossRef]
25. Finn, D.; Yu, J.; Penton, C.R. Soil quality shapes the composition of microbial community stress response and core cell metabolism functional genes. *Appl. Soil Ecol.* **2020**, *148*, 103483. [CrossRef]
26. Prezotti, L.C.; Gomes, J.A.; Dadalto, G.G.; Oliveira, J.A. *Manual de Recomentação de Calagem e Adubação Para o Estado do Espírito Santo: 5ª Aproximação*; SEEA; Incaper; CEDAGRO: Vitória, Brazil, 2007; 305p.

27. da Rocha Junior, P.R.; Sturião, W.P.; Nogueira, N.O.; Passos, R.R.; Donagemma, G.K.; Rangel, O.J.P.; Bhattarai, R. Soil quality indicators to evaluate environmental services at different landscape positions and land uses in the Atlantic Forest biome. *Environ. Sustain. Indic.* **2020**, *7*, 100047. [CrossRef]
28. Lima, M.S.d.; Freire, F.J.; Marangon, L.C.; Almeida, B.G.d.; Ribeiro, E.P.; Santos, R.L.d. Forest Soils in An Urban Forest Remnant in Mata De Dois Irmãos, Recife, Pernambuco State, Brazil. *Ciência Florest.* **2018**, *28*, 542–553. [CrossRef]
29. Johan, P.D.; Ahmed, O.H.; Omar, L.; Hasbullah, N.A. Phosphorus transformation in soils following co-application of charcoal and wood ash. *Agronomy* **2021**, *11*, 2010. [CrossRef]
30. Ng, J.F.; Ahmed, O.H.; Omar, L.; Jalloh, M.B.; Kwan, Y.M.; Musah, A.A.; Poong, K.H. Combined Use of Calciprill and Sodium Silicate Improves Chemical Properties of Low-pH Soil. *Agronomy* **2021**, *11*, 2070. [CrossRef]
31. Schneidewind, U.; Niether, W.; Armengot, L.; Schneider, M.; Sauer, D.; Heitkamp, F.; Gerold, G. Carbon Stocks, Litterfall and Pruning Residues in Monoculture and Agroforestry Cacao Production Systems. *Exp. Agric.* **2019**, *55*, 452–470. [CrossRef]
32. Monroe, P.H.M.; Gama-Rodrigues, E.F.; Gama-Rodrigues, A.C.; Marques, J.R.B. Soil carbon stocks and origin under different cacao agroforestry systems in Southern Bahia, Brazil. *Agric. Ecosyst. Environ.* **2016**, *221*, 99–108. [CrossRef]
33. Thomazini, A.; Mendonça, E.; Cardoso, I.; Garbin, M. SOC dynamics and soil quality index of agroforestry systems in the Atlantic rainforest of Brazil. *Geoderma Reg.* **2015**, *5*, 15–24. [CrossRef]
34. Lima-Perim, J.E.; Romagnoli, E.M.; Dini-Andreote, F.; Durrer, A.; Dias, A.C.F.; Andreote, F.D. Linking the composition of bacterial and archaeal communities to characteristics of soil and flora composition in the Atlantic Rainforest. *PLoS ONE* **2016**, *11*, e0146566. [CrossRef]
35. Pacheco Flores de Valgaz, A.; Naranjo-Morán, J.; Reyes Román, G.; Oviedo-Anchundia, J.; Ratti Torres, M.; Barcos-Arias, M. Discovering the Diversity of Arbuscular Mycorrhizal Fungi Associated with Two Cultivation Practices of Theobroma cacao. *Diversity* **2022**, *14*, 651. [CrossRef]
36. Buyer, J.S.; Baligar, V.C.; He, Z.; Arévalo-Gardini, E. Soil microbial communities under cacao agroforestry and cover crop systems in Peru. *Appl. Soil Ecol.* **2017**, *120*, 273–280. [CrossRef]
37. Sousa, R.d.S.d.R.d.; Lima, G.V.S.; Garcias, J.T.; Gomes, G.d.O.; Mateus, J.R.; Madeira, L.D.P.d.S.; Seldin, L.; Rogez, H.L.G.; Marques, J.M. The Microbial Community Structure in the Rhizosphere of *Theobroma cacao* L. and *Euterpe oleracea* Mart. Is Influenced by Agriculture System in the Brazilian Amazon. *Microorganisms* **2024**, *12*, 398. [CrossRef] [PubMed]
38. Danielson, R.E.; Rodrigues, J.L.M. Impacts of land-use change on soil microbial communities and their function in the Amazon Rainforest. *Adv. Agron.* **2022**, *175*, 179–258.
39. Karimi, B.; Terrat, S.; Dequiedt, S.; Saby, N.P.; Horrigue, W.; Lelièvre, M.; Nowak, V.; Jolivet, C.; Arrouays, D.; Wincker, P. Biogeography of soil bacteria and archaea across France. *Sci. Adv.* **2018**, *4*, eaat1808. [CrossRef]
40. Fujita, M.; Sakumoto, T.; Tanatani, K.; Yu, H.; Mori, K.; Kamimura, N.; Masai, E. Iron acquisition system of *Sphingobium* sp. strain SYK-6, a degrader of lignin-derived aromatic compounds. *Sci. Rep.* **2020**, *10*, 12177.
41. Favero, V.O.; de Carvalho, R.H.; Leite, A.B.C.; dos Santos, D.M.T.; de Freitas, K.M.; Boddey, R.M.; Xavier, G.R.; Rumjanek, N.G.; Urquiaga, S. Bradyrhizobium strains from Brazilian tropical soils promote increases in nodulation, growth and nitrogen fixation in mung bean. *Appl. Soil Ecol.* **2022**, *175*, 104461. [CrossRef]
42. Zhou, J.; Jiang, X.; Wei, D.; Zhao, B.; Ma, M.; Chen, S.; Cao, F.; Shen, D.; Guan, D.; Li, J. Consistent effects of nitrogen fertilization on soil bacterial communities in black soils for two crop seasons in China. *Sci. Rep.* **2017**, *7*, 3267. [CrossRef]
43. Sait, M.; Davis, K.E.R.; Janssen, P.H. Effect of pH on isolation and distribution of members of subdivision 1 of the phylum Acidobacteria occurring in soil. *Appl. Environ. Microbiol.* **2006**, *72*, 1852–1857. [CrossRef]
44. Lacerda Júnior, G.V.; Noronha, M.F.; de Sousa, S.T.P.; Cabral, L.; Domingos, D.F.; Sáber, M.L.; de Melo, I.S.; Oliveira, V.M. Potential of semiarid soil from Caatinga biome as a novel source for mining lignocellulose-degrading enzymes. *FEMS Microbiol. Ecol.* **2017**, *93*, fiw248. [CrossRef]
45. Wilhelm, R.C.; Singh, R.; Eltis, L.D.; Mohn, W.W. Bacterial contributions to delignification and lignocellulose degradation in forest soils with metagenomic and quantitative stable isotope probing. *ISME J.* **2019**, *13*, 413–429. [CrossRef]
46. Ogola, H.J.O.; Selvarajan, R.; Tekere, M. Local geomorphological gradients and land use patterns play key role on the soil bacterial community diversity and dynamics in the highly endemic indigenous afrotemperate coastal scarp forest biome. *Front. Microbiol.* **2021**, *12*, 592725. [CrossRef]
47. Fierer, N.; Bradford, M.A.; Jackson, R.B. Toward an ecological classification of soil bacteria. *Ecology* **2007**, *88*, 1354–1364. [CrossRef]
48. Ren, M.; Zhang, Z.; Wang, X.; Zhou, Z.; Chen, D.; Zeng, H.; Zhao, S.; Chen, L.; Hu, Y.; Zhang, C.; et al. Diversity and contributions to nitrogen cycling and carbon fixation of soil salinity shaped microbial communities in Tarim Basin. *Front. Microbiol.* **2018**, *9*, 431. [CrossRef]
49. Hermans, S.M.; Buckley, H.L.; Case, B.S.; Curran-Cournane, F.; Taylor, M.; Lear, G. Bacteria as emerging indicators of soil condition. *Appl. Environ. Microbiol.* **2016**, *83*, AEM.02826-16. [CrossRef]
50. Ren, N.; Wang, Y.; Ye, Y.; Zhao, Y.; Huang, Y.; Fu, W.; Chu, X. Effects of continuous nitrogen fertilizer application on the diversity and composition of rhizosphere soil bacteria. *Front. Microbiol.* **2020**, *11*, 1948. [CrossRef] [PubMed]
51. Krysenko, S.; Wohlleben, W. Role of Carbon, Nitrogen, Phosphate and Sulfur Metabolism in Secondary Metabolism Precursor Supply in *Streptomyces* spp. *Microorganisms* **2024**, *12*, 1571. [CrossRef] [PubMed]

52. Zhang, B.; Wu, X.; Tai, X.; Sun, L.; Wu, M.; Zhang, W.; Chen, X.; Zhang, G.; Chen, T.; Liu, G. Variation in actinobacterial community composition and potential function in different soil ecosystems belonging to the arid Heihe River Basin of Northwest China. *Front. Microbiol.* **2019**, *10*, 2209. [CrossRef] [PubMed]
53. Bao, Y.; Dolfing, J.; Guo, Z.; Chen, R.; Wu, M.; Li, Z.; Lin, X.; Feng, Y. Important ecophysiological roles of non-dominant Actinobacteria in plant residue decomposition, especially in less fertile soils. *Microbiome* **2021**, *9*, 84. [CrossRef] [PubMed]
54. Song, N.; Xu, H.; Yan, Z.; Yang, T.; Wang, C.; Jiang, H.-L. Improved lignin degradation through distinct microbial community in subsurface sediments of one eutrophic lake. *Renew. Energy* **2019**, *138*, 861–869. [CrossRef]
55. Hanrahan-Tan, D.G.; Henderson, L.; Kertesz, M.A.; Lilje, O. The effects of nitrogen and phosphorus on colony growth and zoospore characteristics of soil *Chytridiomycota*. *J. Fungi* **2022**, *8*, 341. [CrossRef] [PubMed]
56. Huang, Y.; Zheng, X.; Pilgaard, B.; Holck, J.; Muschiol, J.; Li, S.; Lange, L. Identification and characterization of GH11 xylanase and GH43 xylosidase from the chytridiomycetous fungus, *Rhizophlyctis rosea*. *Appl. Microbiol. Biotechnol.* **2019**, *103*, 777–791. [CrossRef] [PubMed]
57. Cerqueira, A.; Silva, T.; Nunes, A.; Nunes, D.; Lobato, L.; Veloso, T.; De Paula, S.; Kasuya, M.; Silva, C. Amazon basin pasture soils reveal susceptibility to phytopathogens and lower fungal community dissimilarity than forest. *Appl. Soil Ecol.* **2018**, *131*, 1–11. [CrossRef]
58. Wijayawardene, N.N.; Hyde, K.D.; Lumbsch, H.T.; Liu, J.K.; Maharachchikumbura, S.S.; Ekanayaka, A.H.; Tian, Q.; Phookamsak, R. Outline of ascomycota: 2017. *Fungal Divers.* **2018**, *88*, 167–263. [CrossRef]
59. Yu, Z.; Chen, L.; Pan, S.; Li, Y.; Kuzyakov, Y.; Xu, J.; Brookes, P.; Luo, Y. Feedstock determines biochar-induced soil priming effects by stimulating the activity of specific microorganisms. *Eur. J. Soil Sci.* **2018**, *69*, 521–534. [CrossRef]
60. Murgia, M.; Fiamma, M.; Barac, A.; Deligios, M.; Mazzarello, V.; Paglietti, B.; Cappuccinelli, P.; Al-Qahtani, A.; Squartini, A.; Rubino, S.; et al. Biodiversity of fungi in hot desert sands. *Microbiol. Open* **2019**, *8*, e00595. [CrossRef]
61. Floudas, D.; Gentile, L.; Andersson, E.; Kanellopoulos, S.G.; Tunlid, A.; Persson, P.; Olsson, U. X-Ray Scattering Reveals Two Mechanisms of Cellulose Microfibril Degradation by Filamentous Fungi. *Appl. Environ. Microbiol.* **2022**, *88*, e00995-22. [CrossRef]
62. Ruiz-Dueñas, F.J.; Barrasa, J.M.; Sánchez-García, M.; Camarero, S.; Miyauchi, S.; Serrano, A.; Linde, D.; Babiker, R.; Drula, E.; Ayuso-Fernández, I.; et al. Genomic Analysis Enlightens Agaricales Lifestyle Evolution and Increasing Peroxidase Diversity. *Mol. Biol. Evol.* **2020**, *38*, 1428–1446. [CrossRef]
63. Wang, Y.; Zhou, X. Effects of green manures on rhizosphere fungal community composition of cucumber seedlings. *Curr. Microbiol.* **2023**, *80*, 87. [CrossRef]
64. Zhou, H.; Di, L.; Hua, X.; Deng, T. The Addition of a Small Dose of Cinnamomum camphora Biomass Unexpectedly Enhanced Lignocellulose Degradation during the Compost of Stropharia rugosoannulata Cultivation Materials. *Sustainability* **2023**, *15*, 10483. [CrossRef]
65. Koechli, C.; Campbell, A.N.; Pepe-Ranney, C.; Buckley, D.H. Assessing fungal contributions to cellulose degradation in soil by using high-throughput stable isotope probing. *Soil Biol. Biochem.* **2019**, *130*, 150–158. [CrossRef]
66. Fuentes-Quiroz, A.; Herrera, H.; Alvarado, R.; Sagredo-Saez, C.; Isabel-Mujica, M.; Vohník, M.; Rolli, E. Cultivable Root-Symbiotic Bacteria of a Pioneer Ericaceous Dwarf Shrub Colonizing Volcanic Deposits and Their Potential to Promote host Fitness. *J. Soil Sci. Plant Nutr.* **2024**, *24*, 3355–3363. [CrossRef]
67. Li, W.; Zhang, Y.; Mao, W.; Wang, C.; Yin, S. Functional potential differences between Firmicutes and Proteobacteria in response to manure amendment in a reclaimed soil. *Can. J. Microbiol.* **2020**, *66*, 689–697. [CrossRef] [PubMed]
68. Lundell, T.K.; Mäkelä, M.R.; de Vries, R.P.; Hildén, K.S. Genomics, lifestyles and future prospects of wood-decay and litter-decomposing basidiomycota. In *Advances in Botanical Research*; Elsevier: Amsterdam, The Netherlands, 2014; Volume 70, pp. 329–370.
69. Manici, L.M.; Caputo, F.; De Sabata, D.; Fornasier, F. The enzyme patterns of Ascomycota and Basidiomycota fungi reveal their different functions in soil. *Appl. Soil Ecol.* **2024**, *196*, 105323. [CrossRef]
70. Manici, L.; Caputo, F.; Fornasier, F.; Paletto, A.; Ceotto, E.; De, M. Ascomycota and Basidiomycota fungal phyla as indicators of land use efficiency for soil organic carbon accrual with woody plantations. *Ecol. Indic.* **2024**, *160*, 111796. [CrossRef]
71. Hu, J.; Jin, V.L.; Konkel, J.Y.; Schaeffer, S.M.; Schneider, L.G.; DeBruyn, J.M. Soil health management enhances microbial nitrogen cycling capacity and activity. *Msphere* **2021**, *6*, 10–1128. [CrossRef]
72. Lori, M.; Armengot, L.; Schneider, M.; Schneidewind, U.; Bodenhausen, N.; Mäder, P.; Krause, H.-M. Organic management enhances soil quality and drives microbial community diversity in cocoa production systems. *Sci. Total Environ.* **2022**, *834*, 155223. [CrossRef]
73. Konishi, N.; Ishiyama, K.; Matsuoka, K.; Maru, I.; Hayakawa, T.; Yamaya, T.; Kojima, S. NADH-dependent glutamate synthase plays a crucial role in assimilating ammonium in the *Arabidopsis* root. *Physiol. Plant.* **2014**, *152*, 138–151. [CrossRef]
74. Yang, Y.; Gao, Y.; Wang, S.; Xu, D.; Yu, H.; Wu, L.; Lin, Q.; Hu, Y.; Li, X.; He, Z. The microbial gene diversity along an elevation gradient of the Tibetan grassland. *ISME J.* **2014**, *8*, 430–440. [CrossRef]
75. Cao, L.; Xu, C.; Sun, Y.; Niu, C.; Leng, X.; Hao, B.; Ma, J.; Liu, Z.; Xu, Z.; Yang, C. Genome-wide identification of glutamate synthase gene family and expression patterns analysis in response to carbon and nitrogen treatment in Populus. *Gene* **2023**, *851*, 146996. [CrossRef] [PubMed]
76. Wang, L.; Xiong, X. Long-Term Organic Manure Application Alters Urease Activity and Ureolytic Microflora Structure in Agricultural Soils. *Agronomy* **2022**, *12*, 3018. [CrossRef]

77. Sinsabaugh, R.; Reynolds, H.; Long, T. Rapid assay for amidohydrolase (urease) activity in environmental samples. *Soil Biol. Biochem.* **2000**, *32*, 2095–2097. [CrossRef]
78. Green, J.L.; Bohannan, B.J.; Whitaker, R.J. Microbial biogeography: From taxonomy to traits. *Science* **2008**, *320*, 1039–1043. [CrossRef]
79. Allison, S.D.; Martiny, J.B. Resistance, resilience, and redundancy in microbial communities. *Proc. Natl. Acad. Sci. USA* **2008**, *105*, 11512–11519. [CrossRef]
80. Adamczyk, B.; Sietiö, O.M.; Straková, P.; Prommer, J.; Wild, B.; Hagner, M.; Pihlatie, M.; Fritze, H.; Richter, A.; Heinonsalo, J. Plant roots increase both decomposition and stable organic matter formation in boreal forest soil. *Nat. Commun.* **2019**, *10*, 3982. [CrossRef]
81. Dotaniya, M.; Aparna, K.; Dotaniya, C.; Singh, M.; Regar, K. Role of soil enzymes in sustainable crop production. In *Enzymes in Food Biotechnology*; Elsevier: Amsterdam, The Netherlands, 2019; pp. 569–589.
82. Mariela, F.-P.; Pável, M.-E.I.; Manuel, S.-R.L.; Jesús, F.-G.M.; Reyes, L.-O. Dehydrogenase and mycorrhizal colonization: Tools for monitoring agrosystem soil quality. *Appl. Soil Ecol.* **2016**, *100*, 144–153. [CrossRef]
83. Mierzwa-Hersztek, M.; Wolny-Koładka, K.; Gondek, K.; Gałązka, A.; Gawryjołek, K. Effect of coapplication of biochar and nutrients on microbiocenotic composition, dehydrogenase activity index and chemical properties of sandy soil. *Waste Biomass Valorization* **2020**, *11*, 3911–3923. [CrossRef]
84. Beheshti, M.; Etesami, H.; Alikhani, H.A. Effect of different biochars amendment on soil biological indicators in a calcareous soil. *Environ. Sci. Pollut. Res.* **2018**, *25*, 14752–14761. [CrossRef]
85. Saha, J.; Sengupta, A.; Gupta, K.; Gupta, B. Molecular phylogenetic study and expression analysis of ATP-binding cassette transporter gene family in Oryza sativa in response to salt stress. *Comput. Biol. Chem.* **2015**, *54*, 18–32. [CrossRef]
86. Giuliani, S.E.; Frank, A.M.; Corgliano, D.M.; Seifert, C.; Hauser, L.; Collart, F.R. Environment sensing and response mediated by ABC transporters. *BMC Genom.* **2011**, *12*, S8. [CrossRef] [PubMed]
87. Siqueira, A.F.; Ormeño-Orrillo, E.; Souza, R.C.; Rodrigues, E.P.; Almeida, L.G.P.; Barcellos, F.G.; Batista, J.S.S.; Nakatani, A.S.; Martínez-Romero, E.; Vasconcelos, A.T.R. Comparative genomics of Bradyrhizobium japonicum CPAC 15 and Bradyrhizobium diazoefficiens CPAC 7: Elite model strains for understanding symbiotic performance with soybean. *BMC Genom.* **2014**, *15*, 1–21. [CrossRef] [PubMed]
88. Florez, P.A.V.; Guerra, D.A.C.; Rosero, S.E.V. Capacidad y diversidad de bacterias celulolíticas aisladas de tres hábitats tropicales en Boyacá, Colombia. *Acta Agronómica* **2016**, *65*, 362–367. [CrossRef]
89. Adetunji, A.T.; Lewu, F.B.; Mulidzi, R.; Ncube, B. The Biological Activities of Beta-Glucosidase, Phosphatase and Urease as Soil Quality Indicators: A Review. *J. Soil Sci. Plant Nutr.* **2017**, *17*, 794–807. [CrossRef]
90. Stege, P.W.; Messina, G.A.; Bianchi, G.; Olsina, R.A.; Raba, J. Determination of β-glucosidase activity in soils with a bioanalytical sensor modified with multiwalled carbon nanotubes. *Anal. Bioanal. Chem.* **2010**, *397*, 1347–1353. [CrossRef]
91. Teixeira, P.C.; Donagemma, G.K.; Fontana, A.; Teixeira, W.G. *Manual de Métodos de Análise de Solo*; Embrapa: Brasília, Brazil, 2017.
92. Moniz, A.; Jorge, J.; Valadares, J. *Métodos de Análise Química, Mineralógica e Física de Solos do Instituto Agronômico de Campinas*; Instituto Agronômico: Campinas, Brazil, 2009.
93. Costa, P.H.d.O.; Nascimento, S.V.d.; Herrera, H.; Gastauer, M.; Ramos, S.J.; Caldeira, C.F.; Oliveira, G.; Valadares, R.B.d.S. Non-specific interactions of rhizospheric microbial communities support the establishment of Mimosa acutistipula var. ferrea in an Amazon rehabilitating mineland. *Processes* **2021**, *9*, 2079. [CrossRef]
94. Qian, C.; Hettich, R.L. Optimized extraction method to remove humic acid interferences from soil samples prior to microbial proteome measurements. *J. Proteome Res.* **2017**, *16*, 2537–2546. [CrossRef]
95. Ogunseitan, O. Direct extraction of proteins from environmental samples. *J. Microbiol. Methods* **1993**, *17*, 273–281. [CrossRef]
96. Nascimento, S.V.d.; Herrera, H.; Costa, P.H.d.O.; Trindade, F.C.; da Costa, I.R.C.; Caldeira, C.F.; Gastauer, M.; Ramos, S.J.; Oliveira, G.; Valadares, R.B.d.S. Molecular Mechanisms Underlying *Mimosa acutistipula* Success in Amazonian Rehabilitating Minelands. *Int. J. Environ. Res. Public Health* **2022**, *19*, 14441. [CrossRef]
97. Oliveira, R.R.; Silva, R.; Nunes, G.L.; Oliveira, G. PIMBA: A PI peline for M eta B arcoding A nalysis. In Advances in Bioinformatics and Computational Biology, In Proceedings of the 14th Brazilian Symposium on Bioinformatics, BSB 2021, Virtual Event, 22–26 November 2021; Springer International Publishing: Berlin/Heidelberg, Germany, 2021; pp. 106–116.
98. Cole, J.R.; Wang, Q.; Fish, J.A.; Chai, B.; McGarrell, D.M.; Sun, Y.; Brown, C.T.; Porras-Alfaro, A.; Kuske, C.R.; Tiedje, J.M. Ribosomal Database Project: Data and tools for high throughput rRNA analysis. *Nucleic Acids Res.* **2014**, *42*, D633–D642. [CrossRef]
99. Nilsson, R.H.; Larsson, K.-H.; Taylor, A.F.S.; Bengtsson-Palme, J.; Jeppesen, T.S.; Schigel, D.; Kennedy, P.; Picard, K.; Glöckner, F.O.; Tedersoo, L. The UNITE database for molecular identification of fungi: Handling dark taxa and parallel taxonomic classifications. *Nucleic Acids Res.* **2019**, *47*, D259–D264. [CrossRef] [PubMed]
100. R Core Team. *R: A Language and Environment for Statistical Computing*; R Foundation for Statistical Computing: Vienna, Austria, 2023. Available online: https://www.R-project.org/ (accessed on 21 May 2024).
101. Parks, D.H.; Tyson, G.W.; Hugenholtz, P.; Beiko, R.G. STAMP: Statistical analysis of taxonomic and functional profiles. *Bioinformatics* **2014**, *30*, 3123–3124. [CrossRef] [PubMed]

International Journal of
Molecular Sciences

MDPI

Article

Difenoconazole Degradation by Novel Microbial Consortium TA01: Metabolic Pathway and Microbial Community Analysis

Tianyue Wang †, Jingyi Sui †, Yi Zhou, Liping Wang, Jia Yang, Feiyu Chen , Xiuming Cui, Ye Yang * and Wenping Zhang *

Key Laboratory of Sustainable Utilization of Panax Notoginseng Resources of Yunnan Province, Faculty of Life Science and Technology, Kunming University of Science and Technology, Kunming 650500, China; wangtianyue26@163.com (T.W.); suijingyi29@163.com (J.S.); zhouyi001016@163.com (Y.Z.); lpwang0301@126.com (L.W.); jia091502@163.com (J.Y.); chenfeiyv2002@163.com (F.C.); cuixm@kust.edu.cn (X.C.)

* Correspondence: yangyekm@163.com or yangye@kust.edu.cn (Y.Y.); wpzhang@kust.edu.cn (W.Z.)

† These authors contributed equally to this work.

Abstract: Difenoconazole, a broad-spectrum systemic fungicide, can effectively prevent and control plant diseases such as rice blast, leaf spot, and black spot caused by *Colletotrichum godetiae*, *Alternaria alternata*, and *Neopestalotiopsis rosae*. However, its residual accumulation in the environment may pose potential toxicity risks to non-target organisms. In this study, a highly efficient DIF-degrading microbial consortium TA01 was enriched from long-term pesticide-contaminated soil by a laboratory-based adaptive evolution strategy. The microbial consortium TA01 was able to degrade 83.87% of 50 mg/L of DIF within 3 days. In addition, three intermediate metabolites were identified using HPLC–MS/MS, and the results indicated that the degradation of DIF by microbial consortium TA01 may involve catalytic reactions such as hydrolysis, dehalogenation, and hydroxylation. High-throughput sequencing results showed that *Pantoea*, *Serratia*, *Ochrobactrum*, and *Bacillus* were the dominant microbial members involved in the degradation process. Finally, bioremediation capacity experiments showed that inoculation with microbial consortium TA01 was able to accelerate the degradation of DIF in the water–sediment system. The findings of this study not only enrich the microbial resources available for DIF degradation but also offer new potential strategies for in situ remediation of DIF contamination.

Keywords: difenoconazole; biodegradation; microbial consortium; degradation pathway; bioremediation

1. Introduction

Difenoconazole is a broad-spectrum triazole fungicide which can effectively prevent and control a variety of plant diseases by inhibiting fungal mycelial growth and conidium formation. Its targets cover the major pathogen groups, such as Ascomycetes, Ascomycetes, and Hemipterans, and are as follows: strong killing activity against sensitive strains of *Colletotrichum godetiae* [1]; significant inhibition of black spot disease of ootheca [2] and black stem disease of quinoa [3] caused by *Alternaria caulina* [1]; significant inhibition of black spot disease of auburn caused by *Alternaria alternata* [2] and black stem disease of quinoa caused by *Alternaria caulina* [3]; and significant prevention of root rot of strawberry caused by *Neopestalotiopsis rosae*, which could recover up to 70% of the yield loss [4]. The agent also controls pea powdery mildew (*Erysiphe pisi*) by interfering with conidiogenesis [5], and controls apple blotch (*Diplocarpon coronariae*) by inhibiting mycelial expansion [6] and potato brown leaf spot (*Phaeosphaeriopsis obtusispora*) [7]. Its broad spectrum is further reflected in the significant inhibition of cross-species pathogens such as *Fusarium* spp., *Didymella segeticola*, and *Mucor xinjiangensis* [8–10]. Field data have shown that this agent has a stable effect in the control of major diseases such as rice blast (*Magnaporthe oryzae*) and maize leaf spot (*Magnaporthiopsis maydis*) [11,12], which emphasizes its key role in the integrated management system of crop diseases [11].

In contrast, the environmental residues of DIF pose a multidimensional risk to non-target organisms and ecosystems. In terms of the mechanisms of toxicity, the compound can cause cross-species toxicity by inducing oxidative stress, interfering with metabolic pathways and activating apoptotic signaling. For example, in zebrafish, it has caused motor dysfunction by inhibiting neurotransmitter levels, key enzyme activities, and neurodevelopmental gene expression [13]; in carp, it has caused a significant increase in intestinal oxidative stress and apoptosis [14], as well as triggering organ damage such as glomerular edema and gill tissue necrosis [15]; and in mammals it manifests as disruption of endocrine function—inhibition of human 3β-HSD1 enzyme activity leads to reduced luteinizing hormone synthesis [16]—and induces ER stress and inflammatory responses in bovine mammary epithelial cells [17]. At the ecological level, DIF has significantly altered the structure of the gut flora of soil organisms (e.g., *Enchytraeus crypticus*), reduced the abundance of beneficial bacteria [18], and inhibited the functioning of the honeybee gut microbiome [19], disrupting the microbial ecological balance of soil organisms. In terms of environmental residues, it has been detected in the surface water of the Yangtze River Delta estuary in China at a concentration of 0.412 ± 0.424 μg/L [20], and has shown a trend towards enrichment in nearshore sediments [21]. In terms of agricultural safety, this agent not only inhibits photosynthesis and interferes with glutathione metabolism, hormone signaling, and other key pathways in wheat [22], but also reduces soluble sugar content, increases acidity, and induces oxidative damage in strawberries, which can significantly deteriorate the flavor quality of the fruit [23]. The above evidence suggests that the environmental behavior of DIF has created a "biotoxicity-ecological imbalance-food safety" chain of risks, and that there is an urgent need to weigh the benefits of control against the potential hazards in application management.

To address the problem of persistent residues of DIF in the environment, a variety of degradation technologies with potential applications have been developed. Among the physicochemical methods based on the adsorption principle, biochar showed 65.3% adsorption of DIF [24], while chitosan biopolymer also demonstrated significant pesticide adsorption capacity through surface functional group complexation [25]. In the field of bioremediation, although the earthworm-mediated bioavailability process can achieve a degradation rate of 42.0% [24], there is still room for improvement in its efficiency. It is noteworthy that microbial degradation technology has become the focus of research because of its high efficiency and environmental friendliness. *Pseudomonas putida* A-3 strain degraded DIF up to 75.98% under optimized conditions [26] and even more groundbreaking is the synergistic bacterial flora system constructed by *Klebsiella*, *Pseudomonas* and *Citrobacter*, which not only promotes plant growth and development but also achieves 85.83–96.59% degradation efficiency of 50 mg/kg of contaminated soil within 45 days in the soil–plant system [27]. These studies indicate that a joint remediation strategy based on microbial community function regulation, which realizes the efficient degradation of pesticides and ecological restoration function at the same time, may become an important direction for future environmental management.

In this study, the construction and mechanisms of microbial degradation systems were investigated systematically in response to the lack of information regarding microbial degradation mechanisms and resources for DIF. An efficient DIF-degrading bacterial population was firstly enriched and screened by a multi-gradient pressure domestication method, and a single factor was used in order to obtain the optimal degradation conditions. On this basis, 16S rRNA high-throughput sequencing was applied to analyze the dynamic succession of microbial consortium structure during the degradation process. The degradation intermediates were traced by high performance liquid chromatography–tandem mass spectrometry (HPLC-MS/MS) to elucidate the metabolic pathway of DIF degradation by bacteria. Finally, the actual remediation efficacy of the microbial consortium on DIF contamination in a composite water–sediment system was assessed by bioremediation simulation experiments. This study provides a theoretical basis and practical foundation for revealing the law of DIF degradation by microbial consortia and developing green bioremediation technology.

2. Results and Discussion

2.1. Determination of the Ability of Microbial Consortium to Degrade DIF

In this study, four microbial consortia—TA01, TB01, TC01, and TD01—were screened from four DIF-contaminated soils by enrichment culture. Dynamic monitoring showed (Figure 1) that all four microbial consortia showed some DIF degradation ability in MSM media with 50 mg/L of DIF, with the highest degradation efficiency belonging to microbial consortium TA01 and a relatively slow degradation rate shown by microbial consortium TD01. In addition, the degradation rate for microbial consortium TA01 was 83.87% after 5 days of incubation, which was significantly higher than that for the other microbial consortia (Consortium B: 82.84%, Consortium C: 76.27%, and Consortium D: 73.18%); therefore, microbial consortium TA01 was established as the core flora resource for the subsequent study.

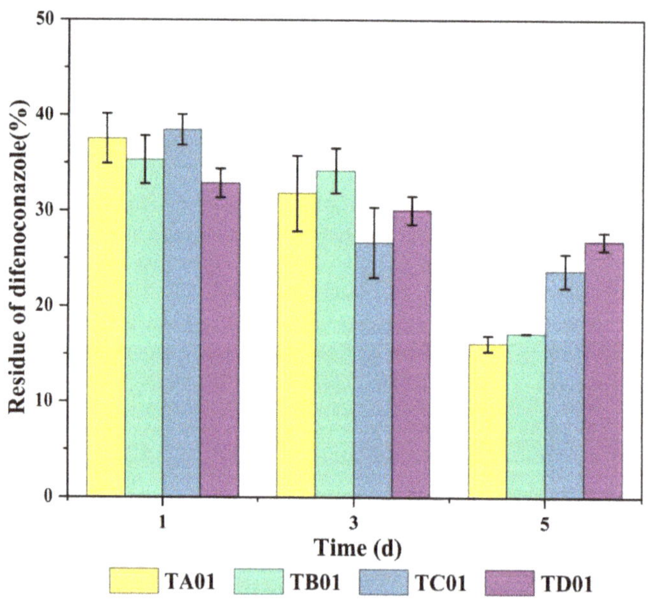

Figure 1. Degradation effect of different microbial consortia on difenoconazole. The bar represents the means of three identical replicates (N = 3) and the error line represents the standard deviation from the mean.

Compared with the reported degradation rate of 62.57% by *Pseudomonas putida* A-3 on 50 mg/L of DIF in 7 d under optimal conditions [27], the present study's microbial consortium TA01 was able to achieve 83.87% degradation in 5 d at the same concentration. In addition, two strains of bacteria—strain T-1 (*Phyllobacterium* sp.) and strain T-2 (*Aeromonas* sp.)—were reported to be able to achieve 96.32% when cultured for fourteen consecutive days (concentration of 20 mg/L) [28]. Although the final degradation rate of these two strains appeared to be higher, when analyzed in the time dimension, microbial consortium TA01 (mainly composed of *Pantoea*, *Serratia*, *Ochrobactrum*, and *Bacillus*) was able to reach a higher degradation rate in a shorter period of time, reflecting a more efficient degradation ability. It can be seen that a variety of microorganisms within each microbial consortium can cooperate with each other and act on DIF from different pathways at the same time, thus accelerating the degradation process and improving degradation efficiency; whereas, due to the single metabolic pathway of a single bacterium, the degradation rate tends to be slower and the degradation effect is not significant.

2.2. Curves of Growth and Degradation Relationship of Microbial Consortium TA01

In view of the fact that microbial consortium TA01 has a superior degradation ability to the other three groups, this study further investigated the growth and degradation relationship. In order to investigate the relationship between growth and degradation for microbial consortia during the degradation of DIF, the growth and degradation rate of microbial consortium TA01 during the degradation of DIF were also monitored. The results, as shown in Figure 2, showed that microbial consortium TA01 was able to rapidly degrade DIF and was able to utilize DIF as the only source of carbon to participate in the metabolic process of the microbial consortium, which could be converted into the nutrients and energy materials required for its growth and metabolism. As can be seen from Figure 2, the microbial consortium TA01 did not have an obvious delay period and achieved rapid degradation of DIF on the first day, with a degradation rate of 76.73%. The DIF degradation rate did not vary much with the continuous growth of microbial consortia and reached 78.31% on day 7. However, there was only weak degradation of DIF in the control group (25.37%).

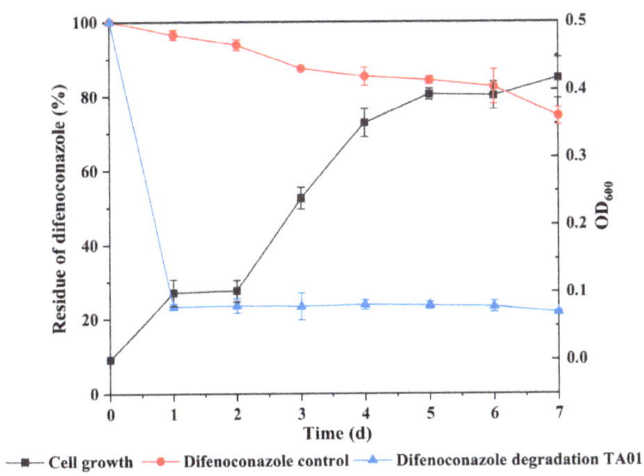

Figure 2. Degradation of difenoconazole during the growth of microbial consortium TA01. Dots indicate the means of three identical replicate samples (N = 3) and error lines indicate the standard deviation from the mean.

For the phenomenon that occurs in Figure 2, this study suggests the following possible causes: At the beginning, the concentration of DIF was sufficient, and the microbial consortium rapidly utilized it as a carbon source for degradation; however, the substrate residue decreased significantly after 2 days, and the degradation reaction was limited by the amount of substrate available, which led to the slowing down of the degradation rate, although the change in degradation rate was not obvious. At the same time, after an acclimatization period, the microbial consortium tends to mature in terms of reproduction conditions (e.g., nutrient transformation, metabolic environment) and enters into a logarithmic growth period, showing exponential growth. On the other hand, the microbial consortium may have activated other metabolic pathways, such as utilization of intracellular storage material and synergistic utilization of other carbon and nitrogen sources in the system, in parallel with the degradation of DIF. These pathways prioritize the growth requirements of the bacterial population so that growth is not limited by the DIF degradation efficiency, leading to an eventual segregation phenomenon of a smooth degradation rate and an exponential growth rate.

2.3. *Effect of Culture Conditions on the Degradation of DIF by Microbial Consortium TA01*

In this experiment, the effects of pH, inoculum size, and initial concentration of DIF on the degradation of DIF by microbial consortium TA01 were investigated using one-way optimization. The effect of the initial concentration of DIF on the degradation of DIF by microbial consortium TA01 is shown in Figure 3a. When the initial concentration of DIF was 50 mg/L, microbial consortium TA01 had the highest DIF degradation rate, which amounted to 74.06% at 3 d. The degradation rate of microbial consortium TA01 was 74.06%. However, the degradation rate kept decreasing when the concentration of DIF was higher than 50 mg/L.

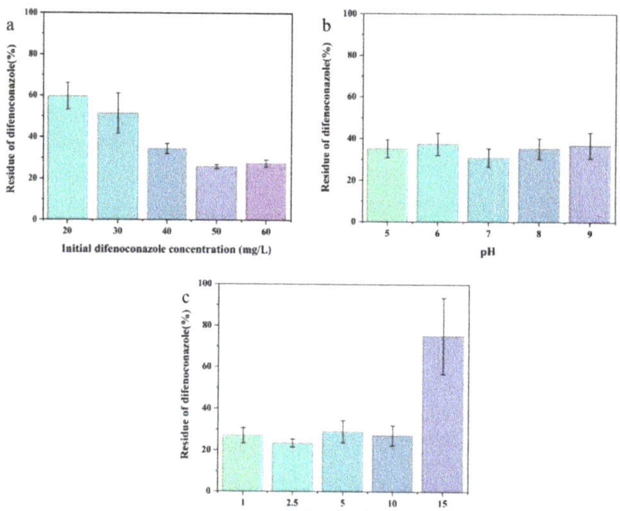

Figure 3. Effect of initial concentration (**a**), pH (**b**), and inoculum (**c**) of difenoconazole on the degradation of difenoconazole. The bar represents the means of three identical replicates ($N = 3$) and the error line represents the standard deviation from the mean.

This may be due to the fact that at low initial concentrations, DIF can induce the expression of biodegradation-related genes to produce a large number of enzymes, thus promoting the metabolism of DIF. However, high concentrations of DIF not only inhibit the growth of the strain but also have a toxic effect on the enzymes involved in microbial degradation, thereby inhibiting enzymatic reactions. This is consistent with the findings of previous studies such as by Tan et al., who found that the degradation rate of chloramphenicol by *Aeromonas media* SZW3 increased gradually in the range of 5–10 mg/L and decreased significantly when above 20 mg/L [29]. In addition, the degradation rate gradually decreased in the range of 20–100 mg/L for strain T-1 (*Phyllobacterium* sp.) and strain T-2 (*Aeromonas* sp.), as studied by Chen et al., who also studied DIF [28]. It can be seen that microbial consortium TA01 (mainly composed of *Pantoea*, *Serratia*, *Ochrobactrum*, and *Bacillus*) in the present study can withstand a higher degradation concentration than the single strain with good degradation.

The degradation of DIF is critically important. The pH affects the activity of microbial enzymes, which in turn affects the growth and metabolic processes of the bacteria. As can be seen in Figure 3b, the DIF degradation efficiency was shown to be high between pH 6 and 7. At pH 7, the degradation rate of microbial consortium TA01 was 69.04%. Too high or too low pH affects the growth and enzyme activity of the bacteria. Many experimental studies have shown that microorganisms generally prefer neutral environments, e.g., strain T-1 (*Phyllobacterium* sp.), strain T-2 (*Aeromonas* sp.) [28], and *Pseudomonas putida* A-3 [26]. This may be due to the fact that a neutral environment is a mild degrading environment where microorganisms can adapt more quickly and thus degrade pesticides better.

The effect of inoculum amount on the degradation of DIF by microbial consortium TA01 is shown in Figure 3c. The DIF degradation efficiency of microbial consortium TA01 increased gradually with the increase in inoculum amount. When the inoculum amount was 2.5%, microbial consortium TA01 showed the highest degradation efficiency for DIF, and the degradation rate reached 76.6% at 3 d. The degradation efficiency of microbial consortium TA01 was higher than that of microbial consortium TA01 when the inoculum amount was 2.5%. When the inoculum amount was increased to 20%, its degradation efficiency did not show significant increase, probably because the growth of the bacteria was too fast and the amount of substrate was not enough to meet the growth and metabolism of the bacteria. When the inoculum is too low, the concentration of bacteria is too low, the growth delay period is prolonged, the cell activity is low, and the time required to reach the appropriate degradation efficiency is longer.

While strain T-1 (*Phyllobacterium* sp.) was reported to achieve its maximum degradation rate at only 5% inoculum (83.25%), strain T-2 (*Aeromonas* sp.) was reported to achieve maximum degradation rate at 10% inoculum (80.28%) [28], which required more biomass than microbial consortium TA01 (mainly composed of *Pantoea*, *Serratia*, *Ochrobactrum*, and *Bacillus*) required to achieve maximum degradation rate at 2.5% inoculum (76.6%). This can be seen in the strong degradation potential of the bacteria at lower biomass inputs. This may be due to the existence of synergistic effects between microorganisms within the microbial consortium, which allows efficient activation of the expression of degradation-related genes and enzymatic activity at low starting numbers, leading to efficient metabolism of pollutants.

2.4. DIF Degradation Kinetics Analysis

During co-metabolism, primary substrate concentration may appear to inhibit microbial growth and substrate degradation kinetics can be modeled using the Andrews equation [30–32]. As shown in Figure 4, in this study, the following kinetic parameters were obtained by nonlinear fitting of the experimental data with Origin 2024: maximum specific degradation rate q_{max} = 0.5 d^{-1}; half-saturation constant K_s = 17.238 mg/L; inhibition constant K_i = 289.344 mg/L; and model coefficient of determination R^2 = 0.97498. Through the derivation of the equation, the maximum specific degradation rate (S_{max}) corresponding to the initial concentration of DIF was calculated to be 70.63 mg/L, which was the optimal initial concentration for the degradation of DIF by microbial consortium TA01.

As shown in Figure 4, when the concentration of DIF was lower than 70.63 mg/L, the specific DIF degradation rate of microbial consortium TA01 increased rapidly with the increase in the initial concentration; when the initial concentration was more than 70.63 mg/L, the specific degradation rate decreased with the increase in the initial concentration, which showed the substrate inhibition effect. This may be due to the high concentration of DIF, which produces toxicity to the microbial consortium and inhibits its growth.

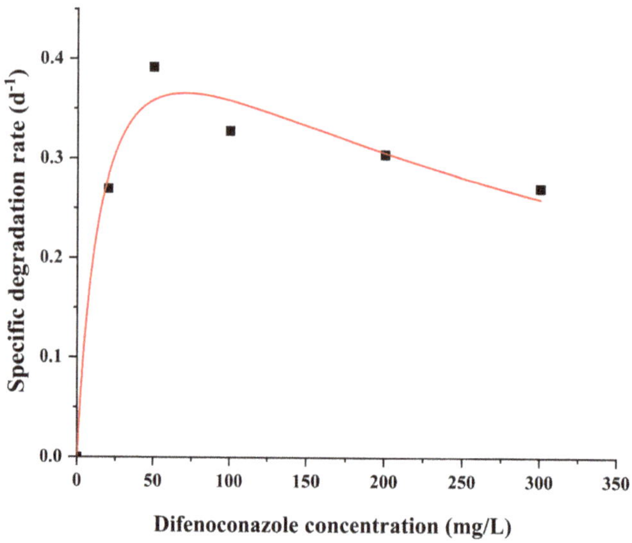

Figure 4. DIF degradation kinetics of TA01 for different initial concentrations: Relationship between the initial difenoconazole concentration and specific degradation rate.

2.5. Community Structure of Degradative Flora and Analysis of Key Degradative Bacteria

In order to analyze the key genera contributing to DIF degradation by microbial consortium TA01, the present study was carried out to investigate the community changes in the process of DIF degradation by microbial consortium TA01 by utilizing 16S rDNA high-throughput sequencing technology. As can be seen from Table 1, the sequence numbers of all sequenced samples range from 80,451 to 87,525. The α-diversity of microbial consortium TA01 showed a phased change: the Chao1 index decreased from an initial 257.05 to 77.12 and then rebounded to 97.00, and the Shannon index decreased from 4.51 to 2.23 and then recovered gradually to 2.78. This trend suggests that high substrate concentrations in the early stages of degradation (0–5 d) lead to a decrease in community richness (enrichment of specific degrading bacteria), whereas the accumulation of metabolic by-products in the later stages (5–7 d) activates secondary metabolic flora and facilitates micro-ecological reconfiguration. The Simpson dominance index rebounded to 0.76 after dropping from 0.89 to 0.60, corroborating the adaptive shift from broad-spectrum competition to specialized metabolism.

Table 1. Bacterial community ordinal number and diversity.

Sample (d)	Sequences	Shannon	Simpson	Chao1
0	87,522	4.51	0.89	257.05
1	85,255	2.23	0.64	87.0
2	85,486	2.31	0.62	87.0
3	87,062	2.17	0.60	94.0
4	82,368	2.42	0.68	86.0
5	84,866	2.77	0.79	77.12
6	83,468	2.84	0.75	111.00
7	80,451	2.78	0.76	97.00

The results of the phylum and genus level analyses of the community structure during DIF degradation by microbial consortium TA01 are shown in Figure 5. At the phylum level (Figure 5a), in the initial stage, the relative abundance of *Proteobacteria* was 60.71%, which coexisted with *Firmicutes* (29.75%), *Actinobacteriota* (3.23%), and *Bacteroidota* (4.7%). As the degradation process progressed, the abundance of *Proteobacteria* continued to increase to 99.20%, while the abundance of other phyla decreased to <0.5%. The relative abundance of *Proteobacteria* increased significantly during the degradation process, in contrast to a sharp decrease in the relative abundance of *Firmicutes*, *Actinobacteriota*, and *Bacteroidota*. It can be seen that this degradation process of *Proteobacteria* showed a significant metabolic advantage. Researchers have found that species of *Proteobacteria* have important applications in industry, agriculture, medicine, health, and environmental protection, especially in industrial and agricultural wastewater treatment, soil remediation, and pollutant degradation [33].

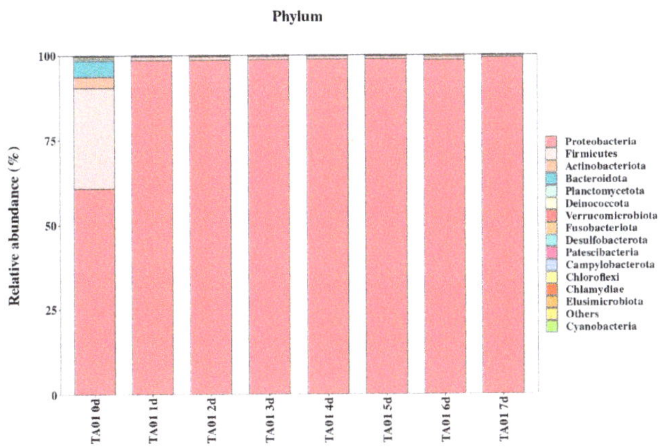

(a)

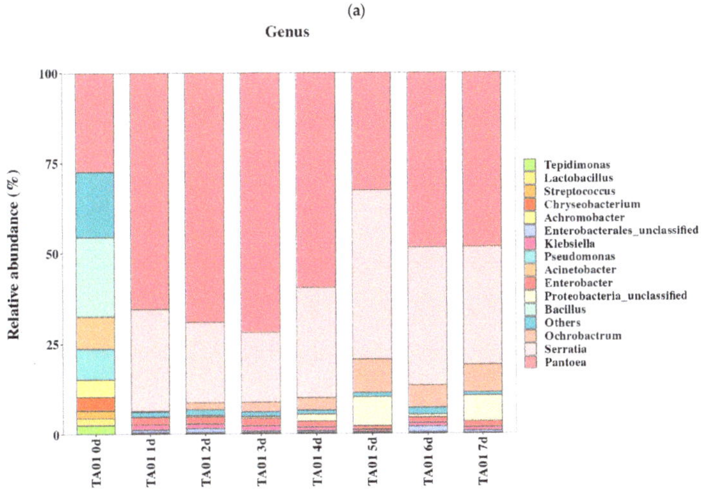

(b)

Figure 5. Evolution of microbial consortia at phylum (**a**) and genus (**b**) level during the degradation of difenoconazole.

In addition, genus level analysis (Figure 5b) revealed the dynamics of the degrading genera in microbial consortium TA01: the initial stage was dominated by *Pantoea* (27.38%), *Bacillus* (21.95%), *Pseudomonas* (8.60%), and *Actinobacteria* (8.86%); in the middle stage of degradation, the abundance of *Pantoea* and *Serratia* increased to 71.79% and 19.45%, respectively; and in the end stage, a stable community was formed, dominated by *Pantoea* (48.33%), *Serratia* (32.49%), and *Ochrobactrum* (7.74%). It is noteworthy that the above genera are reported for the first time in the study of DIF degradation, providing candidate resources for the development of novel degrading strains.

2.6. Characterization of DIF Degradation Products

In this study, three main DIF degradation products from microbial consortium TA01 were found (Table 2). One of them, DIF-TPs 406, with a mass-to-charge ratio of 406.00 for the hydrogenation peak [M+H] and a molecular formula of $C_{19}H_{20}ClN_3O_5$, loses one chlorine atom and adds two oxygen atoms compared to the molecular formula of the DIF parent ($C_{19}H_{17}Cl_2N_3O_3$), a process that occurs by hydroxylation of halogen atoms and hydrolysis to open the ring [28]. DIF-TPs 330, with a mass-to-charge ratio of 330.25 for the hydrogenation peak [M+H] and a molecular formula of $C_{16}H_{13}ClN_3O_3$, should be hydrolyzed for the hydrolysis of an oxygen-containing five-membered heterocyclic ring to open the ring and lose a C_2H_4O molecule. DIF-TPs 290, with a mass-to-charge ratio of 290.25 for the hydrogenation peak [M+H] and a molecular formula of $C_{16}H_{15}ClO_3$, underwent a hydrolysis reaction and lost an N-containing five-membered heterocycle $C_2H_2N_3$ [32].

Table 2. DIF degradation products.

Compounds	Chemical Structures	Molecular Formula	*m/z* [M+H]
DIF-TPs 406	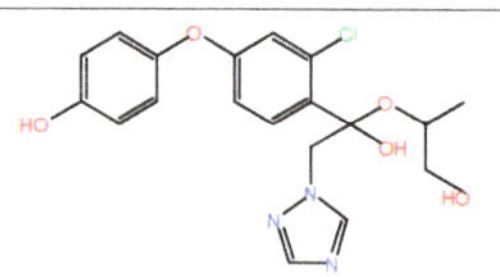	$C_{19}H_{20}ClN_3O_5$	406.00
DIF-TPs 330	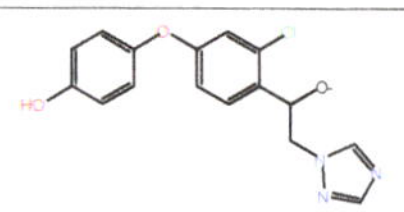	$C_{16}H_{13}ClN_3O_3$	330.25
DIF-TPs 290	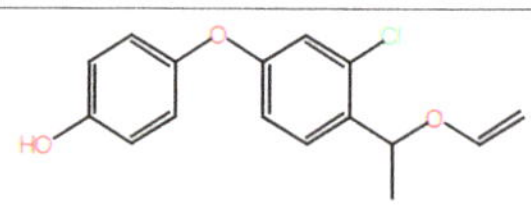	$C_{16}H_{15}ClO_3$	290.25

Based on the structural identification of the degradation metabolites mentioned above, the DIF biodegradation pathway was deduced in this study, as shown in Figure 6. The parent compound was first generated as the intermediate product DIF-TPs 406 (molecular formula $C_{19}H_{20}ClN_3O_5$) by hydrolysis and dehalogenation hydroxylation of the dioxycyclohexane structure; subsequently, the oxygen-containing pentaheterocycle of DIF-TPs 406 was further hydrolyzed by ring-opening and the dehydroxylation of the C_2H_4O group to form DIF-TPs 330 ($C_{16}H_{13}ClN_3O_3$); meanwhile, the nitrogen-containing five-membered heterocyclic ring of DIF-TPs 406 was broken and accompanied by a hydrolysis reaction to generate DIF-TPs 290 ($C_{16}H_{15}ClO_3$). This pathway is highly consistent with the reaction mechanisms proposed by Man et al. [33] and Chen et al. [28], suggesting that the core steps of DIF degradation (e.g., ether bond breaking or heterocyclic ring cleavage) are highly similar in different microbial systems [34].

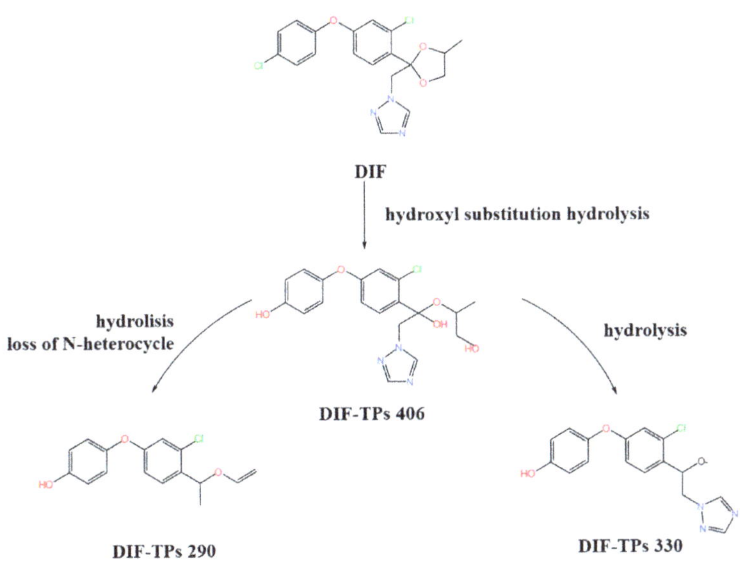

Figure 6. Pathway of difenoconazole degradation by microbial consortium TA01.

2.7. Biodegradation of DIF in Water–Sediment Simulated Pollution Systems

In order to verify the remediation ability of microbial consortium TA01 on the mixed water–sediment system with DIF contamination, four treatment groups were set up in this section to investigate the bioremediation ability of microbial consortium TA01. The results are shown in Figure 7, from which it can be seen that inoculation of microbial consortium TA01 in both sterilized and unsterilized water–sediment systems significantly increased the DIF degradation rate. The DIF degradation rate by microbial consortium TA01 was 75.29% in the 5 d unsterilized water–sediment system, whereas microbial consortium TA01 reached 100% degradation in the sterilized water–sediment system. This indicates that the microbial consortium TA01 is able to exert its degradation ability by growing and multiplying rapidly in the water–sediment system. In contrast, the degradation of phenoxymetrazole in sterilized and unsterilized water–sediment was 68.16% and 94.56% in the treatment group not inoculated with the microbial consortium, respectively. This may be due to the possible presence of some substances with degradation capacity in the water–sediment system.

In addition, in order to better characterize the remediation ability of microbial consortium TA01 on the water–sediment system, this study utilized a first-order kinetic model to fit the data on the degradation of DIF in water–sediment by microbial consortium TA01, and the results are shown in Table 3. The coefficients of determination R^2 for the four treatment groups were 0.97974, 0.98891, 0.88518, and 0.86704, respectively. The degradation rate constants (k) were 0.07067, 0.81563, 0.03094, and 0.30261 d^{-1}, respectively. the degradation half-lives ($t_{1/2}$) were 9.8, 0.9, 22.4, and 2.3 d.

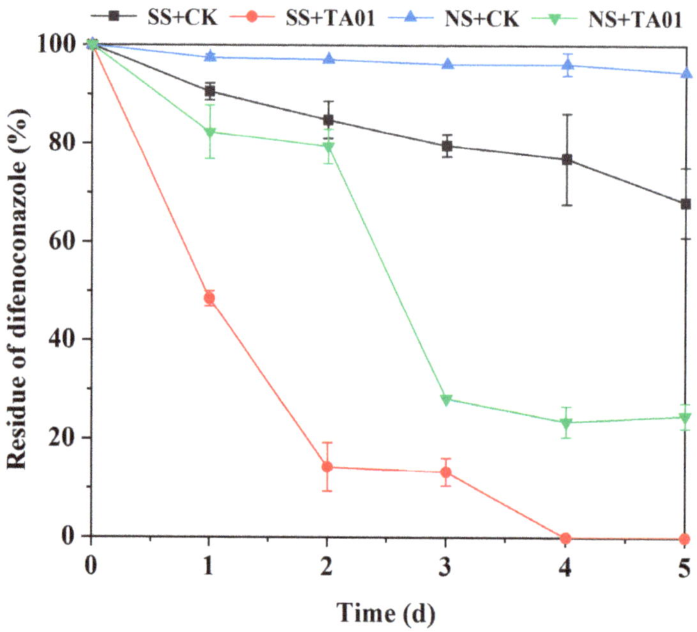

Figure 7. Effect of microbial consortium TA01 on the degradation of difenoconazole in water–sediment pairs.

Table 3. Kinetics of DIF degradation by microbial consortium TA01 in water–sediments.

Treatment	Regression Equation	k (d^{-1})	$t_{1/2}$ (d)	R^2
SS + CK	$C_t = 98.77213e^{-0.07067t}$	0.07067	9.8	0.97974
SS + TA01	$C_t = 100.84707e^{-0.81563t}$	0.81563	0.9	0.98891
NS + CK	$C_t = 99.21675e^{-0.03094t}$	0.03094	22.4	0.88518
NS + TA01	$C_t = 106.36416e^{-0.30261t}$	0.30261	2.3	0.86704

3. Materials and Methods

3.1. Enrichment and Screening of DIF-Degrading Microbial Consortia

Four different soil samples were retrieved from a Chinese herbal medicine plantation area in Kunming, Yunnan Province, and labeled A, B, C, and D. These four soil samples were enriched to create microbial consortia. Enrichment of DIF-degrading microbial consortia was carried out by an incubation procedure described by Bhatt et al. [35]. About 5 g of soil samples were added to an MSM medium containing 50 mg/L of DIF and incubated for 3 d at 30 °C and 200 rpm in a shaker. Meanwhile, 1 mL of the enriched culture was transferred to a fresh MSM medium containing 100 mg/L of DIF after 3 d. Enrichment cultures were serially transferred to fresh enrichment media containing 200, 300, 400, 500, or 600 mg/L of DIF for another 3 d. Four microbial consortia named TA01, TB01, TC01 and TD01 were finally obtained from the soil samples. Among them, microbial consortium TA01 (mainly composed of *Pantoea*, *Serratia*, *Ochrobactrum*, and *Bacillus*) was the one with the highest DIF degradation capacity among the four microbial consortia and, therefore, was selected for subsequent studies on DIF bioremediation.

3.2. Determination of DIF Degradation and Growth for Microbial Consortium TA01

The conserved microbial consortium TA01 was inoculated into an LB medium (30 °C, 200 rpm) for overnight activation, and the bacterial fluid was centrifuged at 3800 rpm for

3 min to collect the bacterial body, the supernatant was discarded and resuspended in sterile water and vortexed for 30–60 s. The washing–centrifugation step was repeated 2 times to remove the medium residue. The seed suspension was prepared by resuspending the final bacterial precipitate in sterile water, transferring it to an inorganic salt medium (MSM) containing 50 mg/L of inoculum with 10% DIF, incubating it at 30 °C and 200 rpm for 7 days under light protection (three replicates were set up for each group). Daily samples were taken at regular intervals on days 0–7 of incubation and were used to determine the biomass of the organisms (a fresh MSM was used as a blank control) and the concentration of DIF (an MSM of uninoculated colonies was used as a blank control), respectively. One milliliter of bacterial biomass was characterized by a UV-2600 ultraviolet spectrophotometer (Shimadzu, Kyoto, Japan) to determine the absorbance at 600 nm (OD_{600}), and the determination of the concentration of DIF was implemented by referring to the method of Zhou et al. [26].

3.3. *Effect of Incubation Conditions on the Degradation of DIF by TA01*

In this experiment, the effect of microbial consortium TA01 on the degradation of difenoconazole by microbial consortium TA01 was investigated in terms of three factors: initial pH (5.0, 6.0, 7.0, 8.0, and 9.0), inoculum (1%, 2.5%, 5%, 10%, and 15%), and the initial concentration of difenoconazole (30, 40, 50, 60, and 70 mg/L). The absence of inoculum was used as a control with three replicates for each factor. Each factor was incubated at 30 °C and 200 rpm. Samples were taken after 3 d of incubation to detect the residual DIF concentration by HPLC.

3.4. *Kinetics of DIF Degradation*

In order to characterize the kinetic parameters during the degradation of DIF by microbial consortium TA01, the dynamic parameters of the degradation process by microbial consortium TA01 under different substrate concentration conditions were nonlinearly fitted. In this experiment, five concentration gradients of 20 mg/L, 50 mg/L, 100 mg/L, 200 mg/L, and 300 mg/L were set as the only carbon sources. The microbial consortium TA01 was inoculated into the inorganic salt medium (MSM) at 10% inoculum, three replicates were set up for each group, and the uninoculated system was used as a blank control and incubated at 30 °C and 200 rpm for 5 days while protected from light. Based on the results, the DIF degradation kinetics of microbial consortium TA01 at different initial concentrations of DIF was calculated using the Andrews equation. Subsequently, the experimental data were processed by Origin 2024 software to calculate the optimal initial concentration of microbial consortia for degradation of DIF by nonlinear least squares curve fitting of the experimental values according to the Andrews equation. The equations are as follows:

$$q = \frac{q_{max} S}{S + K_s + \left(S^2 / K_i\right)} \tag{1}$$

In its formula, S denotes the concentration of DIF (mg/L), q denotes the specific degradation rate (d^{-1}), q_{max} denotes the maximum specific degradation rate (d^{-1}), K_s denotes the half-rate constant (mg/L), and K_i denotes the inhibition coefficient (mg/L).

3.5. *Analysis of the Community Structure and Key Degrading Bacteria of Microbial Consortium TA01*

The microbial consortium TA01 was inoculated into liquid LB medium and incubated at 30 °C and 200 rpm in a constant temperature shaker under light protection for 12 h. The precipitated bacterial bodies were collected by high-speed centrifugation at 3800 rpm for 5 min, washed with sterile water 2–3 times, and then added with an appropriate amount of inorganic salt medium to prepare the bacterial suspension at a certain concentration.

Subsequently, the bacterial suspension was inoculated into an inorganic salt medium containing 50 mg/L of DIF, followed by incubation in a constant temperature shaker at 30 °C and 200 rpm protected from light for 7 d. Samples were taken at 1 d intervals. High-throughput sequencing of 16S rRNA was used to detect the structure and diversity of microbial communities in enrichment culture, and these samples were analyzed by high-throughput sequencing in this study by Hangzhou Lianchuan Biotechnology Co. (Hangzhou, China).

3.6. Identification of DIF Degradation Products

The activated and resuspended bacterial solution was added to MSM containing 50 mg/L DIF and allowed to incubate for 3 d at 30 °C and 200 rpm, while protected from light, and 1 mL samples were taken at 0 h, 6 h, 12 h, 24 h, 36 h, 48 h, and 72 h. MSMs not inoculated with bacterial fluids were used as controls. The DIF-degrading metabolites of microbial consortium TA01 were subsequently identified by HPLC–MS/MS [30]. The detection conditions were as follows: mobile phase—acetonitrile; 0.2% formic acid aqueous solution = 70:30 (*v:v*); column temperature—30 °C; flow rate—0.4 mL/min; injection volume—20 μL; scanning mode—positive ion mode using full scan mode and Produce Ion mode to collect the data; full scanning time of 5 min, scanning range of 50~416 *m*/*z*; temperature of ion source—450 °C; capillary voltage—4000 V [31].

3.7. Bioremediation Potential of Microbial Consortium TA01

Soil taken from the depth of unapplied DIF from the farmland of Kunming University of Science and Technology, Yunnan Province, China, was dried and sieved (2 mm). A sieved soil sample of 5 g was taken into a 250 mL conical flask containing 45 mL of MSM medium and a certain volume of DIF was added to give an initial concentration of 100 mg/L of DIF in the system. Then a 10% bacterial solution (5 mL) was inoculated into this water–sediment remediation system and incubated at 30 °C and 200 rpm for 5 d under light protection, and 1 mL of the sample solution was taken every 1 d for the detection of DIF residues. In this experiment, the unadded bacterial solution was used as a control group and both sterilized and unsterilized treatments were conducted for the water–sediment system, so there were a total of four treatments in this experiment: sterilized system + uninoculated (SS + CK); sterilized system + inoculated (SS + TA01); unsterilized system + uninoculated (NS + CK); and unsterilized system + inoculated (NS + TA01). Three replicates were set up for each treatment.

4. Conclusions

In this study, the microbial consortium TA01, which can utilize DIF as its sole carbon source, was obtained through targeted enrichment screening, and the optimal degradation conditions were determined by one-factor determination. Based on high-throughput sequencing, the core composition of microbial consortium TA01 was *Pantoea*, *Serratia*, *Ochrobactrum*, and *Bacillus*, which is the first report of the synergistic degradation system of multiple genera in the degradation of DIF and provides an important candidate resource for the development of novel composite degrading bacterial agents. This synergistic degradation system is the first to be reported in the study of DIF degradation, and provides an important candidate for the development of new composite degraders. In addition, water–sediment simulation experiments confirmed the significant remediation efficacy of microbial consortium TA01 on DIF contamination in complex environments (with a removal rate of 75.29%), which verified its potential for application in practical ecological remediation.

Author Contributions: Conceptualization, W.Z. and Y.Y.; methodology, T.W., J.S. and Y.Z.; software, L.W.; validation, J.Y. and F.C.; writing—original draft preparation, T.W., J.S. and Y.Z.; writing—review and editing, X.C., Y.Y. and W.Z.; project administration, W.Z.; funding acquisition, Y.Y. All authors have read and agreed to the published version of the manuscript.

Funding: This research was funded by the National Key Research and Development Program of China (No: 2023YFC3503802), the National Natural Science Foundation of China (No: 82360744), Yunnan Fundamental Research Projects (No: 202401AT070316, 202301BE070001-060), and Yunnan Major Scientific and Technological Projects (No: 202202AG050021).

Institutional Review Board Statement: Not applicable.

Informed Consent Statement: Not applicable.

Data Availability Statement: All data are contained within the article.

Acknowledgments: We thank LC-Bio Technologies CO., Ltd., Hangzhou, China, for their support of high-throughput sequencing technology in this study.

Conflicts of Interest: The authors declare no conflicts of interest.

References

1. Peng, K.; Pan, Y.; Tan, T.; Zeng, X.; Lin, M.; Jiang, S.; Zhao, Z.; Tian, F.; Zhao, X. Characterization and fungicide sensitivity of *Colletotrichum godetiae* causing sweet cherry fruit anthracnose in Guizhou, China. *Front. Microbiol.* **2022**, *13*, 923181.
2. Zhou, J.; Wang, T.L.; Miao, Y.H.; Liu, D.H. Identification, biological characteristics, and fungicide screening of pathogen of black spot in *Aconitum carmichaelii*. *Zhongguo Zhong Yao Za Zhi* **2021**, *47*, 1215–1221. (In Chinese)
3. Yin, H.; Zhou, J.; Lv, H.; Qin, N.; Chang, F.J.; Zhao, X. Identification, pathogenicity, and fungicide sensitivity of *Ascochyta caulina* (Teleomorph: *Neocamarosporium calvescens*) associated with black stem on quinoa in China. *Plant Dis.* **2020**, *104*, 2585–2597. [PubMed]
4. Acosta-González, U.; Leyva-Mir, S.G.; Silva-Rojas, H.V.; Rebollar-Alviter, A. Preventive and Curative Effects of Treatments to Manage Strawberry Root and Crown Rot Caused by *Neopestalotiopsis rosae*. *Plant Dis.* **2024**, *108*, 1278–1288. [CrossRef]
5. Maria, N.I.; Khan, R.; Alam, S.S.; Iqbal, O.; Akram, S.; Rajput, N.A.; Younas, M.U.; Qasim, M.; Ali, I.; Elsalahy, H.H.; et al. Unleashing the synergistic effect of promising fungicides: A breakthrough solution for combating powdery mildew in pea plants. *Front. Microbiol.* **2024**, *15*, 1448033.
6. Khodadadi, F.; Martin, P.L.; Donahue, D.J.; Peter, K.A.; Aćimović, S.G. Characterizations of an emerging disease: Apple blotch caused by *Diplocarpon coronariae* (syn. *Marssonina coronaria*) in the mid-atlantic United States. *Plant Dis.* **2022**, *106*, 1803–1817.
7. Park, J.; Kim, S.; Jo, M.; An, S.; Kim, Y.; Yoon, J.; Jeong, M.H.; Kim, E.Y.; Choi, J.; Kim, Y.; et al. Isolation and identification of *Alternaria alternata* from potato plants affected by leaf spot disease in Korea: Selection of Effective Fungicides. *J. Fungi* **2024**, *10*, 53. [CrossRef]
8. Ao, X.; Shi, T.; Yang, W.; Ouyang, H.; Fan, R.; Siddiqui, J.A.; Wu, C.; Lv, Z.; Deng, S.; Chen, X. Biological characterization and in vitro fungicide screening of a new causal agent of walnut leaf spot in Guizhou Province, China. *Front. Microbiol.* **2024**, *15*, 1439487.
9. Song, Y.J.; Zhao, N.L.; Dai, D.R.; Bao, R. Prospects of *Pseudomonas* in microbial fuel, bioremediation, and sustainability. *ChemSusChem* **2025**, *18*, e202401324. [CrossRef]
10. Wang, C.; Hao, D.; Jiao, W.; Li, J.; Yuan, J.; Ma, Y.; Wang, X.; Xu, A.; Wang, M.; Wang, Y. Identification and fungicide sensitivity of *Fusarium* spp. associated with root rot of *scutellaria baicalensis* in Shanxi Province, China. *Phytopathology* **2024**, *114*, 1533–1541.
11. Degani, O.; Dor, S.; Chen, A.; Orlov-Levin, V.; Stolov-Yosef, A.; Regev, D.; Rabinovitz, O. Molecular tracking and remote sensing to ealuate new chemical treatments against the maize late wilt disease causal agent, magnaporthiopsis maydis. *J. Fungi* **2020**, *6*, 54. [CrossRef]
12. Khodadadi, F.; González, J.B.; Martin, P.L.; Giroux, E.; Bilodeau, G.J.; Peter, K.A.; Doyle, V.P.; Aćimović, S.G. Identification and characterization of Colletotrichum species causing apple bitter rot in New York and description of *C. noveboracense* sp. nov. *Sci. Rep.* **2020**, *10*, 11043.
13. Yang, Q.; Deng, P.; Xing, D.; Liu, H.; Shi, F.; Hu, L.; Zou, X.; Nie, H.; Zuo, J.; Zhuang, Z.; et al. Developmental neurotoxicity of difenoconazole in zebrafish embryos. *Toxics* **2023**, *11*, 353. [CrossRef]
14. Wu, X.; Han, H.; Xie, K.; He, N.; Yang, Z.; Jin, X.; Ma, S.; Dong, J. Difenoconazole disrupts carp intestinal physical barrier and causes inflammatory response via triggering oxidative stress and apoptosis. *Pestic. Biochem. Physiol.* **2023**, *194*, 105507. [CrossRef] [PubMed]

15. Nataraj, B.; Hemalatha, D.; Malafaia, G.; Maharajan, K.; Ramesh, M. "Fishcide" effect of the fungicide difenoconazole in freshwater fish (*Labeo rohita*): A multi-endpoint approach. *Sci. Total Environ.* **2022**, *857 Pt 2*, 159425. [CrossRef]
16. Zhai, Y.; Wang, S.; Zhang, B.; Tang, Y.; Wang, H.; Li, J.; Hu, Z.; Wang, Y.; Li, H.; Ge, R.S. The analysis of pesticides and fungicides in the inhibition of human and rat placental 3β-hydroxysteroid dehydrogenase activity: Mode of inhibition and mechanism. *Toxicol. Lett.* **2023**, *379*, 76–86. [CrossRef] [PubMed]
17. Na, M.J.; Lee, W.Y.; Park, H.J. Difenoconazole induced damage of bovine mammary epithelial cells via ER stress and inflammatory response. *Cells* **2024**, *13*, 1715. [CrossRef]
18. Qin, G.; Zhang, Q.; Zhang, Z.; Chen, Y.; Zhu, J.; Yang, Y.; Peijnenburg, W.J.G.M.; Qian, H. Understanding the ecological effects of the fungicide difenoconazole on soil and *Enchytraeus crypticus* gut microbiome. *Environ. Pollut.* **2023**, *326*, 121518. [CrossRef]
19. Han, W.; Ye, Z.; Gu, Y.; Zhong, Y.; Gao, J.; Zhao, S.; Wang, S. Gut microbiota composition and gene expression changes induced in the *Apis cerana* exposed to acetamiprid and difenoconazole at environmentally realistic concentrations alone or combined. *Front. Physiol.* **2023**, *14*, 1174236. [CrossRef]
20. Ganaie, M.I.; Jan, I.; Mayer, A.N.; Dar, A.A.; Mayer, I.A.; Ahmed, P.; Sofi, J.A. Health risk assessment of pesticide residues in drinking water of upper jhelum region in Kashmir Valley-India by GC-MS/MS. *Int. J. Anal. Chem.* **2023**, *2023*, 6802782. [CrossRef]
21. Zhang, S.; Wang, F.; Wang, R.; Cai, M. Spatial assessment of triazole organic compounds in surface water from the coastal estuaries to the East China sea. *Environ. Pollut.* **2023**, *320*, 121024. [CrossRef] [PubMed]
22. Li, J.; Tian, Z.; Han, A.; Li, J.; Luo, A.; Liu, R.; Zhang, Z. Integrative physiological, critical plant endogenous hormones, and transcriptomic analyses reveal the difenoconazole stress response mechanism in wheat (*Triticum aestivum* L.). *Pestic. Biochem. Physiol.* **2023**, *197*, 105688. [CrossRef]
23. Liu, Y.; Liu, R.; Deng, Y.; Zheng, M.; Yu, S.; Nie, Y.; Li, J.Q.; Pan, C.; Zhou, Z.; Diao, J. Insights into the mechanism of flavor loss in strawberries induced by two fungicides integrating transcriptome and metabolome Analysis. *J. Agric. Food Chem.* **2023**, *71*, 3906–3919. [CrossRef] [PubMed]
24. Chen, Y.; Feng, X.; Liu, X.; Zhang, L.; Mao, L.; Zhu, L.; Zheng, Y. Bioavailability assessment of difenoconazole to earthworms (*Eisenia fetida*) in soil by oleic acid-embedded cellulose acetate membrane. *Sci. Total Environ.* **2023**, *905*, 167276. [CrossRef]
25. Altun, Ş.; Kadak, A.E.; Küçükgülmez, A.; Gülnaz, O.; Çelik, M. Explanation of difenoconazole removal by chitosan with Langmuir adsorption isotherm and kinetic modeling. *Toxicol. Res.* **2022**, *39*, 127–133. [CrossRef] [PubMed]
26. Zhou, Y.; Wang, L.; Sui, J.; Chen, F.; Wang, T.; Yang, J.; Chen, S.H.; Cui, X.; Yang, Y.; Zhang, W. Pathway elucidation and key enzymatic processes in the biodegradation of difenoconazole by *Pseudomonas putida* A-3. *J. Agric. Food Chem.* **2025**, *73*, 4770–4786. [CrossRef]
27. Luong, T.T.; Nguyen, T.H.T.; Nguyen, T.D.; Le, V.T.; Pham, T.H.T.; Ho, T.T.; Nguyen, N.L. Degradation of triazole fungicides by plant growth-promoting bacteria from contaminated agricultural soil. *J. Microbiol. Biotechnol.* **2024**, *34*, 56–64. [CrossRef]
28. Chen, X.; Peng, S.; Liu, M.; Wang, L.; Pang, K.; Zhang, L.; Cui, Z.; Liu, A. Highly efficient in-situ cleaner degradation of difenoconazole by two novel dominant strains: Microflora diversity, monoclonal isolation, growth factor optimization, intermediates, and pathways. *Chemosphere* **2023**, *310*, 136863. [CrossRef]
29. Tan, Z.; Yang, X.; Chen, L.; Liu, Y.; Xu, H.J.; Li, Y.; Gong, B. Biodegradation Mechanism of Chloramphenicol by *Aeromonas Media* SZW3 and Genome Analysis. *Bioresour. Technol.* **2022**, *344 Pt B*, 126280. [CrossRef]
30. Huang, Y.; Chen, W.J.; Li, J.; Ghorab, M.A.; Alansary, N.; El-Hefny, D.E.; El-Sayyad, G.S.; Mishra, S.; Zhang, X.; Bhatt, P.; et al. Novel mechanism and degradation kinetics of allethrin using *Bacillus megaterium* strain HLJ7 in contaminated soil/water environments. *Environ. Res.* **2022**, *214 Pt 3*, 113940. [CrossRef]
31. Guo, Y.; Huang, Y.; Pang, S.; Zhou, T.; Lin, Z.; Yu, H.; Zhang, G.; Bhatt, P.; Chen, S. Novel mechanism and kinetics of tetramethrin degradation using an indigenous *Gordonia Cholesterolivorans* A16. *Int. J. Mol. Sci.* **2021**, *22*, 9242. [PubMed]
32. Zhang, W.; Li, J.; Zhang, Y.; Wu, X.; Zhou, Z.; Huang, Y.; Zhao, Y.; Mishra, S.; Bhatt, P.; Chen, S. Characterization of a novel glyphosate-degrading bacterial species, *Chryseobacterium* sp. Y16C, and evaluation of its effects on microbial communities in glyphosate-contaminated soil. *J. Hazard. Mater.* **2022**, *432*, 128689. [PubMed]
33. Sivasamy, S.; Rajangam, S.; Kanagasabai, T.; Bisht, D.; Prabhakaran, R.; Dhandayuthapani, S. Biocatalytic Potential of Pseudomonas Species in the Degradation of Polycyclic Aromatic Hydrocarbons. *J. Basic Microbiol.* **2025**, *65*, e2400448.
34. Man, Y.; Stenrød, M.; Wu, C.; Almvik, M.; Holten, R.; Clarke, J.L.; Yuan, S.; Wu, X.; Xu, J.; Dong, F.; et al. Degradation of difenoconazole in water and soil: Kinetics, degradation pathways, transformation products identification and ecotoxicity assessment. *J. Hazard. Mater.* **2021**, *418*, 126303.
35. Bhatt, P.; Rene, E.R.; Kumar, A.J.; Zhang, W.; Chen, S. Binding interaction of allethrin with esterase: Bioremediation potential and mechanism. *Bioresour. Technol.* **2020**, *315*, 123845. [CrossRef]

International Journal of
Molecular Sciences

MDPI

Review

Microbial β C-S Lyases: Enzymes with Multifaceted Roles in Flavor Generation

Mathieu Schwartz [1,*], Nicolas Poirier [1], Jade Moreno [1], Alena Proskura [1,2], Mélanie Lelièvre [1], Jean-Marie Heydel [1] and Fabrice Neiers [1]

1 Center for Taste and Feeding Behavior, CNRS, INRAE, Institut Agro, University of Burgundy, F-21000 Dijon, France; fabrice.neiers@u-bourgogne.fr (F.N.)

2 International Research Center "Biotechnologies of the Third Millennium", Faculty of Biotechnologies (BioTech), ITMO University, 191002 Saint-Petersburg, Russia

* Correspondence: mathieu.schwartz@inrae.fr

Abstract: β C-S lyases (β-CSLs; EC 4.4.1.8) are enzymes catalyzing the dissociation of β carbon–sulfur bonds of cysteine S-conjugates to produce odorant metabolites with a free thiol group. These enzymes are increasingly studied for their role in flavor generation in a variety of food products, whether these processes occur directly in plants, by microbial β-CSLs during fermentation, or in the mouth under the action of the oral microbiota. Microbial β-CSLs react with sulfur aroma precursors present in beverages, vegetables, fruits, or aromatic herbs like hop but also potentially with some precursors formed through Maillard reactions in cooked foods such as meat or coffee. β-CSLs from microorganisms like yeasts and lactic acid bacteria have been studied for their role in the release of polyfunctional thiols in wine and beer during fermentation. In addition, β-CSLs from microorganisms of the human oral cavity were shown to metabolize similar precursors and to produce aroma in the mouth with an impact on retro-olfaction. This review summarizes the current knowledge on β-CSLs involved in flavor generation with a focus on enzymes from microbial species present either in the fermentative processes or in the oral cavity. This paper highlights the importance of this enzyme family in the food continuum, from production to consumption, and offers new perspectives concerning the utilization of β-CSLs as a flavor enhancer.

Keywords: β C-S lyase; flavor; aromatic thiol; aroma; cysteine S-conjugates; oral microbiota

1. Introduction

The sensory experience encountered during food consumption is considered the main factor in determining the acceptability and consumption of food. Among the various aspects of this experience, flavor perception is recognized as a critical component. A complex interplay of information gathered from the chemical senses, including olfaction, gustation, and somatosensory inputs from the nasal and oral regions, is encompassed by it. The integration of sensory information coming from the chemical composition of the food enables the assessment of its quality. However, it should be noted that the input of these chemical signals to taste and olfactory receptors can be influenced by the environment surrounding these receptors [1–3]. Investigations have evidenced the role played by enzymatic processes in modifying flavor compounds, subsequently impacting how these flavors are perceived [4–8]. Recently, it was shown that the metabolic activity of odorant compounds shapes the brain's representations of the olfactory world [9]. As a result, the enjoyment of food and proper nutrition can be hindered by age-related salivary disorders, or health status such as obesity [10,11]. In addition to human oronasal metabolism, recent scientific inquiries have evidenced the intriguing influence exerted by the oral microbiota on the perception of taste and food preferences [12–15]. Importantly, specific microorganisms residing in the oral microbiota possess the capability to generate aromatic compounds within the mouth, actively participating in the food flavor [16–18]. These findings suggest that variations in the composition of these oral microbes could potentially contribute to differences in flavor perception. While these concepts might be perceived as groundbreaking in the field of chemoperception, it is important to note that some of the molecular mechanisms at play have been well-established in the field of food microbiology.

Food product fermentation has a profound impact on the sensory properties of the final products. Microorganisms have been harnessed by humans for millennia, initially through brewing and baking, and later in the production of bread, cheese, beer, wine, and yogurt. Fermentation is a biological process that not only transforms raw ingredients into an array of products but also adds depth and complexity to the global food flavor [19,20]. Microorganisms like bacteria and yeasts participate in fermentation, converting sugars and other compounds into a spectrum of flavor compounds. Fermentation intensifies flavors by adding new and complex profiles, which are valuable for their uniqueness and complexity. For example, microbial succession during the fermentation of sufu (fermented tofu) was shown to be important for the generation of a diversity of umami peptides [21]. Microbial metabolism can generate new flavor compounds from aromatic precursors that are non-odorant before being metabolized as well as reduce the bitterness of other compounds consequently acting on the flavor in a general manner [22,23]. Advances in food metabolomics, i.e., the large-scale study of metabolites within a food product, have evidenced the diverse array of metabolites generated by microorganisms during fermentation and proved to be valuable for food quality analysis [24]. It was shown that similarities exist in the enzyme families responsible for flavor generation both in the fermentation processes and in the mouth [12]. Microbial enzymes responsible for the generation of these new flavor compounds mainly include glycosidases and carbon–sulfur lyases. Glycosidases have been recently reviewed by Muradova and coworkers, showing the importance of this enzyme family in the generation of aromatic aglycone from non-aromatic glycoconjugates present in a variety of food products including plant-based milks, fruit juices, wines, and teas [25]. Additionally, glycosidase activities were measured in the human saliva and shown to impact the perception of the metabolized aromatic aglycone [18]. A similar observation was brought for another microbial enzyme family, namely, β C-S lyases (β-CSL) [16,26].

β-CSLs catalyze the dissociation of β carbon–sulfur bonds, leading to the release of sulfur aromatic compounds [26–28]. Sulfur compounds play a pivotal role as flavor components, often characterized by their remarkably low detection thresholds, and they frequently leave a distinct sensory mark on food products [29]. β-CSLs from plant or microorganisms act on specific aroma precursors found in foods such as onion, garlic, and fruits but also wine and beer to generate sulfur aromatic compounds [27,30,31]. These enzymes are present in plant-based foods and fermentative microorganisms but are also found within the bacteria of the oral cavity, making them significant contributors to flavor at various points along the food production and consumption continuum. For example, specific yeasts harboring enhanced β-CSL activities have been designed for the improved release of sulfur aroma in wine and beer processes [31,32]. Orthologous enzymes found in oral anaerobe bacteria such as *Fusobacterium nucleatum*, were shown to be able to generate sulfur aroma compounds in the mouth during food consumption [26,33,34]. Gaining insights into the biochemistry of β-CSLs appears interesting for both the biotechnological aspects of flavor improvement as well as our understanding of the molecular mechanisms occurring in the oral cavity and impacting flavor perception. This review focuses on the advances made in the field of β-CSLs in relationship with flavor production, encompassing the sulfur compounds produced in plant-based products, in fermented food microorganisms, as well as in the oral cavity.

2. β C-S Lyases' Roles and Structures in Microorganisms

2.1. Mechanism and Role in the Sulfur Amino Acid Metabolism

β-CSL, also known as cystathionine β-lyases (EC 4.4.1.8) belong to the larger family of carbon–sulfur lyases, which encompass numerous enzymes capable of dissociating carbon–sulfur bonds such as cystathionine γ-lyase, methionine γ-lyase, or cysteine desulfurase [35,36]. β-CSLs catalyze the dissociation of bonds between the atoms C_β and S_γ. In mammals, including humans, β-CSLs have roles in the metabolism of naturally occurring sulfur and selenium-containing compounds, xenobiotics, and anticancer agents [35]. In bacteria, they are encoded by *metC*, *malY*, or *patB* genes. The *metC* gene encodes cystathionine β-lyase involved in the methionine biosynthetic pathway [37]. The *malY* gene encodes a bifunctional enzyme showing both cystathionine β-lyase activity and a negative activity on regulon maltose [38]. The *patB* gene encodes an enzyme with both cystathionine β-lyase and cysteine desulfhydrase activities [39]. The corresponding enzymes of these genes use the pyridoxal 5′-phosphate (PLP) as cofactor to catalyze β-elimination reactions with L-cysteine or L-cysteine derivatives (such as L-cystathionine, S-methyl-L-cysteine, or longer alkylated cysteines) to form the corresponding thiol compounds with pyruvate and ammonia as end-products (Figure 1). The cofactor PLP, which is the active form of its precursor vitamin B6, is covalently attached to the active site lysine residue conserved in this enzyme family [39].

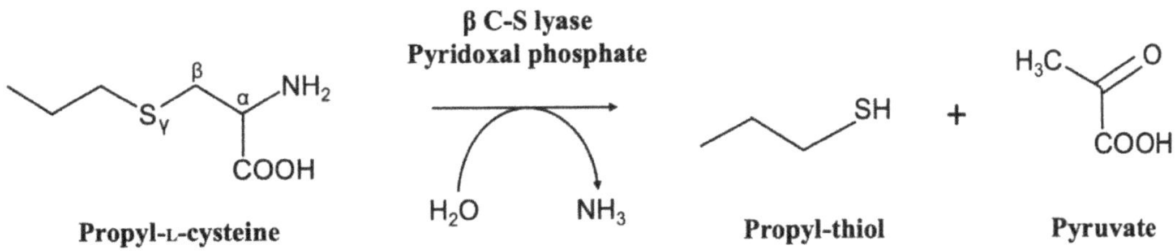

Figure 1. Enzymatic reaction catalyzed by β-CSLs with the example substrate propyl-L-cysteine.

PLP participates in the enzymatic mechanism by switching from a form where it is bound to a conserved lysine residue (named external aldimine) to a form where PLP is then attached to the amino acid substrate (named internal aldimine) [28]. At the end of the reaction, the β-CSL is again in its external aldimine form and can start a new cycle. β-CSLs were shown to be involved in the metabolism of sulfur amino acid and notably in the transsulfuration pathway. This pathway enables the interconversion between cysteine and methionine in both ways. In humans, the reverse transsulfuration pathway allows the conversion of methionine to cysteine. In bacteria such as *E. coli* or *B. subtilis*, β-CSLs play a role in the forward transsulfuration pathway enabling the conversion of cysteine to methionine. Bacterial β-CSLs react with the intermediate compound cystathionine to form homocysteine, which is a precursor of methionine (further formed by methionine synthase). In addition, several CSL were shown to have a wider spectrum of activity and can also catalyze the β-elimination of other amino acids such as L-cysteine, L-cystine, or longer alkylated cysteines such as S-methyl-, S-ethyl-, or S-propyl-L-cysteine (Figure 1) [37,40,41].

2.2. *Production of Volatile Sulfur Compounds by Oral Bacteria β-CSL*

In addition to their role in sulfur amino acid metabolism, some oral bacterial β-CSLs release volatile sulfur compounds (VSCs). Hydrogen sulfide (H_2S) is a notable VSC contributing to the malodor associated with periodontal diseases. Many oral bacteria enzymes were described to produce H_2S, some of them being β-CSLs with a desulfhydrase activity using cysteine as substrate. *Treponema denticola*, an oral pathogen, is involved in periodontitis, and the β-CSL also named cystalysin is considered the main virulence factor of this disease. H_2S is responsible for the lysis of the cell membrane of erythrocytes (haemolysis), participating in the virulence of the infection [42]. The biochemical characterization of cystalysin from *Treponema denticola* [43–45] showed that this enzyme uses the classical PLP cofactor and cysteine to produce H_2S. The H_2S release in the mouth is also responsible for halitosis (bad breath) [46]. Other oral bacteria can produce VSCs using β-CSLs. For example, a C-S lyase from *Fusobacterium varium* was purified and its characterization was performed with various alkyl-cysteine conjugates, all being metabolized [47]. In *Fusobacterium nucleatum*, one of the main producers of sulfur volatiles in the mouth, several sulfur-producing enzymes have been described [33]. A recent study showed that the gene *fn0625* encodes a β-CSL enzyme able to produce sulfur volatile compounds from substrates including L-cysteine, L-cystine, S-ethyl-L-cysteine, S-methyl-L-cysteine, L-cystathionine, and DL-lanthionine, with a preference for L-cystine and L-cystathionine [34]. Additionally, the subspecies *F. nucleatum animalis* presents a β-CSL called FnaPatB1 with high activity towards cysteine conjugates harboring hydrophobic substituents, with this high activity being likely due to an elongated active site loop [26].

2.3. Generation of Odorant Thiol in the Human Sweat by Bacterial β-CSL

Some bacterial β-CSLs are involved in the production of odorant thiols responsible for sweat odors. For example, *Corynebacteria* present in the axillary region of the human skin express β-CSLs [48,49]. A recent study reported the involvement of a particular class of β-CSL exhibiting a specific activity for thio-alcohol precursors in *Staphylococcus hominis* [28]. These enzymes are a particular class of β-CSL (C-T lyase) whose genes have been horizontally transferred approximately 60 million years ago and became part of an exclusive monophyletic cluster of odor-forming staphylococci. Notably, introducing this enzyme alone into staphylococci that do not produce odor results in the acquisition of odor production, indicating that the C-T lyase is both essential and adequate for thioalcohol formation. By examining the structure of the C-T lyase alongside related enzymes, it becomes apparent that the adaptation to thioalcohol precursors has developed by altering the binding site, creating a restricted hydrophobic pocket with a preference for branched aliphatic thioalcohol ligands. The ancient acquisition of this enzyme and its subsequent evolution in specificity for thioalcohol precursors suggest that the production of body odor in humans is a longstanding biological process [28]. Similar mechanisms may exist in other bacteria and notably the ones of the oral cavity, as some β-CSLs having enhanced activity with branched cysteine derivatives were also identified in anaerobes such as *Fusobacterium nucleatum* [26].

2.4. Other Biological Functions of Bacterial β-CSL

In *Escherichia coli*, the *malY* gene encodes a β-CSL able to catalyze classical β-elimination with cysteinyl substrates but also shows the property to act as a repressor of the maltose regulon [50]. The maltose regulon in *E. coli* consists of several genes involved in the metabolism and uptake of maltose and maltodextrins. MalT is the gene activator of the maltose regulon and expression of *malY* represses all MalT-dependent genes. MalY has a cystathionase activity that is not required for its repressing activity on the maltose system. Analysis of MalY crystal structure revealed that amino acids important for repression activity are located far away from the active site and suggests that direct protein–protein interactions are involved between MalY and MalT [37]. Recently, the cryoelectron microscopy structure of the complex formed between MalY and MalT showed that MalY inactivates MalT by blocking its oligomerization and by strengthening ADP-mediated MalT autoinhibition [51]. These results suggest that β-CSLs could have alternative unexplored biological roles involving protein–protein interactions, in addition to their usual C-S lyase activity. They could also have additional catalytic activities. In *Lactobacillus sakei*, Kato and Oikawa identified and characterized Ls-MalY as a bifunctional enzyme harboring both β C-S lyase activity and racemase activity on various amino acids [52]. The PLP cofactor was found to be necessary for both activities occurring in the same active site, suggesting catalytic promiscuity.

2.5. Tridimensional Structure of Microbial β-CSLs

β-CSLs whose three-dimensional structures have been determined to date (all by X-ray crystallography) adopt the type-I fold among the PLP-dependent aminotransferases family. This fold consists in homodimeric or homotetrameric assemblies, although β-CSLs determined to date revealed all a homodimeric fold except for the one identified in *E. coli* (Table 1, Figure 2). The monomer of β-CSLs is characterized by a three-domain fold, exemplified by the structure of *E. coli* CBL (EcCBL, pdb 1CL1) [37]. The N-terminal domain, from residues 1 to 60, is made of three α-helices and one β-strand. This domain is part of the active site of the neighboring subunit, and in the case of tetrameric CBL such as *E. coli* CBL can favor tetramer interaction. The second domain, from residues 61 to 256, has an α/β-structure with a central β-sheet and the conserved Lys210 covalently bound to PLP. The C-terminal domain (from residues 257 to 395) is made of four helices surrounding an antiparallel four-stranded β-sheet. β-CSL enzymes with dimeric structure share this fold globally but with some differences notably in the dimerization interface, which despite some structural homology make them fall in a different subfamily within the type-I PLP enzymes [38].

Table 1. Solved X-ray crystallographic structures of microbial β-CSLs available in the PDB. NP means no publication associated with the PDB deposition.

Name	Organism	PDB Code	Resolution (Å)	Oligomerization	References
BaCBL	*Bacillus anthracis*	3T32	2.00	Homodimer	NP
BsPatB	*Bacillus subtilis*	6QP3	2.30	Homodimer	[39]
CdCBL	*Clostridium difficile*	4DGT	1.55	Homodimer	NP
CdCBL	*Corynebacterium diphtheriae*	3FDB	1.99	Homodimer	NP
EcCBL	*Escherichia coli*	1CL1	1.83	Homotetramer	[37,53]
EcMalY		1D2F	2.50	Homodimer	[38]
FnaPatB1	*Fusobacterium nucleatum subsp. animalis*	7QUG	2.60	Homodimer	[26]
KpCBL	*Klebsiella pneumonia*	8DUY	1.90	Homodimer	NP
LDB CSL	*Lactobacillus delbrueckii subs. bulgaricus*	3DZZ	1.61	Homodimer	[27]
ShPatB	*Staphylococcus hominis*	6QP2	1.60	Homodimer	[28]
Cystalysin	*Treponema denticola*	1C7N	1.90	Homodimer	[42]

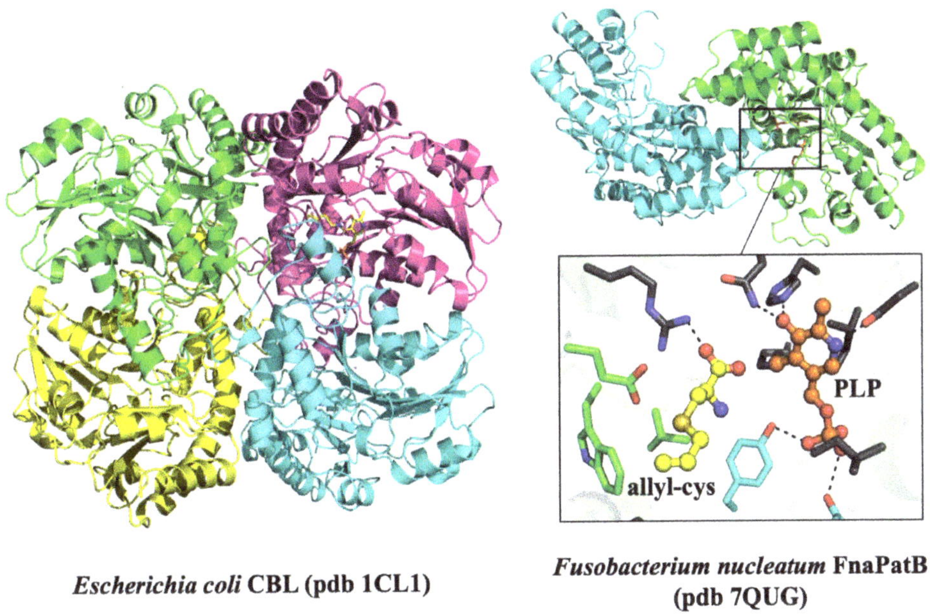

Figure 2. 3D structures of β-CSLs from bacteria: heterotetrameric CBL from *E. coli* and heterodimeric FnaPatB1 from *F. nucleatum*, with an enhanced view of its active site bound to S-allyl-L-cysteine and PLP.

For both oligomeric structures of β-CSL, the active site is made of residues forming a buried pocket where both the PLP cofactor and the amino acid substrate can bind (Figure 2). The PLP cofactor is stabilized deep inside the structure by several hydrogen bonds with conserved amino acids, in addition to a covalent link with the Lys residue involved in the catalytic mechanism, conserved among all PLP-dependent enzymes. The site of the amino acid substrate is located very near from the PLP site but is more accessible to the solvent. It is located in the vicinity of the variable region loop located 20 residues starting from the N-terminal extremity. In the case of FnaPatB1 and ShPatB, this loop is elongated and was suggested to promote favorable hydrophobic contact to stabilize cysteine conjugates with hydrophobic substituents [23,25].

3. Substrates of Microbial β C-S Lyases in Food: Cysteine S-Conjugates As Sulfurous Aroma Precursors

Volatile sulfur compounds have low detection thresholds and play a substantial role in creating the distinctive fragrance of animal products such as cheese and meat [54], plant foods such as vegetables belonging notably to *Brassica* and *Allium* categories, as well as mushrooms [29] and fruits [55]. This sets them apart as a particularly distinct category of plant metabolites, intriguing not only from a flavor chemistry perspective but also in terms of analytical considerations, given their wide variety, low odor thresholds, susceptibility to instability, and reactivity. Sulfur compounds, when considered as metabolic byproducts, exhibit diverse chemical configurations. These compounds are found in food and can be generated or modified through enzymatic and chemical reactions. A category of sulfur aroma precursors stands for its interest in flavor generation: cysteine S-conjugates, which are substrates of β-CSL.

Cysteine S-conjugates have been historically studied for their occurrence in *Allium* species (including onion, garlic, and leeks) and are responsible for the characteristic odor of garlic and onions [56]. Cysteine conjugates represent a significant category of precursors for sulfur-based aromas [16]. They contain a cysteine group connected to an organic group through a carbon–sulfur bond. These compounds lack noticeable odors because they have low volatility (considering their soluble nature as amino acids). However, when β-CSLs metabolize these compounds (either in the plant or by microorganisms), they transform into molecules with a free thiol group and distinct aroma characteristics. These precursors are actually present in various plant-based foods additionally to *Allium* species such as *Brassica* (including cauliflower and broccoli), asparagus, bell peppers, and fruits (including passion fruit, guava, and durian). They are also present in beverages like tea, coffee, wine, sake, and beer due to their presence both in hop and in some malts [16,57–64]. Cysteine conjugates can be formed during cooking through Maillard reactions between cysteine and sugars. As such they are expected in a variety of cooked food including cooked meat and coffee [65,66]. For a recent review on sulfur flavor compounds and cysteine conjugates, the reader is referred to the article from Bonnaffoux and coworkers [67]. Cysteine conjugates exhibit a wide chemical diversity at their S-substituent including aliphatic (saturated or unsaturated), aromatic, or S-oxide groups, and consequently this diversity also occurs on the odorant metabolites formed from these precursors (Table 2). The flavor descriptors of the thiol metabolites range from garlic and onion (e.g., for allyl-thiol, propyl-thiol) to catty, boxtree (e.g., 4-mercapto-4-methylpentan-2-one), meaty (2-methyl-3-furanthiol), or coffee (e.g., 2-furfurylthiol) aromas.

In plants, cysteine conjugates are products of the glutathione pathway. As an example, the synthesis of the flavor precursor S-3-(hexan-1-ol)-L-cysteine (cys-3MH) has been described previously [68]. The synthesis of the glutathione conjugate of 3MH occurs after the conjugation of glutathione to trans-2-hexanal with the involvement of glutathione transferase, and subsequent action of γ-glutamyl transferase and carboxypeptidase, to yield the corresponding cysteine conjugate. This pathway was shown to be modulated by environmental factors like ultraviolet irradiation and water deficit, suggesting that flavor precursor concentrations could be modulated by the geographical region, the specific climate, and the culture condition in which plants were grown. In this context, it was observed that hop cultivars from different growing regions showed different concentrations of S-4-(4-methylpentan-2-one)-L-cysteine (4MMP), mainly due to the usage of copper sulfate as a fungicide that decreases the aromatic thiol potential of hops [69]. Another factor influencing the concentration of cysteine conjugates in grapes is the development of the noble-rot *Botrytis cinerea* on the berries. *B. cinerea* was shown to boost the production of the 3-sulfanyl-hexanol precursor in a model of *Vitis vinifera* cell culture 1000-fold [70].

Concerning malts, evidence was shown of enzymatic and chemical interconversions between 3-sulfanyl-hexanol conjugates during mashing [62]. In addition to cysteine S-conjugates, other sulfur precursors exist as glutathione, γ-glutamyl-cysteine, and cysteine-glycine states, while their biotransformation by β-CSLs is, to our knowledge, not well documented. Importantly, some precursors have been historically shown to be metabolized directly by plant enzymes. The best example is given by the β-CSL named alliinase and its substrate S-allyl-L-cysteine. They are separated in the plant cell but they yield the corresponding metabolite allyl-thiol when mixed together when fresh garlic or onion are cut [71]. Other edible plants or mushrooms were shown to contain C-S lyases important for flavor generation such as seeds of *Parkia speciosa* Hassk [72] and *Lentinula edodes* [73,74].

Table 2. Aroma precursors biotransformed by β C-S lyases from fermentative microorganisms.

Substrate	Metabolite	Organism/Enzyme	Food Product	References
S-4-(4-methylpentan-2-one)-L-cysteine	4-mercapto-4-methylpentan-2-one	β-CSL from *S. cerevisiae, E. limosum*	Sauvignon, Semillon, Chardonnay, Riesling wine	[59,75]
S-4-(4-methylpentan-2-ol)-L-cysteine	4-mercapto-4-methylpentan-2-ol			
S-3-(hexan-1-ol)-L-cysteine	3-mercaptohexan-1-ol			
S-2-(3-methylbutanol)-L-cysteine	2-mercapto-3-methylbutanol	β-CSL from *S. cerevisiae, P. kluyveri*	Lager beer	[76,77]
S-1-(3-pentanone)-L-cysteine	1-mercapto-3-pentanone			
S-3-(3-methylbutanol)-L-cysteine	3-mercapto-3-methylbutanol			
S-3-(3-methylbutanol)-L-cysteine	3-mercapto-3-methylbutanol	β-CSL from *E. limosum*	Passion juice	[78]
S-3-(hexan-1-ol)-L-cysteine	3-mercaptohexan-1-ol			
S-furfuryl-L-cysteine	furfurylthiol	STR3 and CYS3 from *S. cerevisiae*	Baijiu, coffee	[65,66,79]
L-cysteine and benzaldehyde	benzenemethanethiol	Yeast β-CSL	Baijiu Daqu	[80]

4. Importance of β C-S Lyases in Microorganisms Used for Fermented Food Production

4.1. Wine

In the context of wine, specific sulfur compounds are a result of the metabolism of cysteine conjugates by microorganisms having β-CSLs utilized in the fermentation process. Tominaga et al. identified some of these compounds in Sauvignon white wine in 1998, including 4-mercapto-4-methylpentan-2-one, 4-mercapto-4-methylpentan-2-ol, and 3-mercaptohexan-1-ol [59]. These compounds are associated with the typical passion/boxtree aromatic notes of Sauvignon wines and originate from the hydrolysis of nonvolatile cysteinyl precursors present in grape must after being metabolized by yeast. The generation of thiols is facilitated by β-CSLs, which catalyze the dissociation of C-S bonds. Precursors have been identified in various grape varieties such as Semillon, Chardonnay, and Riesling [75]. The corresponding flavor compounds are noteworthy as they contribute to enhancing aroma parameters during wine consumption, including complexity, intensity, and persistence [81]. Some compounds, like S-3-(hexan-1-ol)-L-cysteine, are produced in grapevines through the catabolism of glutathione precursors like S-3-(hexan-1-ol)-glutathione [60], synthesized in grapevines by detoxification systems such as glutathione transferases [22].

4.2. Beer

Polyfunctional thiols derived from cysteine-conjugated precursors, such as 3-methyl-2-buten-1-thiol, 2-mercapto-3-methylbutanol, and 3-mercapto-3-methylbutanol, have been identified in Lager beer [76]. These precursors are naturally present in various hop varieties [61] and are likely synthesized by enzymes from the glutathione transferase family [82], similar to the process in grapevines. It has been demonstrated that introducing a commercial aminotransferase with C-S lyase activity (such as tryptophanase from *E. coli*) to a hop solution can induce the formation of free thiols [61]. In beer, their formation occurs during alcoholic fermentation, driven by the enzymatic action of C-S lyases from *S. cerevisiae* or other yeasts. In this context, a *Pichia kluyveri* strain has been patented for improving thiol levels during beer fermentation [83]. Additionally, Belda et al. reported the identification of several *Saccharomyces* and non-*Saccharomyces* yeast strains, such as *Torulaspora delbrueckii*, *Meyerozyma guilliermondi*, and *Kluyveromyces marxianus*, capable of enhanced thiol release through increased lyase activity. These properties are conferred by the presence of a full-length *IRC7* gene, encoding an active β-CSL, contrary to most yeast species that harbor only a truncated version of the gene [32]. While the release of polyfunctional thiols is desirable for improving beer flavor, it has been shown that the presence of polyphenols and an acidic environment can hamper the activity of C-S lyase during fermentation [84]. Current research aims to address these challenges and determine optimized processes for increasing thiol compounds in beer [62]. Alternative methods include adding fermented beer powder made from Sauvignon Blanc grape skins rich in precursors (e.g., Phantasm™ Powder).

4.3. Baijiu

Baijiu is a traditional alcoholic beverage produced from the solid-state fermentation of cereals. It is associated with more than 2000 aroma compounds, with the most important being the sulfur aroma compounds such as 2-furfuryl thiol and benzenemethanethiol [85]. These two sulfur aroma compounds are produced from their corresponding cysteinylated precursors, which are metabolized by yeast β-CSL. 2-Furfuryl thiol is responsible for the typical sesame flavor of baijiu. This aroma compound is produced in two steps: (i) furfural and cysteine react together to form the conjugate, and (ii) the conjugate is transformed in 2-furfuryl thiol under the action of yeast STR3 and IRC7 enzymes [79]. A similar mechanism was observed in Baijiu Daqu for benzenemethanethiol, which is produced by yeast from cysteine and benzaldehyde as precursors, followed by the action of yeast β-CSLs [80].

4.4. Cheese

Volatile sulfur compounds are important for the flavor of cheese, with methionine as the main aroma precursor. In this regard, two compounds stand for their importance to cheese flavor generation, i.e., methanethiol and dimethyldisulfide. This reaction driving the cheese flavor is mainly catalyzed by methionine γ-lyases (MGLs) in an array of lactic acid bacteria [54] but can also be catalyzed by β-CSLs [86]. A β-CSL was isolated from the strain *Lactococcus lactis* subsp. *cremoris* B78 used for the production of Gouda cheese [87]. This enzyme is able to produce methanethiol from methionine due to a γ-elimination reaction. In *Lactobacillus helveticus* CNRZ 32, overexpression of a β-CSL resulted in higher VSC production from methionine as substrate in a model system, suggesting a similar mechanism [88].

4.5. Genes and Corresponding β-CSLs of Interest in Thiol Release Improvement

As stated above, several yeasts and bacteria are used for their CSL activity to improve the flavor of foods. The main yeast genes responsible for β-CSL activity are *IRC7* [89] and *STR3* [77]. These genes encode cystathionine β-lyases and pyridoxal-5′-phosphate-dependent enzymes catalyzing the dissociation of the C-S bond in various substrates, including L-cystathionine, L-cysteine, L-cystine, and precursors of thiol aroma compounds such as 4-sulfanyl-4-methylpentan-2-one and 3-sulfanylhexanol [69]. Enhancing the release of varietal thiols to improve wine flavor can be achieved by engineering strains for the overexpression of these genes [77,89]. Concerning the enzyme encoded by the *IRC7* gene, studies have shown that two forms exist, the full-length IRC7p retaining full activity and a shorter form that is inactive. Few *Saccharomyces* strains have the fully active form, while most of the lab strains including *Saccharomyces cerevisiae* S288C have an inactive form of the *IRC7* gene that is deprived of the last 38 base pairs [31]. Consequently, the yeast strains that harbor only the truncated form have a reduced potential for thiol release, confirming IRC7p as an interesting target for thiol release [32]. STR3p is a β-CSL with a peroxisomal targeting peptide, and an engineered strain having an extra copy of the *STR3* gene under the PGK1 promoter proved to efficiently improve the thiol release [77]. Additionally, two other genes in yeast were found to improve the release of varietal thiols when overexpressed, i.e., *GLO1* and *CYS3* [89]. However, these two genes do not encode true β-CSLs. *GLO1* encodes glyoxalase and *CYS3* encodes a cystathionine γ-lyase [90]. Overexpression of these genes could indirectly improve the thiol release by modifying the yeast metabolism.

The presence of β-CSLs in lactic bacteria, e.g., *Lactobacillus* species [90], suggests potential applications for improving the flavor of specific red wines undergoing malolactic fermentation, as well as other food products. In the subspecies *bulgaricus* of *Lactobacillus*, a β-CSL (LDB) was identified, with broad substrate specificity toward sulfur-containing amino acids [27]. Its characterization revealed that LDB was able to cleave cysteinylated substrate precursors into the corresponding flavor-contributing thiols, with a catalytic efficiency higher than L-cystathionine. *Lactobacillus plantarum* was also suggested to be a useful tool to enhance thiol release in fermented beverages, although the corresponding β-CSL was not studied yet. Takase and coworkers demonstrated that this strain can improve the 3-sulfanyl hexanol release from its precursors, including its glutathione conjugate form, in fermented grape juices [91]. Previously, the introduction of tryptophanase from *E. coli* to a hop solution induced the formation of free thiols [61]. More recently, Clérat and coworkers discussed in more detail the possibility of using exogenous β-CSLs to improve the release of aromas in wine or beer [92]. Using *Lactobacillus delbrueckii* LDB as an example they show a method to follow the activity of the enzyme in presence of various competing cysteine S-conjugate analogs. Due to constraints in terms of pH, LDB has the potential for aroma increase in beer processes but it is not suitable for winemaking, suggesting the need for further research to identify new targets suitable for the wine processes. Additionally, international regulations regarding the use of exogenous β-CSLs for food production are currently lacking and more studies are warranted. More generally, and according to the Food and Agriculture Organization (FAO), the use of enzymes in food for flavor enhancement involves ensuring non-toxicity and GRAS status, assessing allergenicity, and sourcing from acceptable microbial, plant, or animal origins. Purity and consistent enzymatic activity are crucial, alongside regulatory approval and proper labeling. Enzymes must be specific to their substrates, stable under processing conditions, and effective in enhancing the desired flavor profile without off-flavors. In particular, selected β-CSLs should be specific toward small compound precursors and should not react with the cysteine residues of food proteins, in order to preserve them from alteration or aggregation [93].

5. Involvement of Oral Bacteria β-CSLs in the In-Mouth VSC Release and Perception

5.1. Studies on Pure Compounds

The role of the oral microbiota and its relationships with chemoperception is an increasingly studied topic of research [12]. The oral cavity is colonized by over 700 microbial species, including bacteria, fungi, and viruses [94]. A pioneer work in 2008 led by Starkenmann and coworkers showed that cysteine S-conjugates, the same family of precursors metabolized by microbial enzymes during fermentation, are transformed into their corresponding thiol aromatic compounds under the action of human saliva (Table 3) [16]. This process occurs within a few seconds after the cysteine conjugate precursors are put into the mouth, suggesting that enzymatic mechanisms are compatible with the time of flavor perception. In the same study, Starkenmann et al. showed that the enzymatic reactions of the transformation of cysteine conjugates into thiols are mediated by the oral microbiota. The compounds studied included S-(R/S)-3-(1-hexanol)-L-cysteine, S-(1-propyl)-L-cysteine, and S-((R/S)-2-heptyl)-L-cysteine. Starkenmann and coworkers showed that anaerobic bacteria in saliva, notably *Fusobacterium nucleatum*, are mainly involved in the enzymatic degradation of these precursors, while the appearance of the corresponding thiols increases through time [16]. In humans, odorant thiols are perceived at very low detection thresholds with the involvement of the olfactory receptor OR2T11 [95]. Importantly, *Fusobacterium* spp. possesses an extensive enzymatic repertoire facilitating the metabolism of various sulfur compounds, including cysteine conjugates [33]. More recently, it was shown that human saliva is able to release allyl-thiol, allyl-sulfide, and allyl-disulfide from the precursor compound S-allyl-L-cysteine. Centrifugation of saliva leading to the removal of the saliva pellet containing microbial cells resulted in no metabolization of precursors. Interestingly, the biotransformation was shown to be upregulated by adding the cofactor PLP, while adding a transaminase inhibitor, L-cycloserine, resulted in a drastic decrease in metabolization. Both observations suggest the involvement of β-CSLs from oral bacteria. A subsequent comparative genomic approach identified FnaPatB1 as a putative β-CSL in *F. nucleatum*, which after recombinant production showed high activity with various cysteine S-conjugates precursors and produced the same metabolites as human saliva (Table 3) [26].

Table 3. Aroma precursors biotransformed by β C-S lyases from saliva/oral microbiota.

Substrate	Metabolite	Organism/Enzyme	Foods/Compounds	References
S-3-(hexan-1-ol)-L-cysteine	3-mercaptohexan-1-ol	Human saliva β-CSL, *F. nucleatum* β-CSL	Pure compounds	[16]
S-(1-propyl)-L-cysteine	1-propanethiol			
S-((R/S)-2-Heptyl)-L-cysteine	2-heptanethiol			
S-methyl-L-cysteine-oxide (Methiin)	methanethiol, dimethyl disulfide, dimethyl trisulfide	Human saliva β-CSL	*Allium* and *Brassica* vegetables	[57,58,96]
S-allyl-L-cysteine	allyl-thiol, diallyl-sulfide, diallyl-disulfide	Human saliva β-CSL, FnaPatB1 from *F. nucleatum*	Pure compounds	[26]

5.2. Studies on Food Products

In addition to studies on pure compounds, studies on real foods showed the importance of oral β-CSL activity for perception. As previously mentioned, cysteine conjugates exist in various foods [55,67], and all could be potential substrates for oral microbiota β-CSLs with an impact on flavor perception. Two studies led by Frank and coworkers showed the ability of human saliva to metabolize S-Methyl-Cysteine-Sulfoxide (SMCSO) present in raw cabbage and broccoli into its corresponding metabolites methanethiol and dimethyl-disulfide using proton transfer reaction mass spectrometry (PTR-MS). Differences in the rates of C-S lyase activity present in human saliva and associated sulfur volatile production have been shown, with more than 10-fold differences between individuals [96]. A garlic odor was positively correlated with the release of sulfur aroma compounds. Furthermore, the accumulation of sulfur volatiles originating from SMCSO during more than twenty consecutive mouthfuls was observed, which suggests that these reactions have the potential to influence sensory perception during the act of eating.

In a second ex vivo study, volatile production from Brassica vegetables and associations with liking in an adult/child cohort were explored [58]. Production of volatiles was measured within minutes after raw cauliflower was incubated with fresh human saliva, in accordance with the in vivo approach employed during the first study. Substantial variations were observed in the rate at which sulfur volatiles were generated among different individuals, with no notable distinctions noted between various age groups. Notably, strong positive associations were identified in volatile production between adults and children paired together, indicating shared characteristics in saliva composition and oral microbiome activity. Noteworthy negative correlations were identified between the quantity of sulfur volatile production within the mouth and children's preference for raw cauliflower. Saliva was suggested to be able to trap thiols, probably through salivary proteins [16]. Additionally, several salivary parameters including the salivary antioxidant capacity could be responsible for the modulation of thiol release as it was shown for other reactions in saliva, with a final impact on flavor perception [10].

6. Conclusions

β-CSLs are enzymes with important functions in flavor at multiple points along the food continuum. On the one hand, gaining insights into β-CSLs from fermentative microorganisms appears necessary for improving the thiol release in an array of fermented food. On the other hand, the characterization of this enzyme family in oral microorganisms is also very interesting for increasing our understanding of the molecular mechanisms occurring in the mouth. As thiols are associated with the bad taste of vegetables, strategies to inhibit the C-S lyase activity to limit the oral production of thiol could be valuable for increasing the consumption of green vegetables, especially in children. Molecules from food, especially plant molecules, are potent inhibitors of C-S lyases, and the molecular tools from biochemistry and structural biology can help to identify such inhibitors [26]. In the same way, gaining insights into the inhibition of these enzymes could leverage a problem in the usage of these enzymes as flavor enhancers, because they are inhibited by plant molecules such as polyphenols [84]. While the functional and structural characterization of β-CSLs both from oral bacteria or fermentative microorganisms is still important for gaining knowledge on fundamental biochemical processes, the characterization of their activity in a physiologically/biotechnologically relevant context is even more challenging. However, this challenge is necessary to have a more exhaustive view of the enzymatic processes in the context of (i) an oral environment that modulates the enzymatic activities (salivary proteins, pH, redox status) and (ii) a fermentative process including several steps with different biochemical conditions, whether the β-CSL is present in the microorganism used for fermentation or added exogenously.

Author Contributions: Writing—original draft preparation, M.S. and F.N.; writing—review and editing, M.S., N.P., J.M., A.P., M.L., J.-M.H. and F.N.; funding acquisition, M.S. All authors have read and agreed to the published version of the manuscript.

Funding: This research was funded by the 'Agence Nationale de la Recherche' (ANR-JCJC FLAMME, grant number ANR-22-CE21-0001).

Institutional Review Board Statement: Not applicable.

Informed Consent Statement: Not applicable.

Data Availability Statement: No new data were created or analyzed in this study.

Conflicts of Interest: The authors declare no conflicts of interest.

References

1. Heydel, J.-M.; Coelho, A.; Thiebaud, N.; Legendre, A.; Bon, A.-M.L.; Faure, P.; Neiers, F.; Artur, Y.; Golebiowski, J.; Briand, L. Odorant-Binding Proteins and Xenobiotic Metabolizing Enzymes: Implications in Olfactory Perireceptor Events: Odorant-Binding Proteins and Metabolizing Enzymes. *Anat. Rec.* **2013**, *296*, 1333–1345. [CrossRef]
2. Heydel, J.M.; Hanser, H.I.; Faure, P.; Neiers, F. Odorant Metabolizing Enzymes in the Peripheral Olfactory Process. In *Flavour*; Wiley: Hoboken, NJ, USA, 2017; pp. 34–56.
3. Boichot, V.; Muradova, M.; Nivet, C.; Proskura, A.; Heydel, J.-M.; Canivenc-Lavier, M.-C.; Canon, F.; Neiers, F.; Schwartz, M. The Role of Perireceptor Events in FLavor Perception. *Front. Food Sci. Technol.* **2022**, *9*, 989291. [CrossRef]
4. Munoz-Gonzalez, C.; Brule, M.; Martin, C.; Feron, G.; Canon, F. Molecular Mechanisms of Aroma Persistence: From Noncovalent Interactions between Aroma Compounds and the Oral Mucosa to Metabolization of Aroma Compounds by Saliva and Oral Cells. *Food Chem.* **2021**, *373*, 131467. [CrossRef]
5. Ployon, S.; Brule, M.; Andriot, I.; Morzel, M.; Canon, F. Understanding Retention and Metabolization of Aroma Compounds Using an in Vitro Model of Oral Mucosa. *Food Chem.* **2020**, *318*, 126468. [CrossRef]
6. Schwartz, M.; Neiers, F.; Charles, J.P.; Heydel, J.M.; Munoz-Gonzalez, C.; Feron, G.; Canon, F. Oral Enzymatic Detoxification System: Insights Obtained from Proteome Analysis to Understand Its Potential Impact on Aroma Metabolization. *Compr. Rev. Food Sci. Food Saf.* **2021**, *20*, 5516–5547. [CrossRef] [PubMed]
7. Robert-Hazotte, A.; Faure, P.; Ménétrier, F.; Folia, M.; Schwartz, M.; Le Quéré, J.-L.; Neiers, F.; Thomas-Danguin, T.; Heydel, J.-M. Nasal Odorant Competitive Metabolism Is Involved in the Human Olfactory Process. *J. Agric. Food Chem.* **2022**, *70*, 8385–8394. [CrossRef]
8. Boichot, V.; Menetrier, F.; Saliou, J.-M.; Lirussi, F.; Canon, F.; Folia, M.; Heydel, J.-M.; Hummel, T.; Menzel, S.; Steinke, M.; et al. Characterization of Human Oxidoreductases Involved in Aldehyde Odorant Metabolism. *Sci. Rep.* **2023**, *13*, 4876. [CrossRef] [PubMed]
9. Qian, W.W.; Wei, J.N.; Sanchez-Lengeling, B.; Lee, B.K.; Luo, Y.; Vlot, M.; Dechering, K.; Peng, J.; Gerkin, R.C.; Wiltschko, A.B. Metabolic Activity Organizes Olfactory Representations. *eLife* **2023**, *12*, e82502. [CrossRef]
10. Schwartz, M.; Neiers, F.; Feron, G.; Canon, F. The Relationship Between Salivary Redox, Diet, and Food Flavor Perception. *Front. Nutr.* **2020**, *7*, 612735. [CrossRef] [PubMed]
11. Piombino, P.; Genovese, A.; Esposito, S.; Moio, L.; Cutolo, P.P.; Chambery, A.; Severino, V.; Moneta, E.; Smith, D.P.; Owens, S.M.; et al. Saliva from Obese Individuals Suppresses the Release of Aroma Compounds from Wine. *PLoS ONE* **2014**, *9*, e85611. [CrossRef]
12. Schwartz, M.; Canon, F.; Feron, G.; Neiers, F.; Gamero, A. Impact of Oral Microbiota on Flavor Perception: From Food Processing to In-Mouth Metabolization. *Foods* **2021**, *10*, 2006. [CrossRef] [PubMed]
13. López-Dávalos, P.C.; Requena, T.; Pozo-Bayón, M.Á.; Muñoz-González, C. Decreased Retronasal Olfaction and Taste Perception in Obesity Are Related to Saliva Biochemical and Microbiota Composition. *Food Res. Int.* **2023**, *167*, 112660. [CrossRef] [PubMed]
14. Duarte-Coimbra, S.; Forcina, G.; Pérez-Pardal, L.; Beja-Pereira, A. Characterization of Tongue Dorsum Microbiome in Wine Tasters. *Food Res. Int.* **2023**, *163*, 112259. [CrossRef] [PubMed]
15. Xi, Y.; Yu, M.; Li, X.; Zeng, X.; Li, J. The Coming Future: The Role of the Oral–Microbiota–Brain Axis in Aroma Release and Perception. *Comp. Rev. Food Sci. Food Safe* **2024**, *23*, e13303. [CrossRef] [PubMed]
16. Starkenmann, C.; Le Calve, B.; Niclass, Y.; Cayeux, I.; Beccucci, S.; Troccaz, M. Olfactory Perception of Cysteine-S-Conjugates from Fruits and Vegetables. *J. Agric. Food Chem.* **2008**, *56*, 9575–9580. [CrossRef] [PubMed]
17. Munoz-Gonzalez, C.; Cueva, C.; Angeles Pozo-Bayon, M.; Victoria Moreno-Arribas, M. Ability of Human Oral Microbiota to Produce Wine Odorant Aglycones from Odourless Grape Glycosidic Aroma Precursors. *Food Chem.* **2015**, *187*, 112–119. [CrossRef] [PubMed]
18. Parker, M.; Onetto, C.; Hixson, J.; Bilogrevic, E.; Schueth, L.; Pisaniello, L.; Borneman, A.; Herderich, M.; de Barros Lopes, M.; Francis, L. Factors Contributing to Interindividual Variation in Retronasal Odor Perception from Aroma Glycosides: The Role of Odorant Sensory Detection Threshold, Oral Microbiota, and Hydrolysis in Saliva. *J. Agric. Food Chem.* **2020**, *68*, 10299–10309. [CrossRef] [PubMed]
19. Hirst, M.B.; Richter, C.L. Review of Aroma Formation through Metabolic Pathways of *Saccharomyces cerevisiae* in Beverage Fermentations. *Am. J. Enol. Vitic.* **2016**, *67*, 361–370. [CrossRef]
20. Dzialo, M.C.; Park, R.; Steensels, J.; Lievens, B.; Verstrepen, K.J. Physiology, Ecology and Industrial Applications of Aroma Formation in Yeast. *FEMS Microbiol. Rev.* **2017**, *41*, S95–S128. [CrossRef]
21. Mao, J.; Zhou, Z.; Yang, H. Microbial Succession and Its Effect on the Formation of Umami Peptides during Sufu Fermentation. *Front. Microbiol.* **2023**, *14*, 1181588. [CrossRef]
22. Ferreira, V.; Lopez, R. The Actual and Potential Aroma of Winemaking Grapes. *Biomolecules* **2019**, *9*, 818. [CrossRef] [PubMed]
23. Shin, K.-C.; Nam, H.-K.; Oh, D.-K. Hydrolysis of Flavanone Glycosides by β-Glucosidase from *Pyrococcus furiosus* and Its Application to the Production of Flavanone Aglycones from Citrus Extracts. *J. Agric. Food Chem.* **2013**, *61*, 11532–11540. [CrossRef] [PubMed]
24. Li, S.; Tian, Y.; Jiang, P.; Lin, Y.; Liu, X.; Yang, H. Recent Advances in the Application of Metabolomics for Food Safety Control and Food Quality Analyses. *Crit. Rev. Food Sci. Nutr.* **2021**, *61*, 1448–1469. [CrossRef]

25. Muradova, M.; Proskura, A.; Canon, F.; Aleksandrova, I.; Schwartz, M.; Heydel, J.-M.; Baranenko, D.; Nadtochii, L.; Neiers, F. Unlocking Flavor Potential Using Microbial β-Glucosidases in Food Processing. *Foods* **2023**, *12*, 4484. [CrossRef] [PubMed]
26. Neiers, F.; Gourrat, K.; Canon, F.; Schwartz, M. Metabolism of Cysteine Conjugates and Production of Flavor Sulfur Compounds by a Carbon–Sulfur Lyase from the Oral Anaerobe *Fusobacterium nucleatum*. *J. Agric. Food Chem.* **2022**, *70*, 9969–9979. [CrossRef] [PubMed]
27. Allegrini, A.; Astegno, A.; La Verde, V.; Dominici, P. Characterization of C-S Lyase from *Lactobacillus delbrueckii* Subsp. *Bulgaricus* ATCC BAA-365 and Its Potential Role in Food Flavour Applications. *J. Biochem.* **2017**, *161*, 349–360. [CrossRef]
28. Rudden, M.; Herman, R.; Rose, M.; Bawdon, D.; Cox, D.S.; Dodson, E.; Holden, M.T.G.; Wilkinson, A.J.; James, A.G.; Thomas, G.H. The Molecular Basis of Thioalcohol Production in Human Body Odour. *Sci. Rep.* **2020**, *10*, 12500. [CrossRef] [PubMed]
29. Marcinkowska, M.A.; Jeleń, H.H. Role of Sulfur Compounds in Vegetable and Mushroom Aroma. *Molecules* **2022**, *27*, 6116. [CrossRef]
30. Chin, H.-W.; Lindsay, R.C. Mechanisms of Formation of Volatile Sulfur Compounds Following the Action of Cysteine Sulfoxide Lyases. *J. Agric. Food Chem.* **1994**, *42*, 1529–1536. [CrossRef]
31. Roncoroni, M.; Santiago, M.; Hooks, D.O.; Moroney, S.; Harsch, M.J.; Lee, S.A.; Richards, K.D.; Nicolau, L.; Gardner, R.C. The Yeast IRC7 Gene Encodes a Beta-Lyase Responsible for Production of the Varietal Thiol 4-Mercapto-4-Methylpentan-2-One in Wine. *Food Microbiol.* **2011**, *28*, 926–935. [CrossRef]
32. Belda, I.; Ruiz, J.; Navascues, E.; Marquina, D.; Santos, A. Improvement of Aromatic Thiol Release through the Selection of Yeasts with Increased Beta-Lyase Activity. *Int. J. Food Microbiol.* **2016**, *225*, 1–8. [CrossRef] [PubMed]
33. Basic, A.; Blomqvist, M.; Dahlen, G.; Svensater, G. The Proteins of Fusobacterium Spp. Involved in Hydrogen Sulfide Production from L-Cysteine. *BMC Microbiol.* **2017**, *17*, 61. [CrossRef] [PubMed]
34. Yoshida, Y.; Ito, S.; Kamo, M.; Kezuka, Y.; Tamura, H.; Kunimatsu, K.; Kato, H. Production of Hydrogen Sulfide by Two Enzymes Associated with Biosynthesis of Homocysteine and Lanthionine in *Fusobacterium nucleatum* Subsp. *Nucleatum* ATCC 25586. *Microbiology* **2010**, *156*, 2260–2269. [CrossRef] [PubMed]
35. Cooper, A.J.; Krasnikov, B.F.; Niatsetskaya, Z.V.; Pinto, J.T.; Callery, P.S.; Villar, M.T.; Artigues, A.; Bruschi, S.A. Cysteine S-Conjugate Beta-Lyases: Important Roles in the Metabolism of Naturally Occurring Sulfur and Selenium-Containing Compounds, Xenobiotics and Anticancer Agents. *Amino Acids* **2011**, *41*, 7–27. [CrossRef] [PubMed]
36. Alexander, F.W.; Sandmeier, E.; Mehta, P.K.; Christen, P. Evolutionary Relationships among Pyridoxal-5′-phosphate-dependent Enzymes: Regio-specific α, β and γ Families. *Eur. J. Biochem.* **1994**, *219*, 953–960. [CrossRef] [PubMed]
37. Clausen, T.; Huber, R.; Laber, B.; Pohlenz, H.-D.; Messerschmidt, A. Crystal Structure of the Pyridoxal-5′-Phosphate Dependent Cystathionine β-Lyase fromEscherichia Coliat 1.83 Å. *J. Mol. Biol.* **1996**, *262*, 202–224. [CrossRef] [PubMed]
38. Clausen, T.; Schlegel, A.; Peist, R.; Schneider, E.; Steegborn, C.; Chang, Y.-S.; Haase, A.; Bourenkov, G.P.; Bartunik, H.D.; Boos, W. X-Ray Structure of MalY from Escherichia Coli: A Pyridoxal 5′-Phosphate-Dependent Enzyme Acting as a Modulator in Mal Gene Expression. *EMBO J.* **2000**, *19*, 831–842. [CrossRef] [PubMed]
39. Auger, S.; Gomez, M.P.; Danchin, A.; Martin-Verstraete, I. The PatB Protein of Bacillus Subtilis Is a C-S-Lyase. *Biochimie* **2005**, *87*, 231–238. [CrossRef] [PubMed]
40. Aitken, S.M.; Lodha, P.H.; Morneau, D.J.K. The Enzymes of the Transsulfuration Pathways: Active-Site Characterizations. *Biochim. Biophys. Acta BBA Proteins Proteom.* **2011**, *1814*, 1511–1517. [CrossRef]
41. Auger, S.; Yuen, W.H.; Danchin, A.; Martin-Verstraete, I. The metIC Operon Involved in Methionine Biosynthesis in Bacillus Subtilis Is Controlled by Transcription Antitermination. *Microbiology* **2002**, *148*, 507–518. [CrossRef]
42. Krupka, H.I.; Huber, R.; Holt, S.C.; Clausen, T. Crystal Structure of Cystalysin from Treponema Denticola: A Pyridoxal 5′-phosphate-dependent Protein Acting as a Haemolytic Enzyme. *EMBO J.* **2000**, *19*, 3168–3178. [CrossRef]
43. Chu, L.; Ebersole, J.L.; Kurzban, G.P.; Holt, S.C. Cystalysin, a 46-Kilodalton Cysteine Desulfhydrase from Treponema Denticola, with Hemolytic and Hemoxidative Activities. *Infect. Immun.* **1997**, *65*, 3231–3238. [CrossRef] [PubMed]
44. Chu, L.; Ebersole, J.L.; Kurzban, G.P.; Holt, S.C. Cystalysin, a 46-kDa L-Cysteine Desulfhydrase from Treponema Denticola: Biochemical and Biophysical Characterization. *Clin. Infect. Dis.* **1999**, *28*, 442–450. [CrossRef] [PubMed]
45. Bertoldi, M.; Cellini, B.; Clausen, T.; Voltattorni, C.B. Spectroscopic and Kinetic Analyses Reveal the Pyridoxal 5′-Phosphate Binding Mode and the Catalytic Features of *Treponema denticola* Cystalysin. *Biochemistry* **2002**, *41*, 9153–9164. [CrossRef] [PubMed]
46. Krespi, Y.P.; Shrime, M.G.; Kacker, A. The Relationship between Oral Malodor and Volatile Sulfur Compound-Producing Bacteria. *Otolaryngol. Head. Neck Surg.* **2006**, *135*, 671–676. [CrossRef] [PubMed]
47. Tomisawa, H.; Suzuki, S.; Ichihara, S.; Fukazawa, H.; Tateishi, M. Purification and Characterization of C-S Lyase from Fusobacterium Varium. A C-S Cleavage Enzyme of Cysteine Conjugates and Some S-Containing Amino Acids. *J. Biol. Chem.* **1984**, *259*, 2588–2593. [CrossRef] [PubMed]
48. Starkenmann, C.; Niclass, Y.; Troccaz, M.; Clark, A.J. Identification of the Precursor of (S)-3-Methyl-3-Sulfanylhexan-1-Ol, the Sulfury Malodour of Human Axilla Sweat. *Chem. Biodivers.* **2005**, *2*, 705–716. [CrossRef] [PubMed]
49. Natsch, A.; Schmid, J.; Flachsmann, F. Identification of Odoriferous Sulfanylalkanols in Human Axilla Secretions and Their Formation through Cleavage of Cysteine Precursors by a C–S Lyase Isolated from Axilla Bacteria. *Chem. Biodivers.* **2004**, *1*, 1058–1072. [CrossRef] [PubMed]
50. Zdych, E.; Peist, R.; Reidl, J.; Boos, W. MalY of Escherichia Coli Is an Enzyme with the Activity of a Beta C-S Lyase (Cystathionase). *J. Bacteriol.* **1995**, *177*, 5035–5039. [CrossRef]

51. Wu, Y.; Sun, Y.; Richet, E.; Han, Z.; Chai, J. Structural Basis for Negative Regulation of the *Escherichia coli* Maltose System. *Nat. Commun.* **2023**, *14*, 4925. [CrossRef]
52. Kato, S.; Oikawa, T. A Novel Bifunctional Amino Acid Racemase With Multiple Substrate Specificity, MalY From *Lactobacillus sakei* LT-13: Genome-Based Identification and Enzymological Characterization. *Front. Microbiol.* **2018**, *9*, 403. [CrossRef] [PubMed]
53. Clausen, T.; Huber, R.; Messerschmidt, A.; Pohlenz, H.-D.; Laber, B. Slow-Binding Inhibition of *Escherichia coli* Cystathionine β-Lyase by L-Aminoethoxyvinylglycine: A Kinetic and X-ray Study. *Biochemistry* **1997**, *36*, 12633–12643. [CrossRef]
54. Landaud, S.; Helinck, S.; Bonnarme, P. Formation of Volatile Sulfur Compounds and Metabolism of Methionine and Other Sulfur Compounds in Fermented Food. *Appl. Microbiol. Biotechnol.* **2008**, *77*, 1191–1205. [CrossRef] [PubMed]
55. Cannon, R.J.; Ho, C.T. Volatile Sulfur Compounds in Tropical Fruits. *J. Food Drug Anal.* **2018**, *26*, 445–468. [CrossRef] [PubMed]
56. Li, J.; Dadmohammadi, Y.; Abbaspourrad, A. Flavor Components, Precursors, Formation Mechanisms, Production and Characterization Methods: Garlic, Onion, and Chili Pepper Flavors. *Crit. Rev. Food Sci. Nutr.* **2022**, *62*, 8265–8287. [CrossRef] [PubMed]
57. Yamazaki, Y.; Iwasaki, K.; Mikami, M.; Yagihashi, A. Distribution of Eleven Flavor Precursors, S-Alk(En)Yl-L-Cysteine Derivatives, in Seven Allium Vegetables. *Food Sci. Technol. Res.* **2010**, *17*, 55–62. [CrossRef]
58. Frank, D.; Piyasiri, U.; Archer, N.; Heffernan, J.; Poelman, A.A.M. In-Mouth Volatile Production from Brassica Vegetables (Cauliflower) and Associations with Liking in an Adult/Child Cohort. *J. Agric. Food Chem.* **2021**, *69*, 11646–11655. [CrossRef] [PubMed]
59. Tominaga, T.; des Gachons, C.P.; Dubourdieu, D. A New Type of Flavor Precursors in Vitis Vinifera L Cv Sauvignon Blanc: S-Cysteine Conjugates. *J. Agric. Food Chem.* **1998**, *46*, 5215–5219. [CrossRef]
60. Peyrot Des Gachons, C.; Tominaga, T.; Dubourdieu, D. Sulfur Aroma Precursor Present in S-Glutathione Conjugate Form: Identification of S-3-(Hexan-1-Ol)-Glutathione in Must from *Vitis vinifera* L. Cv. Sauvignon Blanc. *J. Agric. Food Chem.* **2002**, *50*, 4076–4079. [CrossRef]
61. Gros, J.; Tran, T.T.H.; Collin, S. Enzymatic Release of Odourant Polyfunctional Thiols from Cysteine Conjugates in Hop. *J. Inst. Brew.* **2013**, *119*, 221–227. [CrossRef]
62. Chenot, C.; Collin, S.; Suc, L.; Roland, A. Evidence of Enzymatic and Chemical Interconversions of Barley Malt 3-Sulfanylhexanol Conjugates during Mashing. *J. Agric. Food Chem.* **2023**, *71*, 13107–13113. [CrossRef] [PubMed]
63. Starkenmann, C.; Luca, L.; Niclass, Y.; Praz, E.; Roguet, D. Comparison of Volatile Constituents of *Persicaria odorata* (Lour.) Soják (*Polygonum odoratum* Lour.) and *Persicaria hydropiper* L. Spach (*Polygonum hydropiper* L.). *J. Agric. Food Chem.* **2006**, *54*, 3067–3071. [CrossRef]
64. Starkenmann, C.; Niclass, Y.; Escher, S. Volatile Organic Sulfur-Containing Constituents in *Poncirus trifoliata* (L.) Raf. (Rutaceae). *J. Agric. Food Chem.* **2007**, *55*, 4511–4517. [CrossRef]
65. Cerny, C.; Guntz-Dubini, R. Formation of Cysteine-S-Conjugates in the Maillard Reaction of Cysteine and Xylose. *Food Chem.* **2013**, *141*, 1078–1086. [CrossRef]
66. Cerny, C.; Schlichtherle-Cerny, H.; Gibe, R.; Yuan, Y. Furfuryl Alcohol Is a Precursor for Furfurylthiol in Coffee. *Food Chem.* **2021**, *337*, 128008. [CrossRef]
67. Bonnaffoux, H.; Roland, A.; Schneider, R.; Cavelier, F. Spotlight on Release Mechanisms of Volatile Thiols in Beverages. *Food Chem.* **2021**, *339*, 127628. [CrossRef]
68. Kobayashi, H.; Takase, H.; Suzuki, Y.; Tanzawa, F.; Takata, R.; Fujita, K.; Kohno, M.; Mochizuki, M.; Suzuki, S.; Konno, T. Environmental Stress Enhances Biosynthesis of Flavor Precursors, S-3-(Hexan-1-Ol)-Glutathione and S-3-(Hexan-1-Ol)-L-Cysteine, in Grapevine through Glutathione S-Transferase Activation. *J. Exp. Bot.* **2011**, *62*, 1325–1336. [CrossRef] [PubMed]
69. Kishimoto, T.; Kobayashi, M.; Yako, N.; Iida, A.; Wanikawa, A. Comparison of 4-Mercapto-4-Methylpentan-2-One Contents in Hop Cultivars from Different Growing Regions. *J. Agric. Food Chem.* **2008**, *56*, 1051–1057. [CrossRef]
70. Thibon, C.; Cluzet, S.; Mérillon, J.M.; Darriet, P.; Dubourdieu, D. 3-Sulfanylhexanol Precursor Biogenesis in Grapevine Cells: The Stimulating Effect of *Botrytis cinerea*. *J. Agric. Food Chem.* **2011**, *59*, 1344–1351. [CrossRef] [PubMed]
71. Schwimmer, S.; Mazelis, M. Characterization of Alliinase of Allium Cepa (Onion). *Arch. Biochem. Biophys.* **1963**, *100*, 66–73. [CrossRef]
72. Zhang, M.; Batra, R.; Brainta, M.; Huang, D. Purification and Characterisation of a C-S Lyase in Seeds of Parkia Speciosa Hassk. *Food Chem.* **2023**, *404*, 134438. [CrossRef] [PubMed]
73. Lei, X.; Gao, S.; Feng, X.; Huang, Z.; Bian, Y.; Huang, W.; Liu, Y. Effects of GGT and C-S Lyase on the Generation of Endogenous Formaldehyde in Lentinula Edodes at Different Growth Stages. *Molecules* **2019**, *24*, 4203. [CrossRef] [PubMed]
74. Wang, Y.; Bao, D.-P.; Yang, R.-H.; Chen, H.-Y.; Gao, Y.-N.; Li, Y.; Mao, W.-J.; Wu, Y.-Y. Bioinformatics Analyses of C-S Lyases in the Genome of Lentinula Edode. *Mycosystema* **2018**, *37*, 1608–1619. [CrossRef]
75. Peña-Gallego, A.; Hernández-Orte, P.; Cacho, J.; Ferreira, V. S-Cysteinylated and S-Glutathionylated Thiol Precursors in Grapes. A Review. *Food Chem.* **2012**, *131*, 1–13. [CrossRef]
76. Vermeulen, C.; Lejeune, I.; Tran, T.T.; Collin, S. Occurrence of Polyfunctional Thiols in Fresh Lager Beers. *J. Agric. Food Chem.* **2006**, *54*, 5061–5068. [CrossRef] [PubMed]
77. Holt, S.; Cordente, A.G.; Williams, S.J.; Capone, D.L.; Jitjaroen, W.; Menz, I.R.; Curtin, C.; Anderson, P.A. Engineering *Saccharomyces cerevisiae* to Release 3-Mercaptohexan-1-Ol during Fermentation through Overexpression of an S. Cerevisiae Gene, STR3, for Improvement of Wine Aroma. *Appl. Environ. Microbiol.* **2011**, *77*, 3626–3632. [CrossRef]

78. Tominaga, T.; Dubourdieu, D. Identification of Cysteinylated Aroma Precursors of Certain Volatile Thiols in Passion Fruit Juice. *J. Agric. Food Chem.* **2000**, *48*, 2874–2876. [CrossRef] [PubMed]
79. Zha, M.; Sun, B.; Yin, S.; Mehmood, A.; Cheng, L.; Wang, C. Generation of 2-Furfurylthiol by Carbon–Sulfur Lyase from the *Baijiu* Yeast *Saccharomyces cerevisiae* G20. *J. Agric. Food Chem.* **2018**, *66*, 2114–2120. [CrossRef]
80. Zhang, G.; Xiao, P.; Xu, Y.; Li, H.; Li, H.; Sun, J.; Sun, B. Isolation and Characterization of Yeast with Benzenemethanethiol Synthesis Ability Isolated from Baijiu Daqu. *Foods* **2023**, *12*, 2464. [CrossRef]
81. Parker, M.; Capone, D.L.; Francis, I.L.; Herderich, M.J. Aroma Precursors in Grapes and Wine: Flavor Release during Wine Production and Consumption. *J. Agric. Food Chem.* **2018**, *66*, 2281–2286. [CrossRef]
82. Eriksen, R.L.; Padgitt-Cobb, L.K.; Townsend, M.S.; Henning, J.A. Gene Expression for Secondary Metabolite Biosynthesis in Hop (*Humulus lupulus* L.) Leaf Lupulin Glands Exposed to Heat and Low-Water Stress. *Sci. Rep.* **2021**, *11*, 5138. [CrossRef] [PubMed]
83. Holt, S.; Miks, M.H.; de Carvalho, B.T.; Foulquie-Moreno, M.R.; Thevelein, J.M. The Molecular Biology of Fruity and Floral Aromas in Beer and Other Alcoholic Beverages. *FEMS Microbiol. Rev.* **2019**, *43*, 193–222. [CrossRef] [PubMed]
84. Chenot, C.; Willemart, G.; Gros, J.; Collin, S. Ability of Exogenous or Wort Endogenous Enzymes to Release Free Thiols from Hop Cysteinylated and Glutathionylated *S*-Conjugates. *J. Am. Soc. Brew. Chem.* **2023**, *81*, 33–44. [CrossRef]
85. Song, X.; Zhu, L.; Wang, X.; Zheng, F.; Zhao, M.; Liu, Y.; Li, H.; Zhang, F.; Zhang, Y.; Chen, F. Characterization of Key Aroma-Active Sulfur-Containing Compounds in Chinese Laobaigan Baijiu by Gas Chromatography-Olfactometry and Comprehensive Two-Dimensional Gas Chromatography Coupled with Sulfur Chemiluminescence Detection. *Food Chem.* **2019**, *297*, 124959. [CrossRef] [PubMed]
86. Hanniffy, S.B.; Peláez, C.; Martínez-Bartolomé, M.A.; Requena, T.; Martínez-Cuesta, M.C. Key Enzymes Involved in Methionine Catabolism by Cheese Lactic Acid Bacteria. *Int. J. Food Microbiol.* **2009**, *135*, 223–230. [CrossRef] [PubMed]
87. Alting, A.C.; Engels, W.; Van Schalkwijk, S.; Exterkate, F.A. Purification and Characterization of Cystathionine (Beta)-Lyase from *Lactococcus lactis* Subsp. Cremoris B78 and Its Possible Role in Flavor Development in Cheese. *Appl. Environ. Microbiol.* **1995**, *61*, 4037–4042. [CrossRef] [PubMed]
88. Lee, W.-J.; Banavara, D.S.; Hughes, J.E.; Christiansen, J.K.; Steele, J.L.; Broadbent, J.R.; Rankin, S.A. Role of Cystathionine β-Lyase in Catabolism of Amino Acids to Sulfur Volatiles by Genetic Variants of *Lactobacillus helveticus* CNRZ 32. *Appl. Environ. Microbiol.* **2007**, *73*, 3034–3039. [CrossRef] [PubMed]
89. Howell, K.S.; Klein, M.; Swiegers, J.H.; Hayasaka, Y.; Elsey, G.M.; Fleet, G.H.; Høj, P.B.; Pretorius, I.S.; De Barros Lopes, M.A. Genetic Determinants of Volatile-Thiol Release by *Saccharomyces cerevisiae* during Wine Fermentation. *Appl. Environ. Microbiol.* **2005**, *71*, 5420–5426. [CrossRef] [PubMed]
90. Yamagata, S.; D'Andrea, R.J.; Fujisaki, S.; Isaji, M.; Nakamura, K. Cloning and Bacterial Expression of the CYS3 Gene Encoding Cystathionine Gamma-Lyase of *Saccharomyces cerevisiae* and the Physicochemical and Enzymatic Properties of the Protein. *J. Bacteriol.* **1993**, *175*, 4800–4808. [CrossRef]
91. Takase, H.; Sasaki, K.; Kiyomichi, D.; Kobayashi, H.; Matsuo, H.; Takata, R. Impact of Lactobacillus Plantarum on Thiol Precursor Biotransformation Leading to Production of 3-Sulfanylhexan-1-Ol. *Food Chem.* **2018**, *259*, 99–104. [CrossRef]
92. Clérat, L.; Rémond, E.; Schneider, R.; Cavelier, F.; Vivès, E. Exogenous C–S Lyase Enzyme, a Potential Tool To Release Aromas in Wine or Beer? *J. Agric. Food Chem.* **2024**, *72*, 1878–1884. [CrossRef] [PubMed]
93. Zhu, Z.; Pius Bassey, A.; Cao, Y.; Ma, Y.; Huang, M.; Yang, H. Food Protein Aggregation and Its Application. *Food Res. Int.* **2022**, *160*, 111725. [CrossRef] [PubMed]
94. Lamont, R.J.; Koo, H.; Hajishengallis, G. The Oral Microbiota: Dynamic Communities and Host Interactions. *Nat. Rev. Microbiol.* **2018**, *16*, 745–759. [CrossRef] [PubMed]
95. Li, S.; Ahmed, L.; Zhang, R.; Pan, Y.; Matsunami, H.; Burger, J.L.; Block, E.; Batista, V.S.; Zhuang, H. Smelling Sulfur: Copper and Silver Regulate the Response of Human Odorant Receptor OR2T11 to Low-Molecular-Weight Thiols. *J. Am. Chem. Soc.* **2016**, *138*, 13281–13288. [CrossRef]
96. Frank, D.; Piyasiri, U.; Archer, N.; Jenifer, J.; Appelqvist, I. Influence of Saliva on Individual In-Mouth Aroma Release from Raw Cabbage (*Brassica oleracea* Var. Capitata f. *Rubra* L.) and Links to Perception. *Heliyon* **2018**, *4*, e01045. [CrossRef]

International Journal of
Molecular Sciences

MDPI

Review

Wastewater Treatment with Bacterial Representatives of the *Thiothrix* Morphotype

Maria V. Gureeva [1,*], Maria S. Muntyan [2,*], Nikolai V. Ravin [3] and Margarita Yu. Grabovich [1]

1 Department of Biochemistry and Cell Physiology, Voronezh State University, Universitetskaya pl., 1, 394018 Voronezh, Russia; margarita_grabov@mail.ru
2 Belozersky Institute of Physico-Chemical Biology, Lomonosov Moscow State University, Leninskie Gory, 119991 Moscow, Russia
3 Institute of Bioengineering, Research Center of Biotechnology of the Russian Academy of Sciences, Leninsky Prospect, 33-2, 119071 Moscow, Russia; nravin@mail.ru
* Correspondence: maryorl@mail.ru (M.V.G.); muntyan@genebee.msu.ru (M.S.M.)

Abstract: Bacteria of the *Thiothrix* morphotype, comprising the genera *Thiothrix*, *Thiolinea* and *Thiofilum*, are frequently encountered in domestic and industrial wastewater treatment systems, but they are usually not clearly differentiated due to the marked similarity in their morphologies. Methods ranging from light microscopy, FISH and PCR to modern high-throughput sequencing are used to identify them. The development of these bacteria in wastewater treatment systems has both advantages and disadvantages. On the one hand, the explosive growth of these bacteria can lead to activated sludge bulking or clogging of the treatment system's membranes, with a consequent decrease in the water treatment efficiency. On the other hand, members of the *Thiothrix* morphotype can improve the quality of granular sludge and increase the water treatment efficiency. This may be due to their capacity for sulfide oxidation, denitrification combined with the oxidation of reduced sulfur compounds, enhanced biological phosphate removal and possibly denitrifying phosphate removal. The recently obtained pangenome of the genus *Thiothrix* allows the explanation, at the genomic level, of the experimental results of various studies. Moreover, this review summarizes the data on the factors affecting the proliferation of representatives of the *Thiothrix* morphotype.

Keywords: *Thiothrix*; *Thiolinea*; *Thiofilum*; wastewater

1. Introduction

Representatives of filamentous bacteria, similar in morphotype to *Thiothrix*, are attaching filamentous organisms, many of which are capable of forming rosettes, accumulating elemental sulfur inside cells and spreading by gonidia [1]. In running water containing hydrogen sulfide, they form abundant whitish filamentous fouling, visible to the naked eye. In the presence of reduced sulfur compounds, representatives of the *Thiothrix* morphotype are known for their versatile metabolism, which gives them potential for lithoautotrophic and lithoheterotrophic growth [2–5]. Most commonly, they can be found in sulfide springs, deep-sea hydrotherms, irrigation systems and activated sludge from sewage treatment plants [2–5].

Over the last two decades, the taxonomy of the genus *Thiothrix* has been revised several times. In the guidelines for the identification of microorganisms that cause the bulking of activated sludge, filamentous bacteria similar in morphotype to *Thiothrix* were classified as Type 021N, *Thiothrix* I and *Thiothrix* II [6,7]. However, it is worth noting that, even at the time of the publication of these manuals, such classification was shown to be phylogenetically incorrect [8,9]. Subsequently, Boden and Scott [10] proposed that *Thiothrix disciformis* and *Thiothrix eikelboomii*, previously included in Type 021N, should be classified as the genus *Thiolinea*, while *Thiothrix flexilis* should be classified as the genus *Thiofilum*.

Despite the fact that *Thiolinea* and *Thiofilum* were assigned to separate genera 6 years ago, representatives of these genera are often still attributed to the genus *Thiothrix* in works on biological wastewater treatment [11,12]. Moreover, even the more obsolete name Type 021N was used to refer to them for nearly two decades [13–18].

To eliminate such confusion in taxonomy, in the presented review, we combine data on bacteria currently assigned to three genera, *Thiothrix*, *Thiofilum* and *Thiolinea*, based on phylogenetic data, in contrast to their previous assignment to the genus *Thiothrix* only on the basis of the morphotype (Figure 1).

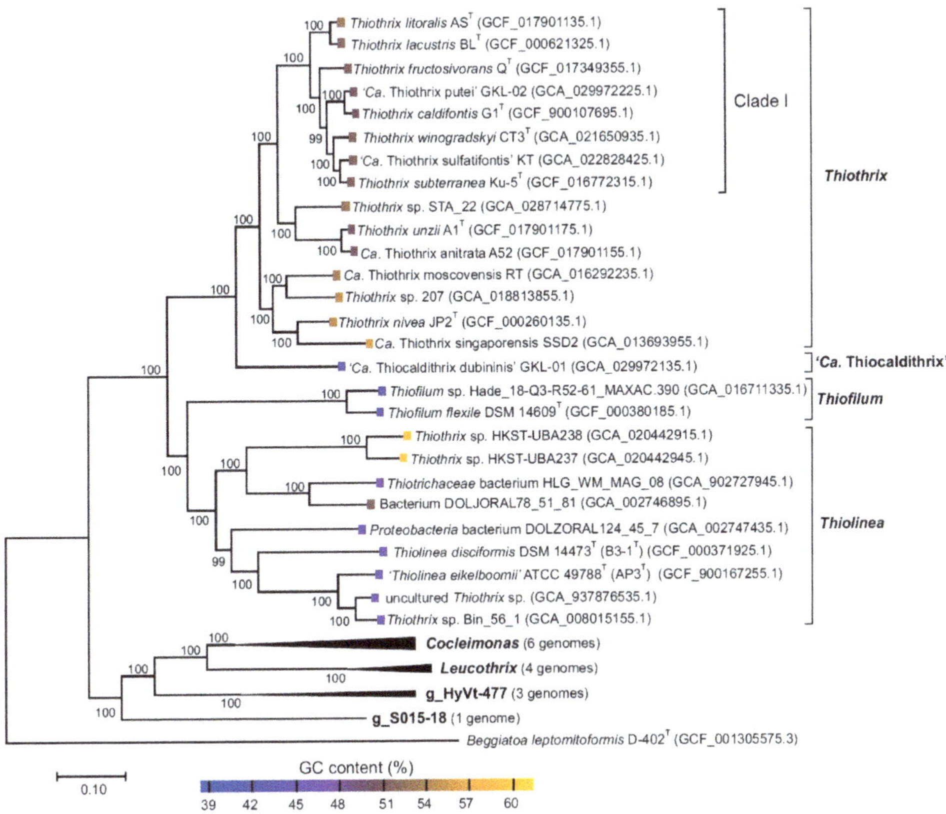

Figure 1. Phylogenetic tree of the family *Thiotrichaceae* based on the concatenated genomic sequences of 120 conserved marker genes (maximum likelihood method). Clade I of the genus *Thiothrix* is marked with a square bracket in the right part of the figure. For tree rooting, the genome of *Beggiatoa leptomitoformis* D-402^T (GCF_001305575.3) was used. Figure was modified from Ravin et al. [19] under the terms of the Creative Commons Attribution 4.0 International License.

The importance of bacteria of the *Thiothrix* morphotype in applied studies is due to the fact that they are often found in domestic [18,20–23] and industrial [24–26] wastewater treatment systems. Biological treatment using complex microbial communities to degrade pollutants into low-toxicity or non-toxic products is the most important method for wastewater treatment [27]. Bacteria of *Thiothrix* morphotype are often found in such microbial communities. These bacteria have been detected in wastewater treatment plants in the pulp and paper [28–31] and textile industries [32], as well as in wastewater from bisphenol A production [33]. In addition, bacteria of the *Thiothrix* morphotype actively proliferate in wastewater from food industries, including the dairy [34], fish canning [35] and potato industries [36].

Bacteria of the *Thiothrix* morphotype are often found in wastewater from pharmaceutical plants. It was shown that *Thiothrix* was one of the dominant populations at the genus level in a decomposition reactor for the wastewater treatment of lincomycin [37], diclofenac and clofibrin acid [23]; spiramycin, oxytetracycline and streptomycin [38]; and tetracycline and sulfamethoxazole [39,40].

The role of bacteria with a morphotype similar to *Thiothrix* in wastewater treatment systems is controversial. On the one hand, a certain change in the environmental parameters can cause their explosive growth, leading to activated sludge bulking and the clogging of membranes, which significantly reduces the efficiency of treatment [26,29,41]. Bulking can be related to the ability of *Thiothrix*-like organisms to form flocks with poor settling characteristics [42], which raises operational problems in biological wastewater treatment systems worldwide [18,25]. At the same time, a number of studies show that, under properly selected conditions, representatives of the *Thiothrix* morphotype, on the contrary, can help to improve the quality of granular sludge and increase the efficiency of wastewater treatment [43,44].

In addition to affecting the characteristics of activated sludge, bacteria of the *Thiothrix* morphotype are capable of performing a number of metabolic processes leading to the removal of pollutants from wastewater. These include sulfide oxidation [45], denitrification coupled with the oxidation of reduced sulfur compounds [23,46], the enhanced biological removal of phosphorus [47] and possibly the denitrifying removal of phosphorus [48,49].

Although the role of *Thiothrix*, *Thiofilum* and *Thiolinea* in the operation of wastewater treatment plants is well known and undisputed, no review has yet been published that discusses the role of these particular bacteria in wastewater treatment systems, rather than filamentous bacteria in general. The aim of this review is to summarize the current information on the types of wastewater treatment plants in which representatives of these three genera occur, the methods for their detection in wastewater and the metabolic processes that they carry out in wastewater treatment, as well as the conditions affecting the development of this group of bacteria. This review is the first to attempt to delineate the available published data on members of the *Thiothrix* morphotype into three genera, *Thiothrix*, *Thiofilum* and *Thiolinea*, according to recent advances in molecular phylogeny. It is also the first review to discuss information on the metabolic processes occurring in members of the genus *Thiothrix* in wastewater treatment plants in the context of its pangenomic data.

2. Types of Wastewater Treatment Plants Where *Thiothrix* Morphotype Can Occur

Microorganisms can be used in biological wastewater treatment in attached form or in suspension form [50]. In the first case, the microorganisms attach to a rigid surface on which they form a biofilm, such as in a rotating biological contactor (RBC); in the second case, the microorganisms are in suspension, as in the conventional activated sludge (CAS) process and in the membrane bioreactor (MBR) [51].

2.1. Activated Sludge

Activated sludge is a community of microorganisms contributing to the biological treatment of wastewater. Although filamentous bacteria are usually present in activated sludge in relatively low numbers, they can cause negative phenomena. Under certain conditions, they proliferate to such an extent that they significantly affect the performance of the wastewater treatment plant, causing sludge bulking and foaming [35]. Wastewater treatment plants in the food [34–36] and pulp and paper industries [28,29] face similar problems.

Most filamentous bacteria in activated sludge belong to the *Thiothrix* morphotype [52]. An analysis of the bacterial populations of 29 industrial wastewater treatment plants using 16S rRNA amplicon sequencing showed that activated sludge bulking was associated with the *Thiothrix* morphotype in 72% of cases [24].

Pure cultures belonging to the genus *Thiothrix* were repeatedly isolated from activated sludge samples [53,54]. Some isolated strains were subsequently described as representatives of new species and genera: *Thiothrix eikelboomii* (later reclassified as a member of the genus *Thiolinea* [10]), *Thiothrix unzii*, *Thiothrix fructosivorans* [8], *Thiothrix disciformis* [9] = *Thiolinea disciformis* [10] and *Thiothrix flexilis* [9] = *Thiofilum flexile* [10]. Based on the genome obtained from the metagenome (MAG) of activated sludge, a new species, *Thiothrix winogradskyi*, was described [55].

Besides pure cultures, new species of *Thiothrix* with *Candidatus* status were described in activated sludge. In this way, '*Candidatus* Thiothrix moskovensis' and '*Ca.* Thiothrix singaporensis' were described based on metagenome-assembled genomes (MAG) obtained from laboratory-scale bioreactors with the enhanced biological removal of phosphorus [56].

2.2. Aerobic Granular Sludge

As a promising alternative to the wastewater treatment process involving activated sludge, aerobic granular sludge (AGS) has received increasing attention in recent decades. In contrast to flocculated activated sludge, AGS is a granular microbial aggregate with a spatially layered structure (anaerobic/microaerobic/aerobic from inside to outside) that promotes the growth of microorganisms with diverse functionalities in one small unit [57].

In treating textile wastewater, AGS was shown to be more effective than conventional activated sludge with *Thiothrix* filaments disappearing from AGS after such treatment, resulting in improved sludge settleability [32]. However, it should be noted that AGS can potentially also face bulking problems due to the proliferation of bacteria similar in morphotype to *Thiothrix*. At the same time, it was shown that in the presence of bacteria of the *Thiothrix* morphotype ($51.4 \pm 8.3\%$ of the total microbial population), stable aerobic granular sludge can be formed, and the morphology of the granules is largely determined by the operation of the reactor and, to a much lesser extent, by the morphologies of individual cells [44]. Moreover, members of the *Thiothrix* morphotype can be among the dominant bacteria in filamentous granular sludge (FGS), which is capable of maintaining relatively stable carbon and phosphorus removal efficiency but is challenging in terms of nitrogen removal [43].

An important factor affecting the granulation and sedimentability of AGS is the food-to-mass ratio (F/M) [58]. As this ratio increases in the first stages of wastewater treatment, the number of bacteria of the *Thiothrix* morphotype increases significantly, which decreases the efficiency of AGS precipitation. However, the gel-like polysaccharides associated with these bacteria effectively help to preserve the granular biomass, which gradually leads to a decrease in the F/M and the restoration of AGS' sedimentability [58]. The C/N ratio also has an important effect on the characteristics of activated sludge and the development of *Thiothrix*. A decrease in this ratio led to increased competition between different microorganisms, including the dominant bacteria from the genus *Defluviicoccus* and the *Thiothrix* morphotype, which reduced the efficiency of the system [59]. Nutrient deficiency, especially nitrogen, also had a negative effect on AGS systems, causing sludge bulking due to the proliferation of the *Thiothrix* morphotype [60].

Incorrectly selected conditions often lead to a significant increase in the size of the granules due to an increase in the number of bacteria similar in morphotype to *Thiothrix* and the disruption of the system [61].

2.3. Membrane Bioreactors

Membrane bioreactors (MBRs) are modern, high-intensity biological treatment facilities. In membrane bioreactors, the separation of activated sludge flocks from treated wastewater is achieved by filtering the sludge mixture through an ultrafiltration or microfiltration membrane, after which microorganisms are removed from the membrane surface by an aeration system [62].

Challenges in operating membrane bioreactors are related to frequent membrane clogging, including the growth of filamentous bacteria [63]. *Thiothrix eikelboomii* (= *Thiolinea eikelboomii*) was reported to decrease the sludge sedimentability, increase the apparent viscosity in aerobic environments and cause dense biofilm formation on the membrane surface [12]. Membrane contamination due to an increase in the number of bacteria of the *Thiothrix* morphotype is more active at low temperatures [41].

Several methods have been proposed to offset the negative effects of bacteria of the *Thiothrix* morphotype in MBR operation. The overgrowth of *Thiothrix*-like organisms was suppressed by using micellial granules to granulate activated sludge [64], by the use of high sludge concentrations in the MBR [63] and by backwashing the membrane with NaClO [65].

2.4. Latest Inventions for Membrane Reactors

It should be especially noted that, for organic pollutant removal, which is difficult when using traditional biodegradation methods, chemical methods with the application of strong oxidizing agents, such as peroxymonosulfate, hydrogen peroxide, etc., have long been used as an alternative. To enhance the action of oxidizing agents, catalysts based on perovskite oxides are used. However, with the growth of industry in recent years, the insufficiency of this variant of the method has been noted, which has stimulated a search for new solutions to improve the efficiency of wastewater treatment.

For this purpose, the latest layered perovskite oxides of the Ruddlesden-Popper structure have recently been synthesized and shown to serve as an effective next-generation catalyst for the activation of peroxymonosulfate, an oxidant used in the oxidative removal of organic pollutants in modern wastewater treatment plants. Unlike conventional perovskites, in Ruddlesden-Popper perovskites, the catalytic role in the decomposition of the pollutant chemical compound is played not by free radical forms of oxygen but by non-free singlet oxygen radicals. For this reason, the newest Ruddlesden-Popper perovskite oxides have demonstrated exceptional efficiency in phenol removal (complete removal from wastewater in 10 min) [66].

The development of new Ruddlesden-Popper perovskite oxides in 2023 showed that they can serve as an effective catalytic component in next-generation composite membranes, increasing the speed and safety of the organic wastewater treatment process. In contrast to conventional perovskite oxides with the formula ABO_3, the Ruddlesden-Popper perovskite oxides of the A_2BO_4 type have improved catalytic properties.

In search of materials with optimal properties, Yang et al. [67] synthesized a La_2CoO_4-δ catalyst with a Ruddlesden-Popper structure, in one of the best variants of which the A-site was 50% replaced by strontium to obtain the $LaSrCoO_4$-δ structure. The use of such a catalyst allowed the complete oxidative decomposition of tetracycline hypochloride in 14 min. The obtained compound embedded in the matrix of a polyacrylonitrile ultrafiltration membrane gave a composite material of extraordinary efficiency in wastewater treatment, the use of which resulted in the 94% degradation of tetracycline hypochloride in only 3 min and a significant reduction in the wastewater's toxicity.

2.5. Biofilms

A biofilm is a set of microorganisms located on the surface of wastewater or parts of treatment facilities. One of the most common types of bioreactors that contain microorganisms in the form of a biofilm is the rotating biological contactor (RBC).

In an RBC, a chain of round plastic discs rotates on a horizontal rod using a motor. This rotation causes the mixing of wastewater volumes, the diffusion of compounds into the biofilm, convection through the pores/medium of the biofilm and the subsequent exchange of products with the reactor and the environment. Oxygen diffusion is obligatory for the oxidation of organic substances. The rotation of the discs creates shear forces that inhibit excess biofilm growth. Excess sludge is transferred to a settling tank [68].

Filamentous organisms can proliferate at the surface–biofilm interface in RBCs. The presence of filaments of the *Thiothrix* morphotype in RBC biofilms has been shown many times, although they are much less common in biofilms than in suspension form. They were detected as part of an RBC biofilm in the simultaneous removal of nitrogen and phosphorus [21]. It was shown that the *Thiothrix* morphotype can be used as an indicator of nitrous and nitric nitrogen's presence in RBCs [69]. In addition, bacteria similar in morphotype to *Thiothrix* were found in the biofilm of a non-aerated reactor for the treatment of sulfide-containing wastewater [70].

3. Methods for Identification of *Thiothrix* in Wastewater Treatment Plants

The significant contribution of the *Thiothrix* morphotype bacteria to the operation of wastewater treatment plants has long been known, leading to the development of methods for their identification using light microscopy (Figure 2). The application of these methods, along with detailed morphological characterization, led to the identification of filamentous bacteria in wastewater treatment plants that were assigned to the genus *Thiothrix* and described as Type 021N, *Thiothrix* I and *Thiothrix* II [6,7]. Therefore, these names were retained in many subsequent articles written by non-taxonomists dealing with microorganisms in wastewater treatment plants [13–18]. The original classification of these bacteria outside of the genus *Thiothrix* was further confirmed by Boden and Scott [10], who proposed placing *Thiothrix flexilis* in the genus *Thiofilum* and *Thiothrix disciformis* and *Thiothrix eikelboomii* in the genus *Thiolinea*.

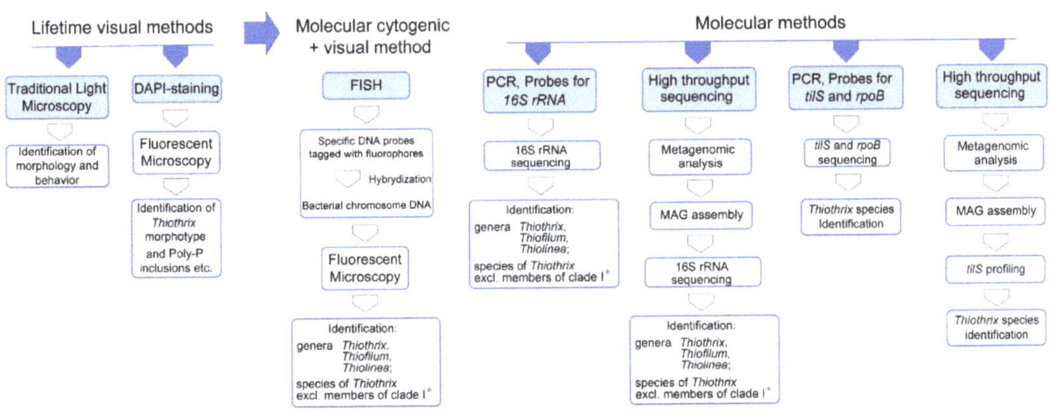

Figure 2. Progress of methods for identification of representatives of *Thiothrix* morphotype from time perspective: light microscopy morphotype identification [6,7]; DAPI staining to confirm bacterial morphotype [71] and identify Poly-P inclusion [47]; FISH for identification of *Thiothrix, Thiofilum* and *Thiolinea* genera and some *Thiothrix* species [31,72,73]; PCR using *tilS* and *rpoB* probes [19] and high-throughput sequencing with *tilS* profiling [19] for *Thiothrix* species identification. The capabilities of the FISH method and the PCR method when using probes for 16SRNA in both methods are current at the time of writing the review and are indicated for the taxonomic groups of *Thiothrix*, *Thiofilum* and *Thiolinea* revised in 2018 [10], with no consideration of the taxonomic status of these groups before 2018. * Not applicable for species identification of *Thiothrix* Clade I, the composition of which is listed in Figure 1, at the date of the review's preparation.

To monitor the population dynamics of *Thiothrix* in wastewater treatment plants, species-specific fluorescence in situ hybridization (FISH) probes are often used, in addition to light microscopy [13,18,26,36,74–77]. Current phylogenetic data suggest that some of the probes used could also be suitable for representatives of the genera *Thiolinea* and *Thiofilum* [18,75,77].

Since the phylogenetic heterogeneity of *Thiothrix* was long known, sets of different probes were developed to identify all known species of the genus known at that time. Their use showed that mixed populations of the *Thiothrix* morphotype are often found in wastewater treatment plants, and bacteria now assigned to the genus *Thiolinea* can also be found in these mixtures [31].

Quantitative PCR has been used less frequently to study *Thiothrix* populations. For Type 021N, primers were developed and used in combination with Sybr Green to quantify these bacteria in laboratory tests [78]. Genus-specific primers were developed for *Thiothrix* 16S rRNA, which showed that, in the wastewater studied, the largest number of bacteria similar in morphotype to *Thiothrix* belonged to the species *Thiothrix eikelboomii* (= *Thiolinea eikelboomii*) [14]. Asvapathanagul et al. also showed that the ratio of *Thiolinea eikelboomii* to the total number of bacteria (TH/TB) (%) was a better predictor of the secondary effluent volume and quality than the *Thiolinea eikelboomii* abundance alone [14].

However, the rapid development of high-throughput sequencing technologies in recent years has led to the increasingly frequent detection of *Thiothrix* by community profiling against 16S rRNA genes [24,52,79,80]. Moreover, representatives of the genera *Thiothrix* [56,81] and *Thiolinea* [11] were identified in activated sludge by metagenomic analysis.

Although it is known that a mixture of several *Thiothrix* species can occur simultaneously in wastewater treatment plants, the identification of these species has so far been a challenge. The reason for this is that the 16S rRNA phylogenetic marker most commonly used in high-throughput sequencing offers an insufficient resolution to unambiguously assign *Thiothrix* to a particular species. In this regard, to identify *Thiothrix* to the species level, primers for other phylogenetic markers—*tilS* and *rpoB*—were constructed [19]. However, they have not yet been used in the analysis of the microbial communities of wastewater treatment plants, but they have shown good results in the analysis of microbial communities from natural biotopes.

4. Metabolic Processes in Wastewater Treatment Involving *Thiothrix* and Prospects for *Thiothrix*'s Further Application

4.1. Sulfide Removal

As prominent representatives of the group of sulfur-oxidizing bacteria, members of the *Thiothrix* morphotype have the metabolic potential to remove toxic sulfide from wastewater. The incoming raw sewage, which contains sulfide resulting from the degradation of protein substrates, is the most probable source of *Thiothrix* found in activated sludge from wastewater treatment plants. For example, the same *Thiothrix* phylotype was detected both in influent raw sewage and in activated sludge from Moscow wastewater treatment plants, and its percentage in the influent microbial community was much higher than in activated sludge and treated water (0.79% vs. 0.05% vs. 0.17%); for the list of 16S rRNA gene reads, see Table S1, SM in Ref. [82].

A well-developed system for sulfur metabolism underlies the broad capabilities of representatives of the genus *Thiothrix* to participate in wastewater treatment from sulfur compounds. Genome studies of all known *Thiothrix* species indicate that all of them contain genes for dissimilatory sulfur metabolism (Figure 3a).

The pangenome of the genus *Thiothrix* comprises *sqr* (*sqrF*, *sqrA*) genes encoding sulfide: quinone oxidoreductase, which is involved in the oxidation of hydrogen sulfide to elemental sulfur; *fccAB* genes encoding sulfide-flavocytochrome-*c* dehydrogenase (FCSD); *soxAXBYZ* genes encoding a branched SOX system for the oxidation of thiosulfate to sulfur and sulfate; genes *dsrABEFHNEMKLJONR* encoding the rDSR complex that carries out the oxidation of sulfur to sulfite; *soeABC* genes encoding quinone-dependent sulfite dehydrogenase catalyzing the direct oxidation of sulfite to sulfate; genes *sat* encoding ATP-sulfurylase of the dissimilatory type; and *aprAB* genes encoding APS reductase for indirect sulfite oxidation [5]. A genome analysis of representatives of other *Thiotrichaceae* genera, namely *Thiolinea disciformis*, *Thiolinea eikelboomii* and *Thiofilum flexile*, showed the presence of all genes of dissimilatory sulfur metabolism with the exception of ATP-sulfurylase of the dissimilatory type (*sat*) and APS-reductase (*aprAB*) for indirect sulfite oxidation, as well as the lack of *fccB* genes of the FCSD complex in *Thiofilum flexile* [5].

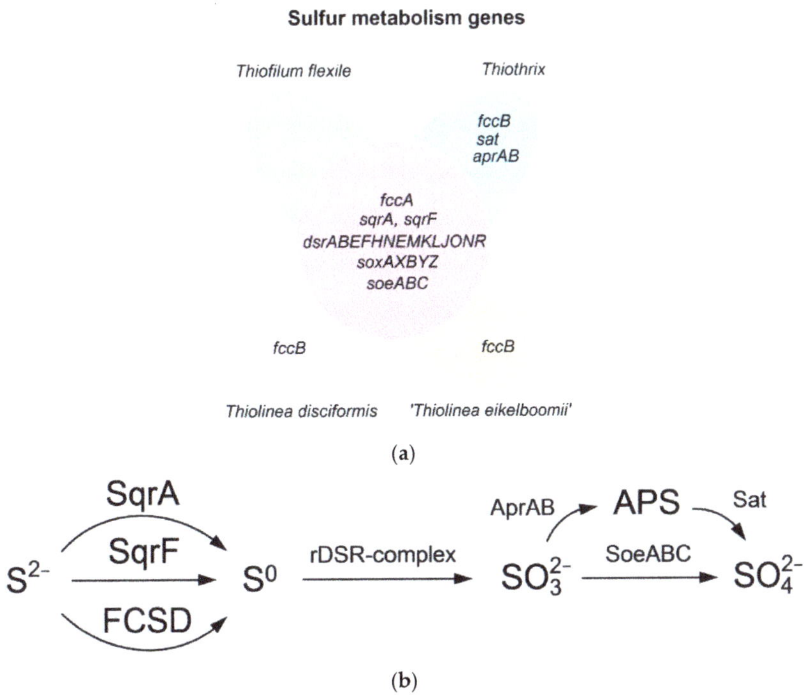

Figure 3. Sulfur metabolism in representatives of *Thiothrix* morphotype. (**a**) EVenn diagram presenting sulfur metabolism genes found in the genomes of the genera *Thiothrix, Thiophilum* and *Thiolinea.* Panel with *Thiothrix* genes is based on pangenome analysis involving 15 genomes of *Thiothrix litoralis* AS^T, *Thiothrix lacustris* BL^T, *Thiothrix fructosivorans* Q^T, '*Ca.* Thiothrix putei', *Thiothrix caldifontis* G1^T, *Thiothrix winogradskyi* CT3^T, '*Ca.* Thiocaldithrix sulfatifontis', *Thiothrix subterranea* Ku-5^T, *Thiothrix* sp. STA_22, *Thiothrix unzii* A1^T, '*Ca.* Thiothrix anitrata' A52, '*Ca.* Thiothrix moscovensis' RT, *Thiothrix* sp. 207, *Thiothrix nivea* JP2^T and '*Ca.* Thiothrix singaporensis' SSD2 and on new 16th genome of '*Ca.* Thiothrix namsaraevi'. The genes in the other three panels are shown based on representative species: *Thiolinea disciformis* DSM 14473^T and '*Thiolinea eikelboomii*' in the case of the genus *Thiolinea*; *Thiofilum flexile* in the case of the genus *Thiofilum.* Figure was prepared using EVenn diagram software (http://www.ehbio.com/test/venn/, accessed on 11 July 2024) [83] and considers data in Refs. [4,5,84]. (**b**) Reactions of sulfur dissimilatory metabolism.

Thus, among the representatives of the *Thiothrix* morphotype, only representatives of the genus *Thiothrix* have the most developed system of oxidative dissimilatory sulfur metabolism (Figure 3b). This suggests that they are the most universal candidates for systems of water treatment from sulfur-containing compounds.

Based on their morphology, microorganisms similar to *Thiothrix* sp. were identified in the sludge beds of upflow anaerobic sludge blanket (UASB) reactors, where the ability of these organisms to oxidize sulfide was demonstrated [45].

When the sulfide concentrations temporarily increased in a fluidized bed reactor used to remove nitrates from a recirculating marine fish culture system, whitish tufts identified as the *Thiothrix* morphotype rapidly developed [85]. In these systems, representatives of the *Thiothrix* morphotype participated in sulfide oxidation using nitrate or oxygen as an electron acceptor [46]. Bacteria similar in morphotype to *Thiothrix* were also found as part of a composite microbial biofilm for the treatment of sulfide-containing wastewater [70] and in a laboratory biodrainage filter treating high concentrations of hydrogen sulfide [74,86].

Moreover, in the latter case, among the nearly full-length 16S rRNA gene sequences of the entire bacterial community, sequences related to *Thiothrix lacustris* accounted for 38% [86].

The ability of *Thiothrix* to oxidize hydrogen sulfide could explain that enrichment by *Thiothrix* improved the removal of sulfamethoxazole (SMX) (sulfur-containing organic matter) and mitigated SMX-induced stress to other microorganisms [40].

In addition to the biological removal of sulfide, the *Thiothrix* morphotype can be involved in other metabolic processes that proceed during the biological treatment of wastewater. *Thiothrix* was shown to be a significant part of the microbial community involved in the processes of nitrogen and phosphorus removal (1.51%), nitrification–denitrification (9.41%) and the aerobic removal of carbon (4.29%) [87].

4.2. Enhanced Biological Removal of Phosphorus

Phosphorus is a key element responsible for the eutrophication of the aquatic environment, which leads to a shift in the equilibrium in the hydroecosystem, subsequently resulting in the development of cyanobacteria, a sharp decrease in the concentration of dissolved oxygen in water, the accumulation of biogenic toxins, the deterioration of water quality, the death of fish and other aquatic organisms and the waterlogging of the water body. In this regard, the removal of phosphorus from wastewater is an important task in preventing the harmful effects of eutrophication in the ecological, economic and social spheres. A biological approach could be among the methods to address this problem. One of the most successful directions of biotechnology is the use of colorless sulfur bacteria of the genus *Thiothrix* in "enhanced biological phosphorus removal" (EBPR) systems. Based on the data of a pangenomic analysis, it can be assumed that the members of the genus *Thiothrix* can behave as phosphorus-accumulating organisms with mixotrophic metabolism during phosphorus removal and, at the same time, use the oxidation of the intracellular sulfur pool as an additional energy source along with poly-β-hydroxy-alkanoates (PHA). A large number of such systems are currently operating successfully around the world. The idea of using biologically safe methods for phosphorus removal based on modified activated sludge systems was implemented in the late 1950s when the "enhanced biological phosphorus removal" (EBPR) system was developed [88]. In an EBPR treatment system, phosphate-accumulating organisms (PAOs) accumulate increased amounts of polyphosphates in their cells and thus biologically remove phosphorus from wastewater [89].

A large number of enzyme proteins are known to be involved in the biological cycle of phosphorus (Figure 4). In bacteria, the fully characterized enzymes that enable the reversible accumulation of phosphorus in the form of polyphosphate (poly-P) are polyphosphate kinases Ppk1 (coupled to ATP dephosphorylation) [90] and Ppk2 (coupled to ATP/GTP dephosphorylation) [91], encoded by the *ppk1* and *ppk2* genes, respectively. It is quite likely that, in addition to this enzyme, there are still unknown mechanisms for poly-P synthesis, such as the Arp complex (*arpABCDEFGH*), which catalyzes poly-P synthesis in archaea and was also shown to have distant homologs in bacteria. Bacterial enzymes that regulate poly-P degradation include exopolyphosphatases Ppx, Epp, GppA (*ppx, epp* and *gppA*); 50/30-nucleotidase SurE (*surE*); poly-AMP phosphotransferase PAP (*pap*); NAD kinase NadK (*nadK*); alkaline phosphatases APases (e.g., *phoX*, *phoD*, *phoA*); and polyphosphate-dependent glucokinase PpgK (*ppgK*). In addition to these enzymes, P_i-transporter Pit, encoded by the *pit* gene (having low substrate affinity), and the ABC-type P_i-transporter Pst, formed by four proteins encoded by the *pstSCAB* genes (having high substrate affinity), are used to transport phosphate molecules into the cell [92]. Closely related to the Pst complex are the two-component signaling proteins PhoR and PhoB (*phoR* and *phoB*) and a cytoplasmic protein PhoU (*phoU*), which regulates Pi entry into the cell under conditions of phosphate deficiency. According to the hypothetical scheme proposed by Gardner et al. [93], the metal-binding protein PhoU performs a negative signaling function, providing the bacterial cell with a mechanism to sense the level of P_i in the environment.

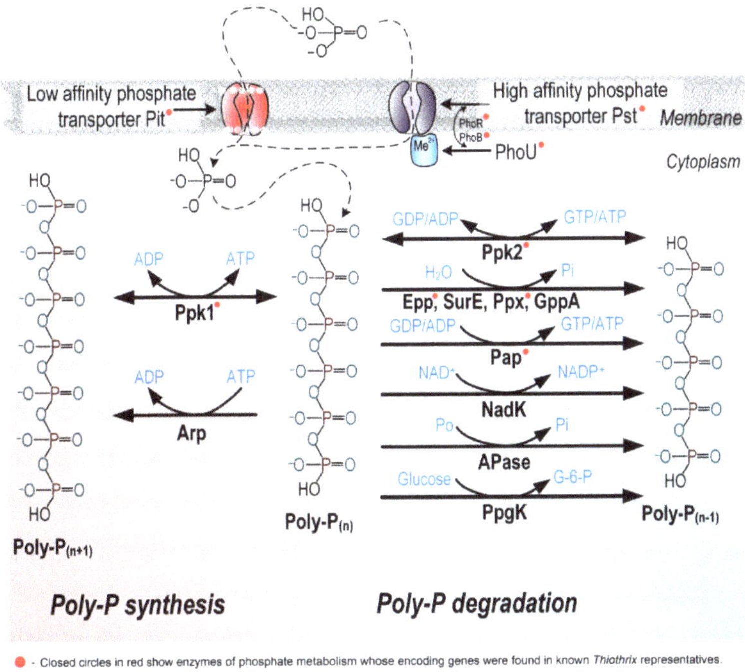

Figure 4. Possible pathways of phosphorus metabolism in bacterial cells. For the depicted names of the enzyme proteins, see the text. The figure was adapted from Roy et al. [92] and Gardner et al. [93] under the terms of the Creative Commons Attribution 4.0 International License. *Thiothrix* enzymes are shown based on the published data [4,11,47,49,56,84,94].

A successfully operating EBPR system requires that a mixed microbial consortium (MMC) be exposed to cyclic anaerobic–aerobic conditions while simultaneously receiving sufficient volatile fatty acids (VFAs) so as to stimulate the metabolism necessary for EBPR activation. Due to this mode of operation, MMC is enriched in phosphorus-accumulating organisms (PAOs).

The physiological group of PAOs comprises representatives of various lines, including '*Ca.* Accumulibacter', *Tetrasphaera* (later '*Ca.* Phosphoribacter' [95]), *Microlunatus phovovorus*, *Dechromonas*, *Thiothrix* and others [96]. '*Ca.* Accumulibacter' and '*Ca.* Phosphoribacter' are considered the main ones [97].

PAO metabolism by '*Ca.* Accumulibacter' is considered classical. These bacteria, under anaerobic conditions, (i) take up VFAs and store them in the form of the intracellular polymer PHA, (ii) use glycogen as a source of reducing equivalents (carbon is also included in PHA) and (iii) hydrolyze internal polyphosphate (poly-P) to form the essential ATP. Under aerobic conditions, PAOs (i) utilize the PHA reserves to grow and replenish the glycogen reserves and (ii) use the ATP generated by oxidative phosphorylation to take up and convert phosphate to poly-P [98] (Figure 5a).

Unlike '*Ca.* Accumulibacter', representatives of *Tetrasphaera* spp. do not possess the ability to store PHA. However, like representatives of '*Ca.* Accumulibacter', they synthesize poly-P inside cells aerobically and use it anaerobically. It is assumed that amino acids and glucose are substrates in the cycle of poly-P transformation in anaerobiosis. During the reaction, these substrates are presumably converted into free amino acids and glycogen, although the role of glycogen is unclear, since it is not found in the activated sludge of *Tetrasphaera* spp. representatives [95] (Figure 5b).

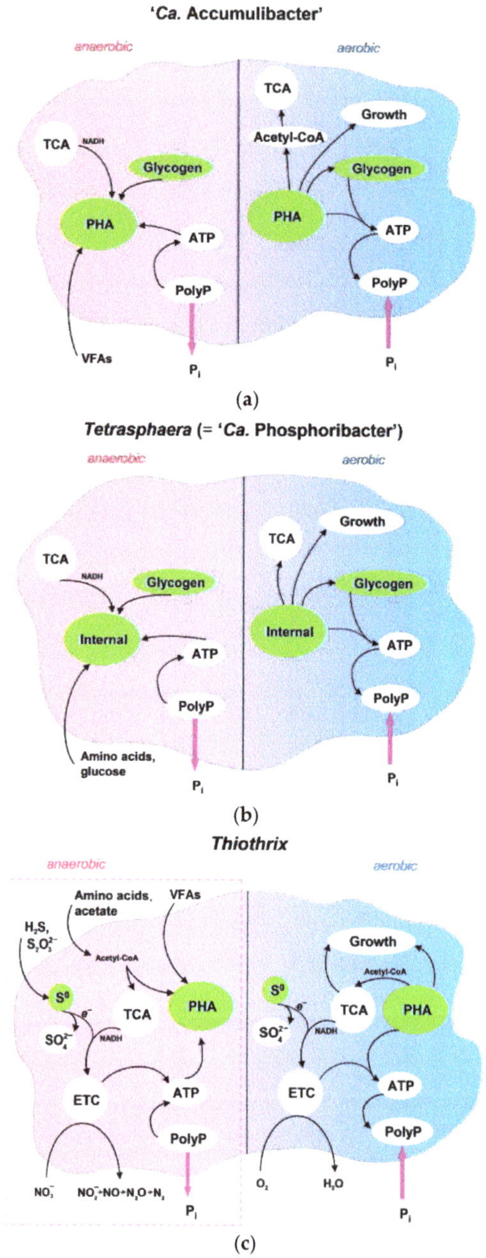

Figure 5. PAO metabolism in (**a**) *'Ca.* Accumulibacter' and (**b**) *Tetrasphaera* (= *'Ca.* Phosphoribacter' [96]) and (**c**) predicted functional potential members of the genus *Thiothrix*. The schemes in panels (**a**,**b**) are based on Nielsen et al. [97] under the terms of the Creative Commons Attribution 4.0 International License; panel (**c**) considers data from published articles [4,5,47,56,94]. In panel (**c**), the metabolic pathways outlined by the dashed line are putative. Internal, internal reserve substances in the cell; PHA, poly-β-hydroxy-alkanoates; S^0, elemental sulfur; VFAs, volatile fatty acids; PolyP, polyphosphate; TCA, tricarboxylic acid cycle; ETC, electron transport chain.

It was shown that members of the genera *Thiothrix* [47] and *Thiolinea* [11] can play a key role in biological phosphorus removal. As shown under aerobic conditions in EBPR systems, in an environment with low organic content but enriched in reduced sulfur compounds, *T. caldifontis* could behave as an efficient phosphate-removing PAO with mixotrophic metabolism. Phosphorus removal from the environment in the aerobic phase proceeds with the oxidation of the intracellular sulfur pool to provide PAO with an additional source of energy [47]. In the same EBPR system, under anaerobic conditions, this species can store carbon in the form of PHA and generate the required energy via the hydrolysis of poly-P stored in the aerobic phase. In the aerobic period, PHA was used as the source of carbon and energy for the growth and synthesis of poly-P (Figure 5c). On the other hand, *Thiothrix* sp. that received glutamate as the only source of nitrogen and carbon was able to store poly-P, probably avoiding PHA synthesis [99].

It should be noted that the growth conditions in EBPR systems differ markedly from those of batch cultivation and often provide preferences for new and rare species. Several representatives of *Thiothrix* were described resulting from the assembly of MAGs obtained from the activated sludge in EBPR systems [11,56,81]. One of the resulting MAGs was closely related to the species *Thiothrix disciformis* (= *Thiolinea disciformis*) [10,11] and two MAGs were described as new species with *Candidatus* status, namely '*Ca.* Thiothrix moscovensis' and '*Ca.* Thiothrix singaporensis' [56].

An important factor determining the function of the representatives of the genus *Thiothrix* as a PAO is the carbon source and the capacity for anaerobic respiration. Due to a rare metabolic feature of '*Ca.* Thiothrix moscovensis' and '*Ca.* Thiothrix singaporensis', namely the presence of FAD-dependent malate: quinone oxidoreductase (MQO) instead of NAD-dependent malate dehydrogenase (MDH), these two species can effectively grow and function as PAOs in EBPR under organotrophic conditions, using acetate as a substrate (Figure 6) [56]. The same metabolic potential is inherent in all representatives of the genera *Thiothrix*, *Thiolinea* and '*Ca.* Thiocaldithrix' since their genomes contain the *mqo* gene instead of the *mdh* gene [4,5]. Presumably, these members of *Thiothrichaceae*, except for species lacking nitrate reductase genes, are better able to overcome the anaerobic phase in the EBPR cycle in the presence of nitrate, allowing them to activate nitrate-mediated respiration (Figure 6). This may be facilitated by the following factors, which, under certain unfavorable conditions, can presumably create advantages for species with MQO instead of MDH or (MDH+MQO), despite the fact that, formally, these two enzymes catalyze the same reaction.

1. MQO is selectively activated 3.4–4-fold during microbial growth on acetate compared to growth on glucose, whereas the corresponding figure for MDH is 1.8 [100,101].

2. In the case of MQO catalysis, but not MDH catalysis, there is no dependence of the reaction progression malate $\rightarrow$ oxaloacetate on the NAD/NADH balance, because the cofactor of MQO is the FAD/FADH pair, while NAD/NADH is involved in the MDH-catalyzed reaction.

3. Moreover, in the case of MQO catalysis, there is no dependence of the reaction on the malate/oxaloacetate balance because the reaction is exergonic ($\Delta G^{\circ\prime} = -18.9$ and -55.0 kJ/mol when the electron acceptor is MQ or UQ, respectively) and proceeds spontaneously. In contrast, the reaction with MDH is endergonic ($\Delta G^{\circ\prime} = +28.6$ kJ/mol) and does not proceed in the forward direction under normal conditions according to the laws of chemistry.

4. It follows from (3) that the simultaneously use of MQO and MDH is disadvantageous under conditions of an energy deficit and a lack of oxygen, because they lead to a futile cycle, where oxaloacetate, reduced from malate by MQO, is converted to malate again due to the reverse reaction catalyzed by MDH. The end result of such a reaction is equivalent to the reaction of noncoupled NADH: quinone oxidoreductase type II (common in *Bacillus* species and many other bacteria), i.e., the production of reduced quinone, but at a higher cost than in the MQO-catalyzed reaction.

5. One possible way to avoid this wastefulness and not waste an "expensive" reducing agent on an unprofitable reaction (see (4)) might be to either block the available MDH or replace MDH with MQO. Switching to MQO catalysis preserves oxaloacetate as an essential substrate for citrate synthase in TCA, as well as for gluconeogenesis and the synthesis of aspartate branch amino acids.

6. The outcome of switching individual microorganisms to MQO maintenance may be the more reliable functioning of TCA under some unfavorable conditions, namely anaerobiosis (respiration with less favorable oxidants compared to oxygen, e.g., nitrate), under carbon-containing substrate limitations, and others.

The electron acceptor from MQO can be menaquinone, which has a more negative redox potential ($E'_0 = -0.075$ V) compared to ubiquinone ($E'_0 = +0.100$ V) and is almost ideally matched to the FAD/FADH pair ($E'_0 = -0.060$ V) (cofactor in MQO) to form a unit in the electron transport chain. It is worth noting that, in contrast to strictly aerobic and strictly anaerobic bacteria, both ubiquinones and menaquinones are synthesized in membranes in facultatively anaerobic bacteria, which include representatives of *Thiothrix*. As shown in *E. coli*, the synthesis of all types of menaquinones increases under anaerobic conditions and exceeds the total amount of ubiquinone three-fold [102,103], which makes menaquinone a more likely electron acceptor from MQO under anaerobic conditions.

Another notable change in the profile of biosynthesis products induced by anaerobiosis in the six *Thiothrix* species studied is a 2.4–6.5-fold increase in the expression of the *narG* gene [104,105] encoding the catalytic subunit of the dissimilatory nitrate reductase NarGHI anchored in the cytoplasmic membrane. It has been shown that the induction of expression by anaerobiosis is accompanied by an increase in the total activity of nitrate reductase in the cells.

Experimental observations on different bacterial species possessing MQO demonstrate good growth rates on acetate and organics at high MQO activity and a significant decrease in growth in MQO deletion mutants in *Corynebacterium glutamicum* [106] and *Pseudomonas aeruginosa* [107], slow growth on fermentable carbon sources and no growth on non-fermentable carbon sources in *Micobacterium stegmatis* [108]. The deletion of *mqo* in *Mycobacterium tuberculosis* leads to stress and impairs the survival of the bacterium [109]; in *Pseudomonas fluorescens*, disruptions in the *mqo* gene led to the loss of the ability to colonize on the root system of tomato [110].

All of the above facts point to an important energetic role of MQO in coupling with nitrate respiration. Although the respiratory chain with Nar enzyme complex as a terminal oxidase and nitrate as an electron acceptor are characterized by lower energetic efficiency as compared to oxygen respiration, the energetic capacity of respiration involving nitrate is comparable to that of the aerobic respiration process (nitrate respiration allows bacteria to transform the energy of substrates into universal energy form with electric potential on the membrane, comparable in magnitude to oxygen respiration). The change in free energy during the oxidation of one molecule of glucose by molecular oxygen ($\Delta G^{\circ\prime} = -2870$ kJ/mol) is of the same order as the oxidation of the same substrate under anaerobic conditions by nitrate reduced to nitrite ($\Delta G^{\circ\prime} = -1770$ kJ/mol). As shown in a pangenomic study, almost all members of the genus *Thiothrix*, with the exception of '*Ca*. Thiothrix anitrata' A52 and '*Ca*. Thiothrix sulfatifontis' KT, possess the capacity for anaerobic respiration using nitrogen-containing compounds [94], which indicates the high potential of *Thiothrix* representatives to persist in bioreactors in the anaerobic phase.

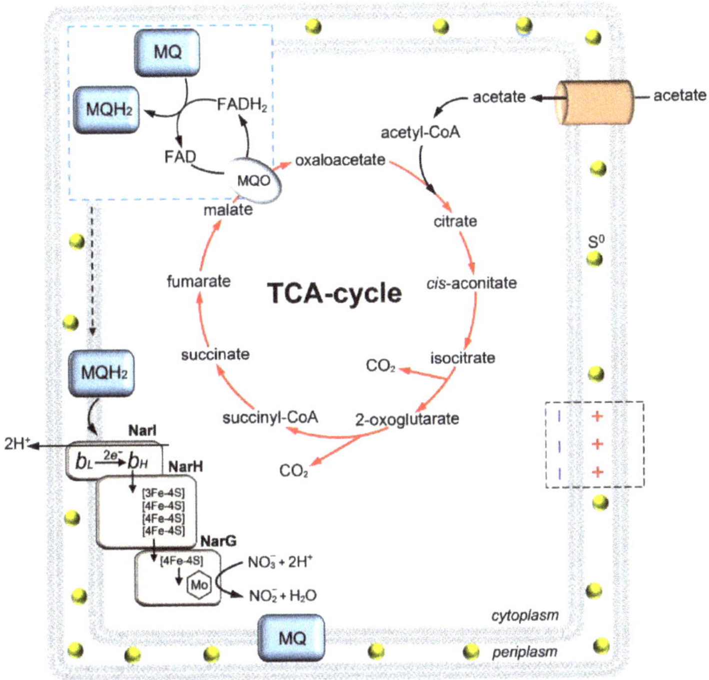

Figure 6. The role of malate: quinone oxidoreductase (MQO) in *Thiothrix* metabolism as determined from experimental and genomic data. MQ and MQH_2—oxidized and reduced forms of menaquinone, respectively; NarI, NarH and NarG—dissimilatory nitrate reductase (Nar) subunits with a cytoplasmically oriented active center; b_L and b_H—low-potential and high-potential cytochrome *b* in the membrane-spanning NarI subunit; [3Fe-4S] and [4Fe-4S]—redox-active iron-sulfur clusters; arrows within Nar subunits indicate the electron pathway between the redox groups. Insets: in the black dashed line on the right side of the figure—the direction of electrical potential at the cytoplasmic membrane, which is charged negatively at the cytoplasmic side and positively at the periplasmic side during the process of nitrogen respiration; in the blue dashed line in the upper left part of the figure—reactions occurring inside the cytoplasmic membrane. Figure is based on Mardanov et al. [56]; Trubitsyn et al. [105]; Guigliarelli et al. [111]; and Martinez-Espinosa et al. [106] under the terms of the Creative Commons Attribution 4.0 International License.

According to Rey-Martínez et al., during the long-term operation of the EBPR system with glutamate as the sole source of carbon and nitrogen, the content of members of the genus *Thiothrix* could be 37%, with *Thiothrix* members able to participate in phosphorus removal by accumulating poly-P [99]. However, another study showed that although *Thiothrix* was one of the dominant PAOs at different stages of EBPR bioreactor operation, it was more abundant with propionic acid as the carbon source than when glutamate was used [112]. The use of glycine as a carbon source had an adverse effect on *Thiothrix* and other PAOs in the EBPR system [96].

It should be noted that the explosive growth of members of the genus *Thiothrix* leading to unfavorable sludge bulking can be observed in EBPR systems under changes in cultivation conditions, such as stopping the use of the nitrification inhibitor allylthiourea, which is used to stimulate the growth of *Tetrasphaera*-related PAOs [113].

The intensive study of *Thiothrix* metagenomes in the last decade has revealed the significant potential of the species of this genus to address the challenge of phosphate removal from the environment. In the two genomes of '*Ca.* Thiothrix moscovensis' and

'*Ca.* Thiothrix singaporensis' derived from laboratory EBPR, as well as in the genomes of *T. lacustris* BL^T, *T. caldifontis* G1^T and *T. nivea* JP2^T, the genes *phoURB* and *pstSACB*, encoding the systems of signaling regulation and the transport of inorganic phosphate into the cell, and the genes *ppk1* and *epp*, whose products catalyze poly-P synthesis and degradation, were identified [56]. Rubio-Rincón et al. documented the visual mass accumulation of poly-P inclusion granules in the cells of *T. caldifontis*, corroborated by data on the presence of gene *ppk2* in its genome [47]. To date, we have detected the presence of multiple genes *ppk2* in the available genomes of the most identified species of the genus *Thiothrix* (in the following, access to only one of several genes in each species is shown), namely *T. unzii* A1^T [QTR55213.1] [94], *T. fructosivorans* Q^T [QTX11709.1] [94], *Thiothrix subterranea* Ku-5^T [WML89006.1] [19], '*Ca.* Thiothrix putei' [WGZ96527.1] [5], '*Ca.* Thiothrix singaporensis' [QLQ34282.1] [114], *T. lacustris* BL^T [WML90248.1] [19], *T. nivea* JP2^T [EIJ34895.1], *T. litoralis* AS^T [QTR45363.1] [94], '*Ca.* Thiothrix anitrata' A52 [QTR51311.1] [94], '*Ca.* Thiocaldithrix sulfatifontis' [UOG93965.1] [55], *T. winogradskyi* CT3^T [UJS26581.1] [55], '*Ca.* Thiothrix sp. Deng01' [WP_324693012.1] [115] and a number of unidentified *Thiothrix* species, as well as the members of the other *Thiotrichaceae* genera, '*Ca.* Thiocaldithrix dubininis' [WGZ92350.1] [5], *Thiofilum flexile* DSM 14,609 (formerly *Thiothrix flexilis*) [ARCL01000001.1] and *Thiolinea eikelboomii* (formerly *Thiothrix eikelboomii*) [SKA88474.1]. According to Matsuura et al., *Thiothrix disciformis* (= *Thiolinea disciformis*) possesses genes for the major systems of phosphate metabolism (*ppk*, *ppx*, *pst* and *pit*), which include two different phosphate transport systems, Pst and Pit, with high and low affinity for phosphate [11]. A metagenomic analysis performed by the research group of Dong et al. also showed that the identified representatives of the genus *Thiothrix* have the genes of phosphate transport, *pit* and *phnCDE* and other genes of phosphate metabolism: *phoURB*, *pst*, *ppk* [49]. Most recently, Chernitsyna et al. reported that the genes encoding the systems of Pi transport into the cell (*pstDCABS*) and the signaling regulation of phosphate transport (*phoURB*), as well as the systems of poly-P synthesis and degradation, namely polyphosphate kinase (*ppk1*) and exopolyphosphatase (*ppx*), were found in the MAG-assembled genome of '*Ca.* Thiothrix namsaraevi', a new member of the genus *Thiothrix* from Lake Baikal [84]. Ravin et al. identified genes pstDCABS, phoURB and *epp*, as well as genes *ppk1* and *pap*, for poly-P synthesis and hydrolysis in three species of *Thiothrix*, *T. subterranea* sp. nov. Ku-5, *T. litoralis* sp. nov. AS and '*Ca.* Thiothrix anitrata' sp. nov. A52, obtained from other geographical locations in Russia—a coal mine, the White Sea littoral and a sulfur spring in the Volgograd region [94]. According to a summarized pangenomic analysis, the core genome of 12 *Thiothrix* species (*T. winogradskyi* CT3^T, *T. lacustris* BL^T, *T. litoralis* AS^T, *T. subterranea* Ku-5^T, *T. caldifontis* G1^T, *T. unzii* A1^T, *T. nivea* JP2^T, *T. fructosivorans* Q^T, '*Ca.* Thiothrix moscovensis' RT, '*Ca.* Thiothrix anitrata', '*Ca.* Thiothrix sulfatifontis' KT, MAG of *Thiothrix* sp. 207) contains genes for the main systems of phosphate metabolism, with the exception of only one of the three genes of the cellular phosphate transport system, *pstB*, which may be absent in some species [4]. As can be seen from the recent publications, almost all representatives of the genus *Thiothrix*, with a few exceptions, have genes for the signal regulation of Pi import into the cell, namely *phoUR* or *phoURB*, encoding a system of two or three corresponding proteins PhoU, PhoR and PhoB, as well as genes of Pi transporters, *pstDCABS*, *pit* and *phnCDE*; polyphosphate degradation and synthesis, *ppk1* and *ppk2*; exopolyphosphatases, *epp* and *ppx*; and poly-AMP phosphotransferase, *pap* (Figure 4).

4.3. Nitrogen Removal

In addition to phosphorus, nitrogen is one of the two major biogenic elements that threaten to upset the equilibrium in water bodies under anthropogenic impacts. While phosphorus and especially phosphates, which pose the greatest problem for water treatment, are mainly responsible for ecological damage to natural inland water reservoirs, nitrogenous compounds are more problematic for marine water reservoirs, although both types of compounds are dangerous for any water body under uncontrolled inputs. Members of the *Thiothrix* morphotype are prominent representatives of sulfur-oxidizing bacteria with metabolic potential to remove toxic nitrates from wastewater. For many species of the genus *Thiothrix*, the ability to denitrify during mixotrophic growth in the presence of reduced sulfur and organic compounds (lactate and acetate) has been experimentally demonstrated [105]. No data are available on the ability of *Thiothrix* pure cultures to denitrify in the absence of reduced sulfur compounds. An analysis of the pangenome of the genus *Thiothrix* showed the presence of denitrification genes in the variable part of the pangenome. All members of the genus, with the exception of '*Ca.* Thiothrix anitrata' and '*Ca.* Thiothrix sulfatifontis' KT, have genes encoding at least one form of nitrate reductase, which reduces nitrate to nitrite (Table 1). Most strains contain genes for nitrite reductase (*nirS*), which reduces nitrite to nitric oxide, and nitric oxide reductase NorBC, which reduces nitric oxide to nitrous oxide. The nitrous oxide reductase NosZ genes were found in only one of the strains involved in the pangenomic analysis [4].

Table 1. Distribution of nitrogen metabolism genes in *Thiothrix* genomes. Genes: *narG*, membrane-bound nitrate reductase; *napAB*, periplasmic nitrate reductase; *nasA*, assimilatory nitrate reductase; *nasB*, assimilatory nitrite reductase; *nirS*, dissimilatory nitrite reductase; *nirBD*, assimilatory nitrite reductase; *norBC*, nitric oxide reductase; *nosZ*, nitric oxide reductase. Table was based on Ravin et al. [5] and Chernitsyna et al. [84] under the terms of the Creative Commons Attribution 4.0 International License.

Strain	Dissimilatory Nitrate Reduction				Assimilatory Nitrate and Nitrite Reduction	
	$NO_3^- \rightarrow NO_2^-$	$NO_2^- \rightarrow NO$	$NO \rightarrow N_2O$	$N_2O \rightarrow N_2$	$NO_3^- \rightarrow NO_2^-$	$NO_2^- \rightarrow NH_3$
MAG *Thiothrix* sp. 207	*narGHIJ*, *napAB*	*nirS*	*norBC*	*nosZ*	*nasA*, *nasD*	*nirBD*
MAG *Thiothrix* sp. STA 22	*narGHIJ*	*nirS*	*norBC*	*nosZ*	*nasA*, *nasD*	*nirBD*
T. litoralis AS[T]	*narGHIJ*	*nirS*	*norBC*	-	*nasA*, *nasD*	*nirBD*
T. fructosivorans Q[T]	*narGHIJ*	*nirS*	*norBC*	-	*nasA*, *nasD*	*nirBD*
T. caldifontis G1[T]	*narGHIJ*	*nirS*	*norBC*	-	*nasA*, *nasD*	*nirBD*
T. winogradskyi CT3[T]	*narGHIJ*	*nirS*	*norBC*	-	*nasA*, *nasD*	*nirBD*
T. unzii A1[T]	*narGHIJ*	*nirS*	*norBC*	-	*nasA*, *nasD*	*nirBD*
'*Ca.* Thiothrix singaporensis' SSD2	*narGHIJ*	*nirS*	*norBC*	-	*nasA*, *nasD*	*nirBD*
'*Ca.* Thiothrix moscovensis' RT	*narGHIJ*	-	*norBC*	-	*nasA*, *nasD*	*nirBD*
'*Ca.* Thiothrix putei' GK-02	*narGHIJ*	-	*norBC*	-	-	-
T. subterranea Ku-5[T]	*narGHIJ*	-	-	-	*nasA*, *nasD*	*nirBD*
T. lacustris BL[T]	*narGHIJ*	-	-	-	*nasA*, *nasD*	*nirBD*
'*Ca.* Thiothrix sulfatifontis' KT	-	-	-	-	*nasA*, *nasD*	*nirBD*
T. nivea JP2[T]	*napAB*	-	-	-	*nasA*, *nasD*	*nirBD*
'*Ca.* Thiothrix anitrata' A52	-	-	-	-	-	-
'*Ca.* Thiothrix namsaraevi'	*narGHIJ*	-	-	-	*nasA*	*nirBD*

In case of uncontrolled inputs due to effluents from animal farms, agricultural factories, industrial companies and the sewage systems of residential complexes, high concentrations of nitrogenous compounds lead to the destructive eutrophication of water bodies and ammonium toxicity to fish and other aquatic fauna as well as flora. Kosgey et al. presented traditional schemes for the biological treatment of wastewater contaminated with nitrogenous compounds [116]. These schemes are based on the use of specialized bacteria with different types of nitrogen metabolism, which allows the sequential conversion of toxic ammonium ions (NH_4^+) to safe dinitrogen molecules (N_2) through a chain consisting of nitrifier bacteria, which oxidize ammonium (AOB) and nitrite (NOB), and denitrifier bacteria, which reduce nitrate at the expense of organic carbon, which acts as an electron donor (Figure 7). In this scheme, representatives of the *Thiothrix* morphotype can play a role in the last stage.

Representatives of the *Thiothrix* morphotype are often found in nitrification–denitrification systems [32,117–120]. Their role in these systems can involve denitrification coupled with the oxidation of reduced sulfur compounds. In this case, sulfides can be used as an electron donor for the reduction of nitrates and nitrites [121]. There are reports that the *Thiothrix* morphotype is capable of nitrate-mediated sulfide oxidation [46]. It was shown that bacteria similar in morphotype to *Thiothrix* began to play the role of denitrifiers in the simultaneous nitrification–denitrification system when the oxygen concentration decreased [122].

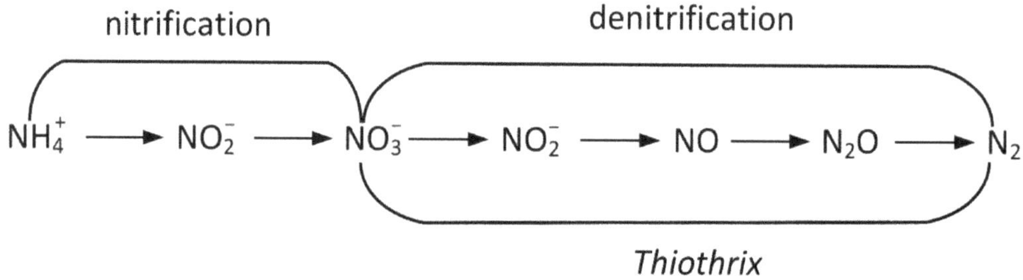

Figure 7. Nitrification–denitrification reactions. In the processes of the transformation of nitrogenous compounds leading to nitrogen removal from wastewater, representatives of the genus *Thiothrix* are able to perform the denitrification step.

Zhao et al. found a large number of bacteria of the *Thiothrix* morphotype in biofilm reactors with simultaneous nitrification–denitrification processes, where pomelo peel, which can biodegrade to form sulfide, served as a carrier [123]. This explains the appearance of dominant representatives of the *Thiothrix* morphotype in the system and the presence of denitrification associated with the oxidation of reduced sulfur compounds.

4.4. Denitrifying Removal of Phosphorus

Two methods of phosphorus removal by denitrifying have been described. One is based on the operation of both denitrification and phosphorus accumulation reactions in the same microorganism. Another variant is based on the separate running of these reactions, where denitrification is carried out by one microorganism and phosphorus accumulation by another [124]. Denitrifying organisms that accumulate phosphate can simultaneously remove nitrogen and phosphorus under rotational anaerobic and anoxic/aerobic conditions [125].

The role of bacteria of the *Thiothrix* morphotype in denitrifying phosphorus removal has been suggested in different studies. The presence of these bacteria in the denitrifying phosphorus removal system was mentioned, but their function as denitrifying or dephosphorizing bacteria was not considered [126]. Other studies indicate a dephosphorizing role for the bacteria of the *Thiothrix* morphotype in such systems. For example, Li et al. found that bacteria similar in morphotype to *Thiothrix* played a crucial role in the accumulation of precipitated phosphate in a reactor with simultaneous nitrification–denitrification and phosphorus removal [127]. Moreover, Ai et al. stated that, in a wastewater treatment system where denitrifying phosphorus removal served as the primary pathway of phosphorus removal, the *Thiothrix* morphotype was associated with phosphorus removal [128].

In addition, bacteria similar in morphotype to *Thiothix* are believed to be capable of denitrifying phosphorus removal [48]. An analysis of metagenomes from a biofilm reactor in which denitrifying phosphorus removal was performed showed that bacteria similar in morphotype to *Thiothrix* had genes for denitrification and *pit* genes for the phosphate transporter [49]. Thus, representatives of the *Thiothrix* morphotype are potentially capable of denitrifying phosphorus removal within a single organism.

5. Factors Affecting *Thiothrix* Proliferation

In wastewater treatment systems, due to the significant role of bacteria similar in morphotype to *Thiothrix*, it is important to consider factors that can lead to their increased growth and methods to control it.

Bacteria of the *Thiothrix* morphotype are able to dominate in activated sludge in the presence of thiosulfate, which can be related to the ability of these bacteria to grow mixotrophically using reduced sulfur compounds as an energy source [26]. Consistent with this phenomenon, the core genome of the genus *Thiothrix* was found to contain genes encoding sulfur metabolism complexes: the SQR and FCSD complexes, participating in the oxidation of sulfide to polysulfide/sulfur; the Sox complex (*soxAXBYZ*), which participates in the branched pathway of thiosulfate oxidation to sulfur and sulfate; and the rDsr system (*dsrABCEFHEMPKJOL*), which participates in the subsequent oxidation of elemental sulfur [4].

It was shown that the presence of sulfide in wastewater treatment systems can evoke the active growth of the *Thiothrix* morphotype [44,47,85]. High levels of dissolved oxygen (DO) are detrimental to bacteria similar in morphotype to *Thiothrix* and inhibit their growth, while, conversely, low DO values favor the growth of these bacteria [21]. This is due to the fact that they belong to the group of colorless sulfur-oxidizing bacteria, which are known to be catalase-negative and prefer growth in narrow microaerobic eco-niches where fewer toxic reactive oxygen species are produced. At low DO levels and low nitrate concentrations, bacteria of the *Thiothrix* morphotype gain a competitive advantage over other microorganisms due to the ability to accumulate nitrates intracellularly, which leads to the explosive growth of these bacteria and activated sludge bulking [129]. This phenomenon is related to the ability of bacteria of the *Thiothrix* morphotype to utilize nitrate as an alternative electron acceptor. The appearance of bisphenol A in wastewater also leads to an increase in the abundance of the *Thiothrix* morphotype [33].

To suppress the growth of *Thiothrix*, the addition of polyaluminum chloride and a decrease in VFAs were tested, but these approaches were ineffective in resolving the problem conclusively [34]. Although a correlation was observed between the propionic acid content and the Type 021N-mediated bulking of sludge, this relationship was, in reality, an indirect effect of the inhibition of flocculating microorganisms, which gave filamentous bacteria a competitive advantage in growth [29].

Several methods have been demonstrated to reduce the abundance of bacteria similar in morphotype to *Thiothrix* in wastewater. These include the use of intermittent fasting [34], moderate oxygen concentrations with low aerobic periods in the surrounding habitat [130] and the addition of 1‰ hydrogen peroxide [80], the backwashing of membrane reactors with NaClO in wastewater systems [65] and even exposure to a magnetic field [79].

Regarding the growth rate of bacteria currently assigned to the genus *Thiolinea*, it was demonstrated to have a statistically significant correlation with the food/mass ratio (daily activated sludge load, defined as the ratio of the daily amount of incoming organic pollutants F to ash-free substance of activated sludge M (F:M)) and the ammonium ion concentration in the primary effluent, as well as an inverse correlation with the dissolved oxygen concentration in the microaerobic selector [14].

6. Conclusions

Thus, in wastewater treatment systems, bacteria of the *Thiothrix* morphotype are more often found in activated sludge or in membrane bioreactors and much more rarely in biofilms. Modifications to the operating parameters of biological treatment systems can lead to the explosive growth of the *Thiothrix* morphotype and result in system disruption. The proliferation of the *Thiothrix* morphotype is enhanced under low concentrations of dissolved oxygen and in the presence of sulfide and bisphenol A and can be suppressed by peroxide, hypochlorite or exposure to a magnetic field. Historically, representatives of this morphotype were detected based on their morphology; later, FISH probes and PCR primers were developed for their detection; and, now, high-throughput sequencing methods are increasingly used for this purpose. *Thiothrix*-like bacteria have a flexible metabolism, which allows them to perform a number of biological wastewater treatment processes, such as sulfide oxidation, denitrification and enhanced biological phosphorus removal. For bacteria belonging to the genus *Thiothrix*, a pangenome has been obtained, which makes it possible to perform the genetic determination of these metabolic processes. Obtaining pangenomes of the genera *Thiofilum* and *Thiolinea* is a task for further research and could lead to a better understanding of the metabolism of these genera and to the synthesis of data on the metabolism of the representatives of the *Thiothrix* morphotype in wastewater treatment systems. Prospects for the further development of wastewater treatment with the participation of representatives of the *Thiothrix* morphotype include the study of the possibility of using representatives of this group in the removal of phosphorus compounds in EBPR, as well as in the treatment of hydrogen sulfide. Namely, due to the knowledge of their metabolic potential, we wish to draw the attention of researchers not only to the possibility of identifying the presence of representatives of the *Thiothrix* morphotype in various wastewater treatment systems, etc., but also to the possibility of the targeted introduction of these organisms in the process of phosphorus and hydrogen sulfide removal.

Author Contributions: Conceptualization, M.Y.G. and N.V.R.; writing—original draft preparation, M.V.G., M.S.M. and M.Y.G.; visualization, M.V.G., M.S.M. and M.Y.G.; writing—review and editing, M.V.G., M.S.M., N.V.R. and M.Y.G.; supervision, M.Y.G.; project administration, N.V.R.; funding acquisition, M.Y.G. All authors have read and agreed to the published version of the manuscript.

Funding: This research was funded by the Russian Science Foundation as an extension of grant number 20-14-00137 (https://rscf.ru/en/project/20-14-00137/, accessed on 15 May 2023 to M.Y.G.) and the Ministry of Science and Higher Education of the Russian Federation (N.V.R., search for *Thiothrix* phylotypes in sewage and activated sludge).

Institutional Review Board Statement: Not applicable.

Informed Consent Statement: Not applicable.

Data Availability Statement: The original contributions presented in the study are included in the article; further inquiries can be directed to the corresponding authors.

Conflicts of Interest: The authors declare no conflicts of interest.

References

1. Larkin, J.M. Isolation of *Thiothrix* in pure culture and observation of a filamentous epiphyte on *Thiothrix*. *Curr. Microbiol.* **1980**, *4*, 155–158. [CrossRef]
2. Larkin, J.M.; Strohl, W.R. *Beggiatoa*, *Thiothrix*, and *Thioploca*. *Annu. Rev. Microbiol.* **1983**, *37*, 341–367. [CrossRef] [PubMed]
3. Unz, R.F.; Head, I.M.; Genus, I. *Thiothrix* Winogradsky 1888, 39^{AL}. In *Bergey's Manual of Systematic Bacteriology: Part B: Gammaproteobacteria*, 2nd ed.; Garrity, G., Brenner, D.J., Krieg, N.R., Staley, J.T.E., Eds.; Springer: Boston, MA, USA, 2005; Volume 2, pp. 131–142.
4. Ravin, N.V.; Rudenko, T.S.; Smolyakov, D.D.; Beletsky, A.V.; Gureeva, M.V.; Samylina, O.S.; Grabovich, M.Y. History of the study of the genus *Thiothrix*: From the first enrichment cultures to pangenomic analysis. *Int. J. Mol. Sci.* **2022**, *23*, 9531. [CrossRef]
5. Ravin, N.V.; Muntyan, M.S.; Smolyakov, D.D.; Rudenko, T.S.; Beletsky, A.V.; Mardanov, A.V.; Grabovich, M.Y. Metagenomics revealed a new genus '*Candidatus* Thiocaldithrix dubininis' gen. nov., sp. nov. and a new species '*Candidatus* Thiothrix putei' sp. nov. in the family *Thiotrichaceae*, some members of which have traits of both Na^+- and H^+-motive energetics. *Int. J. Mol. Sci.* **2023**, *24*, 14199. [CrossRef]
6. Jenkins, D.; Richard, M.G.; Daigger, G.T. *Manual on the Causes and Control of Activated Sludge Bulking and Other Solids Separation Problems*, 3rd ed.; CRC Press: Boca Raton, FL, USA, 2003; 236p. [CrossRef]
7. Eikelboom, D.H. *Process Control of Activated Sludge Plants by Microscopic Investigation*; IWA Publishing: London, UK, 2000; 156p.
8. Howarth, R.; Unz, R.F.; Seviour, E.M.; Seviour, R.J.; Blackall, L.L.; Pickup, R.W.; Jones, J.G.; Yaguchi, J.; Head, I.M. Phylogenetic relationships of filamentous sulfur bacteria (*Thiothrix* spp. and Eikelboom type 021N bacteria) isolated from wastewater-treatment plants and description of *Thiothrix eikelboomii* sp. nov., *Thiothrix unzii* sp. nov., *Thiothrix fructosivorans* sp. nov. and *Thiothrix defluvii* sp. nov. *Int. J. Syst. Bacteriol.* **1999**, *49*, 1817–1827. [CrossRef]
9. Aruga, S.; Kamagata, Y.; Kohno, T.; Hanada, S.; Nakamura, K.; Kanagawa, T. Characterization of filamentous Eikelboom type 021N bacteria and description of *Thiothrix disciformis* sp. nov. and *Thiothrix flexilis* sp. nov. *Int. J. Syst. Evol. Microbiol.* **2002**, *52*, 1309–1316. [CrossRef]
10. Boden, R.; Scott, K.M. Evaluation of the genus *Thiothrix* Winogradsky 1888 (Approved Lists 1980) emend. Aruga et al. 2002: Reclassification of *Thiothrix disciformis* to *Thiolinea disciformis* gen. nov., comb. nov., and of *Thiothrix flexilis* to *Thiofilum flexile* gen. nov., comb nov., with emended description of *Thiothrix*. *Int. J. Syst. Evol. Microbiol.* **2018**, *68*, 2226–2239. [CrossRef] [PubMed]
11. Matsuura, N.; Masakke, Y.; Karthikeyan, S.; Kanazawa, S.; Honda, R.; Yamamoto-Ikemoto, R.; Konstantinidis, K.T. Metagenomic insights into the effect of sulfate on enhanced biological phosphorus removal. *Appl. Microbiol. Biotechnol.* **2021**, *105*, 2181–2193. [CrossRef]
12. Wu, X.; Huang, J.; Lu, Z.; Chen, G.; Wang, J.; Liu, G. *Thiothrix eikelboomii* interferes oxygen transfer in activated sludge. *Water Res.* **2019**, *151*, 134–143. [CrossRef] [PubMed]
13. Guo, J.; Peng, Y.; Yang, X.; Wang, Z.; Zhu, A. Changes in the microbial community structure of filaments and floc formers in response to various carbon sources and feeding patterns. *Appl. Microbiol. Biotechnol.* **2014**, *98*, 7633–7644. [CrossRef]

14. Asvapathanagul, P.; Olson, B.H.; Gedalanga, P.B.; Hashemi, A.; Huang, Z.; La, J. Identification and quantification of *Thiothrix eikelboomii* using qPCR for early detection of bulking incidents in a full-scale water reclamation plant. *Appl. Microbiol. Biotechnol.* **2015**, *99*, 4045–4057. [CrossRef]
15. Guo, J.; Peng, Y.; Wang, Z.; Yuan, Z.; Yang, X.; Wang, S. Control filamentous bulking caused by chlorine-resistant Type 021N bacteria through adding a biocide CTAB. *Water Res.* **2012**, *46*, 6531–6542. [CrossRef]
16. Kocerba-Soroka, W.; Fiałkowska, E.; Pajdak-Stós, A.; Klimek, B.; Kowalska, E.; Drzewicki, A.; Salvadó, H.; Fyda, J. The use of rotifers for limiting filamentous bacteria Type 021N, a bacteria causing activated sludge bulking. *Water Sci. Technol.* **2013**, *67*, 1557–1563. [CrossRef] [PubMed]
17. Gnida, A.; Żabczyński, S.; Surmacz-Górska, J. Filamentous bacteria in the nitrifying activated sludge. *Water Sci. Technol.* **2018**, *77*, 2709–2713. [CrossRef]
18. Deepnarain, N.; Nasr, M.; Kumari, S.; Stenström, T.A.; Reddy, P.; Pillay, K.; Bux, F. Decision tree for identification and prediction of filamentous bulking at full-scale activated sludge wastewater treatment plant. *Process Saf. Environ. Prot.* **2019**, *126*, 25–34. [CrossRef]
19. Ravin, N.V.; Smolyakov, D.D.; Markov, N.D.; Beletsky, A.V.; Mardanov, A.V.; Rudenko, T.S.; Grabovich, M.Y. *tilS* and *rpoB*: New molecular markers for phylogenetic and biodiversity studies of the genus *Thiothrix*. *Microorganisms* **2023**, *11*, 2521. [CrossRef]
20. Zhao, X.; Xie, Y.; Sun, B.; Liu, Y.; Zhu, S.; Li, W.; Zhao, M.; Liu, D. Unraveling microbial characteristics of simultaneous nitrification, denitrification and phosphorus removal in a membrane-aerated biofilm reactor. *Environ. Res.* **2023**, *239*, 117402. [CrossRef]
21. Lee, S.; Basu, S.; Tyler, C.W.; Pitt, P.A. A survey of filamentous organisms at the Deer Island treatment plant. *Environ. Technol.* **2003**, *24*, 855–865. [CrossRef] [PubMed]
22. Lee, H.; Choi, E.; Yun, Z.; Park, Y.K. Microbial structure and community of RBC biofilm removing nitrate and phosphorus from domestic wastewater. *J. Microbiol. Biotechnol.* **2008**, *18*, 1459–1469.
23. Feng, Y.; Guo, M.; Jia, X.; Liu, N.; Li, X.; Li, X.; Song, L.; Wang, X.; Qiu, L.; Yu, Y. Combined effects of electrical current and nonsteroidal antiinflammatory drugs (NSAIDs) on microbial community in a three-dimensional electrode biological aerated filter (3DE-BAF). *Bioresour. Technol.* **2020**, *309*, 123346. [CrossRef]
24. Seguel Suazo, K.; Dobbeleers, T.; Dries, J. Bacterial community and filamentous population of industrial wastewater treatment plants in Belgium. *Appl. Microbiol. Biotechnol.* **2024**, *108*, 43. [CrossRef]
25. Onetto, C.A.; Eales, K.L.; Grbin, P.R. Remediation of *Thiothrix* spp. associated bulking problems by raw wastewater feeding: A full-scale experience. *Syst. Appl. Microbiol.* **2017**, *40*, 396–399. [CrossRef] [PubMed]
26. Nielsen, P.H.; de Muro, M.A.; Nielsen, J.L. Studies on the in situ physiology of *Thiothrix* spp. present in activated sludge. *Environ. Microbiol.* **2000**, *2*, 389–398. [CrossRef]
27. Wu, G.; Yin, Q. Microbial niche nexus sustaining biological wastewater treatment. *NPJ Clean Water* **2020**, *3*, 33. [CrossRef]
28. Fourest, E.; Craperi, D.; Deschamps-Roupert, C.; Pisicchio, J.-L.; Lenon, G. Occurrence and control of filamentous bulking in aerated wastewater treatment plants of the French paper industry. *Water Sci. Technol.* **2004**, *50*, 29–37. [CrossRef] [PubMed]
29. Pelletier, C.; Fitzsimmons, M.A.; Deschênes, S.; Paice, M. Impact of septic compounds and operational conditions on the microbiology of an activated sludge system. *Water Sci. Technol.* **2007**, *55*, 135–142. [CrossRef]
30. Dumonceaux, T.J.; Hill, J.E.; Pelletier, C.P.; Paice, M.G.; Van Kessel, A.G.; Hemmingsen, S.M. Molecular characterization of microbial communities in Canadian pulp and paper activated sludge and quantification of a novel *Thiothrix eikelboomii*-like bulking filament. *Can. J. Microbiol.* **2006**, *52*, 494–500. [CrossRef]
31. Kim, S.B.; Goodfellow, M.; Kelly, J.; Saddler, G.S.; Ward, A.C. Application of oligonucleotide probes for the detection of *Thiothrix* spp. in activated sludge plants treating paper and board mill wastes. *Water Sci. Technol.* **2002**, *46*, 559–564. [CrossRef]
32. Zou, J.; Yang, J.; He, H.; Wang, X.; Mei, R.; Cai, L.; Li, J. Effect of seed sludge type on aerobic granulation, pollutant removal and microbial community in a sequencing batch reactor treating real textile wastewater. *Int. J. Environ. Res. Public Health* **2022**, *19*, 10940. [CrossRef] [PubMed]
33. Ferrer-Polonio, E.; Alvim, C.B.; Fernández-Navarro, J.; Mompó-Curell, R.; Mendoza-Roca, J.A.; Bes-Piá, A.; Alonso-Molina, J.L.; Amorós-Muñoz, I. Influence of bisphenol A occurrence in wastewaters on biomass characteristics and activated sludge process performance. *Sci. Total Environ.* **2021**, *778*, 146355. [CrossRef]
34. Henriet, O.; Meunier, C.; Henry, P.; Mahillon, J. Filamentous bulking caused by *Thiothrix* species is efficiently controlled in full-scale wastewater treatment plants by implementing a sludge densification strategy. *Sci. Rep.* **2017**, *7*, 1430. [CrossRef]
35. Figueroa, M.; Val del Río, A.; Campos, J.L.; Méndez, R.; Mosquera-Corral, A. Filamentous bacteria existence in aerobic granular reactors. *Bioprocess Biosyst. Eng.* **2015**, *38*, 841–851. [CrossRef] [PubMed]
36. van der Waarde, J.; Krooneman, J.; Geurkink, B.; van der Werf, A.; Eikelboom, D.; Beimfohr, C.; Snaidr, J.; Levantesi, C.; Tandoi, V. Molecular monitoring of bulking sludge in industrial wastewater treatment plants. *Water Sci. Technol.* **2002**, *46*, 551–558. [CrossRef]
37. Li, Y.; Zhou, J.; Gong, B.; Wang, Y.; He, Q. Cometabolic degradation of lincomycin in a Sequencing Batch Biofilm Reactor (SBBR) and its microbial community. *Bioresour. Technol.* **2016**, *214*, 589–595. [CrossRef] [PubMed]
38. Wang, R.F.; Wang, Q.; Zhang, H.; Qi, R. [Investigation for filamentous bacteria community diversity in activated sludge under various kinds and concentration conditions of antibiotics]. *Huan Jing Ke Xue* **2015**, *36*, 2239–2244. (In Chinese) [PubMed]
39. Du, B.; Yang, Q.; Li, X.; Yuan, W.; Chen, Y.; Wang, R. Impacts of long-term exposure to tetracycline and sulfamethoxazole on the sludge granules in an anoxic-aerobic wastewater treatment system. *Sci. Total Environ.* **2019**, *684*, 67–77. [CrossRef]

40. Zhang, B.; He, Y.; Shi, W.; Liu, L.; Li, L.; Liu, C.; Lens, P.N.L. Biotransformation of sulfamethoxazole (SMX) by aerobic granular sludge: Removal performance, degradation mechanism and microbial response. *Sci. Total Environ.* **2023**, *858*, 159771. [CrossRef]
41. Ma, Z.; Wen, X.; Zhao, F.; Xia, Y.; Huang, X.; Waite, D.; Guan, J. Effect of temperature variation on membrane fouling and microbial community structure in membrane bioreactor. *Bioresour. Technol.* **2013**, *133*, 462–468. [CrossRef]
42. Richard, M.G.; Shimizu, G.P.; Jenkins, D. The growth physiology of the filamentous organism Type-021N and its significance to activated sludge bulking. *J. WPCF* **1985**, *57*, 1152–1162. Available online: https://www.jstor.org/stable/25042817 (accessed on 19 August 2024).
43. Cheng, W.; Zhang, L.; Xu, W.; Sun, Y.; Wan, J.; Li, H.; Wang, Y. Formation and characteristics of filamentous granular sludge. *Water Sci. Technol.* **2020**, *82*, 364–372. [CrossRef]
44. de Graaff, D.R.; van Loosdrecht, M.C.M.; Pronk, M. Stable granulation of seawater-adapted aerobic granular sludge with filamentous *Thiothrix* bacteria. *Water Res.* **2020**, *175*, 115683. [CrossRef]
45. Garcia, G.P.; Souza, C.L.; Glória, R.M.; Silva, S.Q.; Chernicharo, C.A. Biological oxidation of sulphides by microorganisms present in the scum layer of UASB reactors treating domestic wastewater. *Water Sci. Technol.* **2012**, *66*, 1871–1878. [CrossRef]
46. Cytryn, E.; Minz, D.; Gelfand, I.; Neori, A.; Gieseke, A.; De Beer, D.; Van Rijn, J. Sulfide-oxidizing activity and bacterial community structure in a fluidized bed reactor from a zero-discharge mariculture system. *Environ. Sci. Technol.* **2005**, *39*, 1802–1810. [CrossRef]
47. Rubio-Rincón, F.J.; Welles, L.; Lopez-Vazquez, C.M.; Nierychlo, M.; Abbas, B.; Geleijnse, M.; Nielsen, P.H.; van Loosdrecht, M.C.M.; Brdjanovic, D. Long-term effects of sulphide on the enhanced biological removal of phosphorus: The symbiotic role of *Thiothrix caldifontis*. *Water Res.* **2017**, *116*, 53–64. [CrossRef]
48. Miao, X.N.; Wang, Q.; Guo, K.C.; Liu, W.R.; Shen, Y.L. [Start-up and optimization of denitrifying phosphorus removal in ABR-MBR coupling process]. *Huan Jing Ke Xue* **2020**, *41*, 4150–4160. (In Chinese) [CrossRef]
49. Dong, K.; Qiu, Y.; Wang, X.; Yu, D.; Yu, Z.; Feng, J.; Wang, J.; Gu, R.; Zhao, J. Towards low carbon demand and highly efficient nutrient removal: Establishing denitrifying phosphorus removal in a biofilm-based system. *Bioresour. Technol.* **2023**, *372*, 128658. [CrossRef]
50. Waqas, S.; Bilad, M.R.; Man, Z.; Wibisono, Y.; Jaafar, J.; Indra Mahlia, T.M.; Khan, A.L.; Aslam, M. Recent progress in integrated fixed-film activated sludge process for wastewater treatment: A review. *J. Environ. Manag.* **2020**, *268*, 110718. [CrossRef]
51. Bakar, S.N.H.A.; Hasan, H.A.; Mohammad, A.W.; Abdullah, S.R.S.; Haan, T.Y.; Ngteni, R.; Yusof, K.M.M. A review of moving-bed biofilm reactor technology for palm oil mill effluent treatment. *J. Clean. Prod.* **2018**, *171*, 1532–1545. [CrossRef]
52. Hong, Y.; Yao, J.Q.; Ma, B.; Xu, S.; Zhang, Y.J. [Filamentous sludge microbial community of a SBR reactor based on high-throughput sequencing]. *Huan Jing Ke Xue* **2018**, *39*, 3279–3285. (In Chinese) [CrossRef] [PubMed]
53. Williams, T.M.; Unz, R.F. Filamentous sulfur bacteria of activated sludge: Characterization of *Thiothrix*, *Beggiatoa*, and Eikelboom type 021N strains. *Appl. Environ. Microbiol.* **1985**, *49*, 887–898. [CrossRef]
54. Tandoi, V.; Caravaglio, N.; Di Dio Balsamo, D.; Majone, M.; Tomei, M.C. Isolation and physiological characterization of *Thiothrix* sp. *Water Sci. Technol.* **1994**, *29*, 261–269. [CrossRef]
55. Ravin, N.V.; Rossetti, S.; Beletsky, A.V.; Kadnikov, V.V.; Rudenko, T.S.; Smolyakov, D.D.; Moskvitina, M.I.; Gureeva, M.V.; Mardanov, A.V.; Grabovich, M.Y. Two new species of filamentous sulfur bacteria of the genus *Thiothrix*, *Thiothrix winogradskyi* sp. nov. and '*Candidatus* Thiothrix sulfatifontis' sp. nov. *Microorganisms* **2022**, *10*, 1300. [CrossRef]
56. Mardanov, A.V.; Gruzdev, E.V.; Smolyakov, D.D.; Rudenko, T.S.; Beletsky, A.V.; Gureeva, M.V.; Markov, N.D.; Berestovskaya, Y.Y.; Pimenov, N.V.; Ravin, N.V.; et al. Genomic and metabolic insights into two novel *Thiothrix* species from enhanced biological phosphorus removal systems. *Microorganisms* **2020**, *8*, 2030. [CrossRef] [PubMed]
57. van Dijk, E.J.H.; Pronk, M.; van Loosdrecht, M.C.M. A settling model for full-scale aerobic granular sludge. *Water Res.* **2020**, *186*, 116135. [CrossRef]
58. Wang, G.; Huang, X.; Wang, S.; Yang, F.; Sun, S.; Yan, P.; Chen, Y.; Fang, F.; Guo, J. Effect of food-to-microorganisms ratio on aerobic granular sludge settleability: Microbial community, potential roles and sequential responses of extracellular proteins and polysaccharides. *J. Environ. Manag.* **2023**, *345*, 118814. [CrossRef]
59. Huang, Y.; Zhang, J.; Liu, J.; Gao, X.; Wang, X. Effect of C/N on the microbial interactions of aerobic granular sludge system. *J. Environ. Manag.* **2024**, *349*, 119505. [CrossRef]
60. Xu, P.; Xie, Z.; Shi, L.; Yan, X.; Fu, Z.; Ma, J.; Zhang, W.; Wang, H.; Xu, B.; He, Q. Distinct responses of aerobic granular sludge sequencing batch reactors to nitrogen and phosphorus deficient conditions. *Sci. Total Environ.* **2022**, *834*, 155369. [CrossRef] [PubMed]
61. Zhou, J.H.; Ren, Q.; Xu, X.L.; Fang, J.Y.; Wang, T.; Wang, K.M.; Wang, H.Y. Enhancing stability of aerobic granules by microbial selection pressure using height-adjustable influent strategy. *Water Res.* **2021**, *201*, 117356. [CrossRef]
62. Rahman, T.U.; Roy, H.; Islam, M.R.; Tahmid, M.; Fariha, A.; Mazumder, A.; Tasnim, N.; Pervez, M.N.; Cai, Y.; Naddeo, V.; et al. The advancement in Membrane Bioreactor (MBR) technology toward sustainable industrial wastewater management. *Membranes* **2023**, *13*, 181. [CrossRef]
63. Huang, R.; Geng, M.; Gao, S.; Yin, X.; Tian, J. In-depth insight into improvement of simultaneous nitrification and denitrification/biofouling control by increasing sludge concentration in membrane reactor: Performance, microbial assembly and metagenomic analysis. *Bioresour. Technol.* **2023**, *393*, 130013. [CrossRef]

64. Xiao, X.; Guo, H.; Ma, F.; Zhang, J.; Ma, X.; You, S. New insights into mycelial pellets for aerobic sludge granulation in membrane bioreactor: Bio-functional interactions among metazoans, microbial communities and protein expression. *Water Res.* **2023**, *228*, 119361. [CrossRef]
65. Wang, Z.; Meng, F.; He, X.; Zhou, Z.; Huang, L.N.; Liang, S. Optimisation and performance of NaClO-assisted maintenance cleaning for fouling control in membrane bioreactors. *Water Res.* **2014**, *53*, 1–11. [CrossRef]
66. Yang, L.; Jiao, Y.; Xu, X.; Pan, Y.; Su, C.; Duan, X.; Sun, H.; Liu, S.; Wang, S.; Shao, Z. Superstructures with atomic-level arranged perovskite and oxide layers for advanced oxidation with an enhanced non-free radical pathway. *ACS Sustain. Chem. Eng.* **2022**, *10*, 1899–1909. [CrossRef]
67. Yang, N.; Yu, J.; Zhang, L.; Sun, Y.; Zhang, L.; Jiang, B. Integration of efficient $LaSrCoO_4$ perovskite and polyacrylonitrile membrane to enhance mass transfer for rapid contaminant degradation. *J. Water Process. Eng.* **2023**, *53*, 103804. [CrossRef]
68. Hassard, F.; Biddle, L.; Cartmell, E.; Jefferson, B.; Tyrrel, S.; Stephenson, T. Rotating biological contactors for wastewater treatment—A review. *Process Saf. Environ. Prot.* **2015**, *94*, 285–306. [CrossRef]
69. Galván, A.; de Castro, F. Relationships among filamentous microorganisms in rotating biological contactors. *Indian J. Microbiol.* **2007**, *47*, 15–25. [CrossRef] [PubMed]
70. Ferrera, I.; Massana, R.; Casamayor, E.O.; Balagué, V.; Sánchez, O.; Pedrós-Alió, C.; Mas, J. High-diversity biofilm for the oxidation of sulfide-containing effluents. *Appl. Microbiol. Biotechnol.* **2004**, *64*, 726–734. [CrossRef]
71. Moissl, C.; Rudolph, C.; Huber, R. Natural communities of novel archaea and bacteria with a string-of-pearls-like morphology: Molecular analysis of the bacterial partners. *Appl. Environ. Microbiol.* **2002**, *68*, 933–937. [CrossRef]
72. Wagner, M.; Amann, R.; Kampfer, P.; Assmus, B.; Hartmann, A.; Hutzler, P.; Springer, N.; Schleifer, K. Identification and In-Situ detection of Gram-negative filamentous bacteria in activated sludge. *Syst. Appl. Microbiol.* **1994**, *17*, 405–417.
73. Kanagawa, T.; Kamagata, Y.; Aruga, S.; Kohno, T.; Horn, M.; Wagner, M. Phylogenetic analysis of and oligonucleotide probe development for eikelboom type 021N filamentous bacteria isolated from bulkingactivated sludge. *Appl. Environ. Microbiol.* **2000**, *66*, 5043–5052. [CrossRef]
74. Maestre, J.P.; Rovira, R.; Alvarez-Hornos, F.J.; Fortuny, M.; Lafuente, J.; Gamisans, X.; Gabriel, D. Bacterial community analysis of a gas-phase biotrickling filter for biogas mimics desulfurization through the rRNA approach. *Chemosphere* **2010**, *80*, 872–880. [CrossRef]
75. Lou, I.; de Los Reyes, F.L., 3rd. Substrate uptake tests and quantitative FISH show differences in kinetic growth of bulking and non-bulking activated sludge. *Biotechnol. Bioeng.* **2005**, *92*, 729–739. [CrossRef]
76. Gulez, G.; de Los Reyes, F.L., 3rd. Multiple approaches to assess filamentous bacterial growth in activated sludge under different carbon source conditions. *J. Appl. Microbiol.* **2009**, *106*, 682–691. [CrossRef]
77. Gaval, G.; Pernelle, J.J. Impact of the repetition of oxygen deficiencies on the filamentous bacteria proliferation in activated sludge. *Water Res.* **2003**, *37*, 1991–2000. [CrossRef] [PubMed]
78. Vervaeren, H.; De Wilde, K.; Matthys, J.; Boon, N.; Raskin, L.; Verstraete, W. Quantification of an Eikelboom type 021N bulking event with fluorescence in situ hybridization and real-time PCR. *Appl. Microbiol. Biotechnol.* **2005**, *68*, 695–704. [CrossRef]
79. Chen, R.; Shuai, J.; Xie, Y.; Wang, B.; Hu, X.; Guo, W.; Lyu, W.; Zhou, D.; Mosa, A.; Wang, H. Aerobic granulation and microbial community succession in sequencing batch reactors treating the low strength wastewater: The dual effects of weak magnetic field and exogenous signal molecule. *Chemosphere* **2022**, *309*, 136762. [CrossRef] [PubMed]
80. He, X.M.; Ding, L.L.; Zhang, L.L.; Gu, Z.J.; Ren, H.Q. [Performance, sludge characteristics, and the microbial community dynamics of bulking sludge under different nitrogen and phosphorus imbalances]. *Huan Jing Ke Xue* **2018**, *39*, 1782–1793. (In Chinese) [CrossRef] [PubMed]
81. Arumugam, K.; Bessarab, I.; Haryono, M.A.S.; Liu, X.; Zuniga-Montanez, R.E.; Roy, S.; Qiu, G.; Drautz-Moses, D.I.; Law, Y.Y.; Wuertz, S.; et al. Recovery of complete genomes and non-chromosomal replicons from activated sludge enrichment microbial communities with long read metagenome sequencing. *npj Biofilms Microbiomes* **2021**, *7*, 23. [CrossRef]
82. Begmatov, S.; Beletsky, A.V.; Dorofeev, A.G.; Pimenov, N.V.; Mardanov, A.V.; Ravin, N.V. Metagenomic insights into the wastewater resistome before and after purification at large-scale wastewater treatment plants in the Moscow city. *Sci. Rep.* **2024**, *14*, 6349. [CrossRef]
83. Yang, M.; Chen, T.; Liu, Y.X.; Huang, L. Visualizing set relationships: EVenn's comprehensive approach to Venn diagrams. *iMeta* **2024**, *3*, e184. [CrossRef]
84. Chernitsyna, S.M.; Elovskaya, I.S.; Bukin, S.V.; Bukin, Y.S.; Pogodaeva, T.V.; Kwon, D.A.; Zemskaya, T.I. Genomic and morphological characterization of a new *Thiothrix* species from a sulfide hot spring of the Zmeinaya bay (Northern Baikal, Russia). *Antonie Van Leeuwenhoek* **2024**, *117*, 23. [CrossRef]
85. Cytryn, E.; Minz, D.; Gieseke, A.; van Rijn, J. Transient development of filamentous *Thiothrix* species in a marine sulfide oxidizing, denitrifying fluidized bed reactor. *FEMS Microbiol. Lett.* **2006**, *256*, 22–29. [CrossRef] [PubMed]
86. Maestre, J.P.; Rovira, R.; Gamisans, X.; Kinney, K.A.; Kirisits, M.J.; Lafuente, J.; Gabriel, D. Characterization of the bacterial community in a biotrickling filter treating high loads of H_2S by molecular biology tools. *Water Sci. Technol.* **2009**, *59*, 1331–1337. [CrossRef]
87. Wen, D.D.; Yuan, L.J.; Chen, X.; Wang, Y.; Shen, T.T.; Liu, X.B. [Effect of microbial community structure and metabolites on sludge settling ability under three different switching condition processes]. *Huan Jing Ke Xue* **2018**, *39*, 4644–4652. (In Chinese) [CrossRef]

88. Wentzel, M.C.; Comeau, Y.; Ekama, G.A.; van Loosdrecht, M.C.M.; Brdjanovic, D. Enhanced biological phosphorus removal. In *Biological Wastewater Treatment. Principles, Modelling and Design*; Henze, M., van Loosdrecht, M., Ekama, G., Brdjanovic, D., Eds.; IWA Publishing: London, UK, 2008; pp. 155–220.
89. Zuthi, M.F.; Guo, W.S.; Ngo, H.H.; Nghiem, L.D.; Hai, F.I. Enhanced biological phosphorus removal and its modeling for the activated sludge and membrane bioreactor processes. *Bioresour. Technol.* **2013**, *139*, 363–374. [CrossRef] [PubMed]
90. Ahn, K.; Kornberg, A. Polyphosphate kinase from *Escherichia coli*. Purification and demonstration of a phosphoenzyme intermediate. *J. Biol. Chem.* **1990**, *265*, 11734–11739.
91. Zhang, H.; Ishige, K.; Kornberg, A. A polyphosphate kinase (PPK2) widely conserved in bacteria. *Proc. Natl. Acad. Sci. USA* **2002**, *99*, 16678–16683. [CrossRef] [PubMed]
92. Roy, S.; Guanglei, Q.; Zuniga-Montanez, R.; Williams, R.B.; Wuertz, S. Recent advances in understanding the ecophysiology of enhanced biological phosphorus removal. *Curr. Opin. Biotechnol.* **2021**, *67*, 166–174. [CrossRef]
93. Gardner, S.G.; Johns, K.D.; Tanner, R.; McCleary, W.R. The PhoU Protein from *Escherichia coli* interacts with PhoR, PstB, and metals to form a phosphate-signaling complex at the membrane. *J. Bacteriol.* **2014**, *196*, 1741–1752. [CrossRef]
94. Ravin, N.V.; Rudenko, T.S.; Smolyakov, D.D.; Beletsky, A.V.; Rakitin, A.L.; Markov, N.D.; Fomenkov, A.; Sun, L.; Roberts, R.J.; Novikov, A.A.; et al. Comparative genome analysis of the genus *Thiothrix* involving three novel species, *Thiothrix subterranea* sp. nov. Ku-5, *Thiothrix litoralis* sp. nov. AS and "*Candidatus* Thiothrix anitrata" sp. nov. A52, revealed the conservation of the pathways of dissimilatory sulfur metabolism and variations in the genetic inventory for nitrogen metabolism and autotrophic carbon fixation. *Front. Microbiol.* **2021**, *12*, 760289. [CrossRef]
95. Singleton, C.M.; Petriglieri, F.; Wasmund, K.; Nierychlo, M.; Kondrotaite, Z.; Petersen, J.F.; Peces, M.; Dueholm, M.S.; Wagner, M.; Nielsen, P.H. The novel genus, '*Candidatus* Phosphoribacter', previously identified as *Tetrasphaera*, is the dominant polyphosphate accumulating lineage in EBPR wastewater treatment plants worldwide. *ISME J.* **2022**, *16*, 1605–1616. [CrossRef]
96. Tian, Y.; Chen, H.; Chen, L.; Deng, X.; Hu, Z.; Wang, C.; Wei, C.; Qiu, G.; Wuertz, S. Glycine adversely affects enhanced biological phosphorus removal. *Water Res.* **2022**, *209*, 117894. [CrossRef] [PubMed]
97. Nielsen, P.H.; McIlroy, S.J.; Albertsen, M.; Nierychlo, M. Re-evaluating the microbiology of the enhanced biological phosphorus removal process. *Curr. Opin. Biotechnol.* **2019**, *57*, 111–118. [CrossRef]
98. Coats, E.R.; Deyo, B.; Brower, N.; Brinkman, C.K. Effects of anaerobic HRT and VFA loading on the kinetics and stoichiometry of enhanced biological phosphorus removal. *Water Environ. Res.* **2021**, *93*, 1608–1618. [CrossRef]
99. Rey-Martínez, N.; Badia-Fabregat, M.; Guisasola, A.; Baeza, J.A. Glutamate as sole carbon source for enhanced biological phosphorus removal. *Sci. Total Environ.* **2019**, *657*, 1398–1408. [CrossRef]
100. Molenaar, D.; van der Rest, M.E.; Drysch, A.; Yücel, R. Functions of the membrane-associated and cytoplasmic malate dehydrogenases in the citric acid cycle of *Corynebacterium glutamicum*. *J. Bacteriol.* **2000**, *182*, 6884–6891. [CrossRef] [PubMed]
101. van der Rest, M.E.; Frank, C.; Molenaar, D. Functions of the membrane-associated and cytoplasmic malate dehydrogenases in the citric acid cycle of *Escherichia coli*. *J. Bacteriol.* **2000**, *182*, 6892–6899. [CrossRef] [PubMed]
102. Wissenbach, U.; Kröger, A.; Unden, G. The specific functions of menaquinone and demethylmenaquinone in anaerobic respiration with fumarate, dimethylsulfoxide, trimethylamine N-oxide and nitrate by *Escherichia coli*. *Arch. Microbiol.* **1990**, *154*, 60–66. [CrossRef]
103. Wallace, B.J.; Young, I.G. Role of quinones in electron transport to oxygen and nitrate in *Escherichia coli*. Studies with a ubiA- menA- double quinone mutant. *Biochim. Biophys. Acta* **1977**, *461*, 84–100. [CrossRef]
104. Trubitsyn, I.V.; Andreevskikh, Z.G.; Yurevich, L.I.; Belousova, E.V.; Tutukina, M.N.; Merkel, A.Y.; Dubinina, G.A.; Grabovich, M.Y. Capacity for nitrate respiration as a new aspect of metabolism of the filamentous sulfur bacteria of the genus *Thiothrix*. *Microbiology* **2013**, *82*, 15–21. [CrossRef]
105. Trubitsyn, I.V.; Belousova, E.V.; Tutukina, M.N.; Merkel, A.Y.; Dubinina, G.A.; Grabovich, M.Y. Expansion of ability of denitrification within the filamentous colorless sulfur bacteria of the genus *Thiothrix*. *FEMS Microbiol. Lett.* **2014**, *358*, 72–80. [CrossRef]
106. Martinez-Espinosa, R.M.; Dridge, E.J.; Bonete, M.J.; Butt, J.N.; Butler, C.S.; Sargent, F.; Richardson, D.J. Look on the positive side! The orientation, identification and bioenergetics of 'Archaeal' membrane-bound nitrate reductases. *FEMS Microbiol. Lett.* **2007**, *276*, 129–139. [CrossRef] [PubMed]
107. Kretzschmar, U.; Rückert, A.; Jeoung, J.H.; Görisch, H. Malate:quinone oxidoreductase is essential for growth on ethanol or acetate in *Pseudomonas aeruginosa*. *Microbiology* **2002**, *148*, 3839–3847. [CrossRef] [PubMed]
108. Harold, L.K.; Jinich, A.; Hards, K.; Cordeiro, A.; Keighley, L.M.; Cross, A.; McNeil, M.B.; Rhee, K.; Cook, G.M. Deciphering functional redundancy and energetics of malate oxidation in mycobacteria. *J. Biol. Chem.* **2022**, *298*, 101859. [CrossRef]
109. Kumar, R.; Sharma, P.; Chauhan, A.; Singh, N.; Prajapati, V.M.; Singh, S.K. Malate:quinone oxidoreductase knockout makes *Mycobacterium tuberculosis* susceptible to stress and affects its in vivo survival. *Microbes Infect.* **2024**, *26*, 105215. [CrossRef] [PubMed]
110. Cho, S.T.; Chang, H.H.; Egamberdieva, D.; Kamilova, F.; Lugtenberg, B.; Kuo, C.H. Genome analysis of *Pseudomonas fluorescens* PCL1751: A *Rhizobacterium* that controls root diseases and alleviates salt stress for its plant host. *PLoS ONE* **2015**, *10*, e0140231. [CrossRef]

111. Guigliarelli, B.; Asso, M.; More, C.; Augier, V.; Blasco, F.; Pommier, J.; Giordano, G.; Bertrand, P. EPR and redox characterization of iron-sulfur centers in nitrate reductases A and Z from Escherichia coli: Evidence for a high-potential and a low-potential class and their relevance in the electron-transfer mechanism. *Eur. J. Biochem.* **1992**, *207*, 61–68. [CrossRef]
112. Zhang, C.; Guisasola, A.; Baeza, J.A. Exploring the stability of an A-stage-EBPR system for simultaneous biological removal of organic matter and phosphorus. *Chemosphere* **2023**, *313*, 137576. [CrossRef]
113. Wang, H.; Lin, L.; Zhang, L.; Han, P.; Ju, F. Microbiome assembly mechanism and functional potential in enhanced biological phosphorus removal system enriched with *Tetrasphaera*-related polyphosphate accumulating organisms. *Environ. Res.* **2023**, *233*, 116494. [CrossRef]
114. Arumugam, K.; Bessarab, I.; Haryono, M.A.; Liu, X.; Zuniga-Montanez, R.E.; Roy, S.; Qiu, G.; Drautz-Moses, D.I.; Law, Y.Y.; Wuertz, S.; et al. Analysis procedures for assessing recovery of high quality, complete, closed genomes from Nanopore long read metagenome sequencing. *BioRxiv* **2020**. [CrossRef]
115. Ishige, K.; Zhang, H.; Kornberg, A. Polyphosphate kinase (PPK2), a potent, polyphosphate-driven generator of GTP. *Proc. Natl. Acad. Sci. USA* **2002**, *99*, 16684–16688. [CrossRef]
116. Kosgey, K.; Zungu, P.V.; Bux, F.; Kumari, S. Biological nitrogen removal from low carbon wastewater. *Front. Microbiol.* **2022**, *13*, 968812. [CrossRef]
117. Wu, H.; Zhang, Q.; Chen, X.; Zhu, Y.; Yuan, C.; Zhang, C.; Zhao, T. Efficiency and microbial diversity of aeration solid-phase denitrification process bioaugmented with HN-AD bacteria for the treatment of low C/N wastewater. *Environ. Res.* **2021**, *202*, 111786. [CrossRef]
118. Xia, Z.; Wang, Q.; She, Z.; Gao, M.; Zhao, Y.; Guo, L.; Jin, C. Nitrogen removal pathway and dynamics of microbial community with the increase of salinity in simultaneous nitrification and denitrification process. *Sci. Total Environ.* **2019**, *697*, 134047. [CrossRef]
119. Salcedo Moyano, A.J.; Delforno, T.P.; Subtil, E.L. Simultaneous nitrification-denitrification (SND) using a thermoplastic gel as support: Pollutants removal and microbial community in a pilot-scale biofilm membrane bioreactor. *Environ. Technol.* **2021**, *13*, 4411–4425. [CrossRef]
120. Luan, Y.N.; Yin, Y.; An, Y.; Zhang, F.; Wang, X.; Zhao, F.; Xiao, Y.; Liu, C. Investigation of an intermittently-aerated moving bed biofilm reactor in rural wastewater treatment under low dissolved oxygen and C/N condition. *Bioresour. Technol.* **2022**, *358*, 127405. [CrossRef]
121. Tang, K.; Baskaran, V.; Nemati, M. Bacteria of the sulphur cycle: An overview of microbiology, biokinetics and their role in petroleum and mining industries. *Biochem. Eng. J.* **2009**, *44*, 73–94. [CrossRef]
122. Feng, L.; Jia, R.; Zeng, Z.; Yang, G.; Xu, X. Simultaneous nitrification-denitrification and microbial community profile in an oxygen-limiting intermittent aeration SBBR with biodegradable carriers. *Biodegradation* **2018**, *29*, 473–486. [CrossRef]
123. Zhao, J.; Feng, L.; Yang, G.; Dai, J.; Mu, J. Development of simultaneous nitrification-denitrification (SND) in biofilm reactors with partially coupled a novel biodegradable carrier for nitrogen-rich water purification. *Bioresour. Technol.* **2017**, *243*, 800–809. [CrossRef] [PubMed]
124. Dai, H.; Sun, Y.; Wan, D.; Abbasi, H.N.; Guo, Z.; Geng, H.; Wang, X.; Chen, Y. Simultaneous denitrification and phosphorus removal: A review on the functional strains and activated sludge processes. *Sci. Total Environ.* **2022**, *835*, 155409. [CrossRef] [PubMed]
125. Zheng, C.; Wang, C.; Zhang, M.; Wu, Q.; Chen, M.; Ding, C.; He, T. [Denitrifying phosphate accumulating organisms and its mechanism of nitrogen and phosphorus removal]. *Sheng Wu Gong Cheng Xue Bao* **2023**, *39*, 1009–1025. (In Chinese) [CrossRef] [PubMed]
126. Gao, C.D.; Zhang, N.; Han, H.; Ren, H.; Li, Y.; Hou, C.Y.; Wang, C.D.; Peng, Y.Z. [Microbial diversity of filamentous sludge bulking at low temperature]. *Huan Jing Ke Xue* **2020**, *41*, 3373–3383. (In Chinese) [CrossRef] [PubMed]
127. Li, Y.; Zou, J.; Zhang, L.; Sun, J. Aerobic granular sludge for simultaneous accumulation of mineral phosphorus and removal of nitrogen via nitrite in wastewater. *Bioresour. Technol.* **2014**, *154*, 178–184. [CrossRef]
128. Ai, S.; Du, L.; Nie, Z.; Liu, W.; Kang, H.; Wang, F.; Bian, D. Characterization of a novel micro-pressure double-cycle reactor for low temperature municipal wastewater treatment. *Environ. Technol.* **2023**, *44*, 394–406. [CrossRef] [PubMed]
129. Su, Y.; Zhang, Y.; Zhou, X.; Jiang, M. Effect of nitrate concentration on filamentous bulking under low level of dissolved oxygen in an airlift inner circular anoxic-aerobic incorporate reactor. *J. Environ. Sci.* **2013**, *25*, 1736–1744. [CrossRef]
130. Lei, Z.; Xue, J.; Feng, Y.; Li, Y.Y.; Kong, Z.; Chen, R. Sludge granulation in PN/A enhances nitrogen removal from mainstream anaerobically pretreated wastewater. *Sci. Total Environ.* **2023**, *895*, 165048. [CrossRef]

International Journal of *Molecular Sciences*

MDPI

Review

Biodegradation of Poly(ε-caprolactone): Microorganisms, Enzymes, and Mechanisms

Nikolay Krumov [1], Nikolina Atanasova [1], Ivanka Boyadzhieva [1], Kaloyan Petrov [2] and Penka Petrova [1,*]

[1] Institute of Microbiology, Bulgarian Academy of Sciences, 1113 Sofia, Bulgaria; krumov.1993@gmail.com (N.K.); nikolina@microbio.bas.bg (N.A.); petrovaim@abv.bg (I.B.)
[2] Institute of Chemical Engineering, Bulgarian Academy of Sciences, 1113 Sofia, Bulgaria; kkpetrov@iche.bas.bg
* Correspondence: ppetrova@microbio.bas.bg

Abstract: Poly(ε-caprolactone) (PCL) is a synthetic plastic known for its excellent physicochemical properties and a wide range of applications in packaging, coatings, foaming, and agriculture. In medicine, its versatility allows it to function as a scaffold for drug delivery, sutures, implants, tissue engineering, and 3D printing. In addition to its biocompatibility, PCL's most notable characteristic is its biodegradability. However, this property is affected by temperature, microbial activity, and environmental conditions, which means PCL can sometimes remain in nature for long periods. This review shows that various types of microorganisms can efficiently degrade PCL, including different strains of *Pseudomonas* spp., *Streptomyces* spp., *Alcaligenes faecalis*, and fungi like *Aspergillus oryzae*, *Fusarium* spp., *Rhizopus delemar*, and *Thermomyces lanuginosus*. These microorganisms produce enzymes such as lipases, esterases, and cutinases that break down PCL into smaller molecules that act as substrates. The review also examines the phylogenetic diversity of organisms capable of biodegrading PCL, the biochemical pathways involved in this process, and specific aspects of the genetic framework responsible for the expression of the enzymes that facilitate degradation. Targeted research on microbial PCL biodegradation and its practical applications could significantly aid in reducing and managing plastic waste on a global ecological scale.

Keywords: plastic; biodegradation; bacteria; fungi; esterase; lipase; cutinase

1. Introduction

Poly(ε-caprolactone) (PCL) is a synthetic biodegradable polyester known for its excellent physicochemical properties, thus finding wide applications in packaging, coatings, foaming, and agriculture [1,2]. Its chemical structure closely resembles that of polylactic acid (PLA) and polyhydroxyalkanoates (PHAs) [3,4]. However, unlike these biopolymers, its monomer, ε-caprolactone (ε-CL), is not biosynthesized by microorganisms but must be obtained through chemical synthesis [5]. As a petroleum-derived polymer, the production of PCL relies on non-renewable fossil resources, raising environmental concerns such as greenhouse gas emissions and the depletion of fossil carbon during production (Figure 1). Despite these drawbacks, PCL remains a more environmentally friendly alternative to non-biodegradable plastics such as polyethylene, polypropylene, and polyvinyl chloride, significantly contributing to global plastic pollution [6].

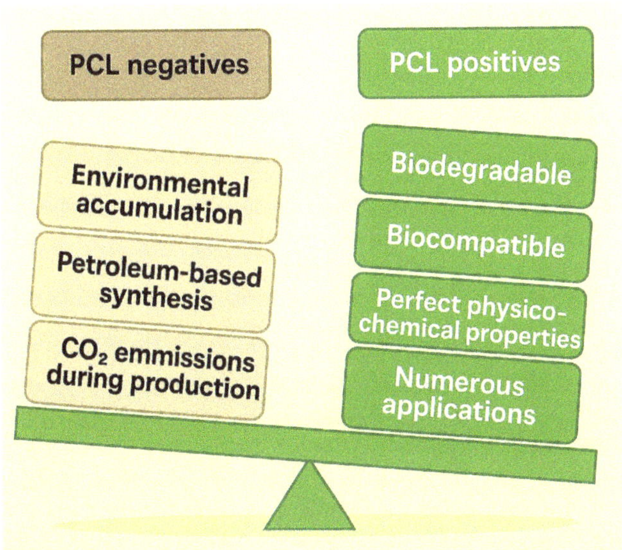

Figure 1. Pros and cons of using PCL.

In 2022, the global market value for PCL reached USD 485.0 million, with projections suggesting it will increase to USD 1.1 billion by 2031, driven by a compound annual growth rate (CAGR) of 9.5% [7]. PCL offers several compelling advantages over other biodegradable polymers. For instance, PLA lacks impact resistance and exhibits poor thermal properties [4], whereas PCL is characterized by its rubber-like texture, high elasticity, and excellent mechanical strength. These properties render PCL highly suitable for applications that require flexibility and structural resilience. Additionally, PCL demonstrates superior solubility in various organic solvents, good compatibility in blends [7], and ease of processing, including techniques such as 3D printing and electrospinning [8,9]. Depending on the processing method, PCL shows slow degradation, which can benefit applications that demand long-term material stability, such as medical implants or agricultural films [10].

One of PCL's most critical environmental attributes is its biodegradability. However, PHA, PCL, and their blends are not always biodegradable or may need conditions only available in industrial-scale composting facilities [9,10]. Efficient polycaprolactone (PCL) breakdown in natural environments relies on specific external factors, including temperature, salinity, microbial activity, and other ecological conditions. When these factors are suboptimal, the polymer's persistent nature can lead to limited biodegradability and microplastic accumulation, posing long-term environmental risks [11,12].

The present review examines the potential of microorganisms as effective biodegraders of PCL, summarizing scientific data on the isolation and enzymatic characterization of various natural isolates from the past decade. The primary search engines used were Web of Science, Scopus, and Google Scholar, with a focus on peer-reviewed studies on PCL biodegradation, enzyme characterization, and biotechnological applications. Approximately two-thirds of the articles were published between 2020 and 2025. About 60% of the articles addressed bacterial strains with PCL-degrading activity, while approximately 30% focused on fungal species. Consequently, the extensive biodiversity of taxonomic groups and their enzymatic activities, genetic basis, and mechanisms of PCL degradation are detailed. Clarifying the specificity of various classes of microorganisms emphasizes some as particularly promising and indicates that PCL is a readily available substrate for them.

2. PCL Synthesis, Structure, Physicochemical Properties, and Applications

2.1. PCL Synthesis Methods

Poly(ε-caprolactone), or (1,7)-polyoxepan-2-one, can be synthesized from oil using various methods [13]. One approach involves the direct condensation of 6-hydroxyhexanoic acid (6-hydroxycaproic acid, 6-HCA), as illustrated in Figure 2 [11,14]. However, the most commonly used technique is the ring-opening polymerization (ROP) of ε-caprolactone (ε-CL) with catalysts (Figure 2). ROP is widely preferred in industrial settings for producing PCL due to its ability to yield polymers with higher molecular weights, narrower molecular weight distributions (lower polydispersity index), and better control over the polymerization process. The ring-opening polymerization (ROP) of ε-caprolactone typically utilizes one of three types of catalysts—anionic, cationic, or coordination–insertion—each producing polymers with distinct structural and molecular properties [15]. Anionic ROP relies on alkali metal salts such as potassium tert-butoxide, potassium isopropoxide, KH, phenyllithium, and lithium diisopropylamide as catalysts, where the ring-opening occurs specifically at the acyl–oxygen bond [13]. The active species in this process are alkoxide anions. A key limitation of anionic ROP is the occurrence of intramolecular transesterification, causing the formation of low-molecular-weight or cyclic polymers unless the reaction is halted beforehand. Nevertheless, this approach enables the controlled synthesis of high-molecular-weight polymers in polar solvents [16].

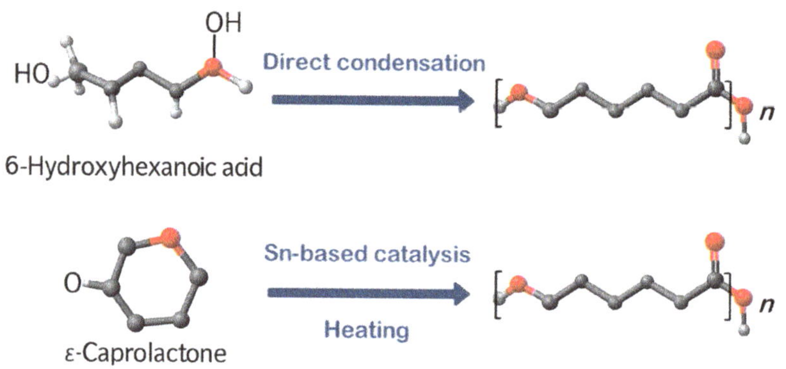

Figure 2. Two methods for synthesizing PCL: either through the direct condensation of 6-hydroxyhexanoic acid or by opening the ε-caprolactone ring in the presence of catalysts. Designation: *n*, variable number of monomers.

In cationic ROP, a positively charged intermediate is formed that is attacked by the monomer's carbonyl oxygen, leading to ring opening via an SN2-like mechanism. This method is often challenging to control and generally results in polymers with low molecular weights or oligomers. Catalysts for this pathway include cationic aluminum complexes, trityl tetrafluoroborate (Ph_3CBF_4), and iron(III) chloride-based systems [17].

The coordination–insertion mechanism, or quasi-anionic ROP, involves coordinating the ε-caprolactone monomer to a metal center via its carbonyl oxygen. This reaction typically employs tin-based catalysts, including stannous (II) ethyl hexanoate, stannous (II) octoate, or stannous triflate [2,13]. The activated monomer then undergoes acyl–oxygen bond cleavage, allowing for the sequential insertion of monomers and the formation of the polymer chain. The end product can be chemically modified or blended with other materials to achieve the desired characteristics.

Sustainable PCL composites can be synthesized by initiating the ring-opening polymerization of ε-caprolactone using the surface hydroxyl groups of exfoliated oxidized biochar derived from hardwood waste biomass. This method enables neat, catalyst-assisted polymerization and produces composites with improved crystallinity, mechanical properties, and degradability compared to pure PCL [18].

Further, the possibility of producing or synthesizing ε-CL from renewable sources has also been investigated. Pyo et al. [5] developed a sustainable synthetic pathway for producing 6-HCA, adipic acid, and ε-CL by integrating biocatalysis and chemical catalysis. Bio-based 6-HCA was obtained from biomass containing glucose, fructose, or sucrose through the intermediate metabolite cyclohexanone, followed by the physicochemical conversion of 6-HCA to ε-CL using an ion exchange catalyst and a molecular sieve [5].

Alternative enzymatic methods include Baeyer–Villiger monooxygenases, such as cyclohexanone monooxygenase from *Acinetobacter calcoaceticus*, which catalyzes the oxidation of cyclohexanone to ε-CL in the presence of oxygen [19]. However, this enzyme is unstable and is inhibited by both the substrate and the product.

Another enzymatic method employs *Candida antarctica* lipase A, which facilitates the direct ring-opening oligomerization of ε-CL with up to 200 mM of cyclohexanol, yielding over 20 g/L of oligo-ε-CL [20]. However, this technique faces economic constraints due to the high cost of petroleum-derived cyclohexanol [5]. Previously, Dong et al. [21] utilized a lipase from Pseudomonas sp. to conduct ring-opening polymerization of lactones, achieving higher molecular weights and better monomer conversion compared to the condensation of the corresponding linear hydroxyesters and the copolymerization of ε-CL with lactide and cyclopentadecanolide [21].

2.2. PCL Structure and Physicochemical Properties

PCL is a linear, unbranched polymer in the polyester family, composed of up to 1000 repeating units of the cyclic ester caprolactone [22]. Its typical aliphatic structure contributes to its hydrophobic nature and solubility in organic solvents [23]. PCL exhibits various physical and mechanical properties depending on its chain length and molecular weight; higher-molecular-weight variants (e.g.,~40,000 Da) are rigid yet flexible. In comparison, lower-molecular-weight types like PCL-300 (10,000 Da) and PCL-150 (5000 Da) tend to be more brittle [24]. Being hydrophobic, the polymer is highly soluble at room temperature in organic solvents such as chloroform, toluene, and benzene. However, it does not dissolve well in acetone and acetonitrile, though solubility improves when heated to 50 °C [25]. It remains insoluble in water, alcohol, and petroleum ether [26].

PCL is a semi-crystalline polyester whose crystallinity decreases as molecular weight increases [27]. Its relatively low glass transition temperature (−60 °C) and melting range (58–64 °C) facilitate easy thermal processing but reduce stability at elevated temperatures—a limitation that can be addressed through blending or cross-linking [28,29]. PCL exhibits excellent thermoformability, making it widely used in 3D printing applications, where heating and melting are common fabrication steps [30]. It also demonstrates shape memory behavior [31,32].

PCL degrades slowly, typically over 2–3 years, and its high toughness is attributed to the presence of amorphous regions in its rubbery phase [13]. Table 1 compares the key physicochemical properties of PCL and other biodegradable polymers.

Table 1. Comparison of key physicochemical properties of PCL and other biodegradable polymers.

Property	PCL	PLA	PHA	Reference
Chemical structure	Aliphatic polyester	Aliphatic polyester	Polyhydroxyalkanoate	[2,4]
Melting temperature (Tm °C)	58–64 °C	130–180 °C	170–180 °C	[6,10]
Glass transition temperature (Tg, °C)	−60 °C	55–60 °C	−10 to 5 °C	[11]
Crystallinity (%)	45–70%	35–40%	30–80%	[4,11]
Degradation rate (in vivo/in vitro)	Slow (months–years)	Moderate (weeks–months)	Fast (depending on microbial strain)	[14]
Biocompatibility	Excellent	Moderate to high	Excellent	[1,2,11]
Solubility in organic solvents	High	Poor	Poor	[25,26]
Processability	Easy (low Tm, flexible)	Requires high temperature	Brittle, limited solubility	[28,29]
Shape memory effect	Present	Absent or weak	Absent	[31,32]
Mechanical behavior	Tough, flexible	Stiff, brittle	Variable, often brittle	[14]

2.3. PCL Applications

Polycaprolactone (PCL) is widely valued across various industries for its distinctive thermoplastic behavior, biocompatibility, and hydrophobic nature. Its applications extend to the biomedical field, cosmetics, and agriculture (Figure 3), where it enhances product functionality, improves delivery mechanisms, and supports sustainable practices.

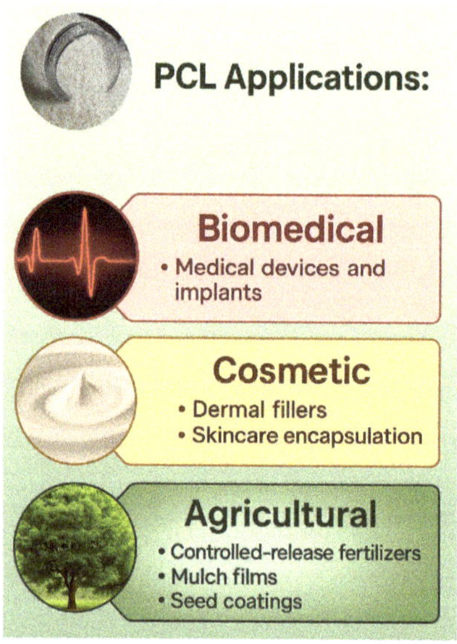

Figure 3. PCL applications.

2.3.1. Biomedical Applications

PCL has become a valuable material in the biomedical field due to its unique combination of biodegradability, ease of processing, and mechanical flexibility [33]. Given its advantageous rheological and viscoelastic behavior, the Food and Drug Administration of the USA (FDA) has approved PCL for use in long-term bioresorbable medical devices and implants, where it serves as a biodegradable alternative to traditional metallic components like screws and plates [34]. Its relatively low melting point enables thermo-responsive behaviors, making it ideal for advanced applications such as self-tightening sutures, biodegradable stents [35], and smart implants that can respond to body temperature and integrate seamlessly into surrounding tissues [36]. These characteristics contribute to developing less invasive surgical procedures and improved healing outcomes. PCL's resistance to water further supports its use in drug delivery systems [7], bone and tissue scaffolds [37], and orthopedic fixation devices [38]. These applications leverage mechanical stability and gradual degradation over extended periods, which aligns with tissue regeneration timelines.

PCL's excellent biocompatibility is a crucial property for various medical applications. When implanted, PCL does not elicit a strong immune or inflammatory response, allowing tissues in direct contact with it to heal wholly and swiftly [36,37]. However, the hydrophobic nature of PCL and its relatively slow degradation can restrict specific applications. As a result, researchers have concentrated on developing PCL-based blends and composites to address these limitations [7,9,12,30]. PCL composites frequently demonstrate improved hydrophilicity, mechanical strength, and tailored degradation rates, enhancing their biomedical potential [28]. For example, PCL/chitosan (CS) blends have demonstrated enhanced hydrophilicity and cytocompatibility. The double-porosity scaffolds created from PCL/CS facilitate cell adhesion and nutrient transport, making them ideal for bone and skin tissue engineering [7]. Lignin–PCL composites have also shown potential in neural tissue regeneration due to their antioxidant properties and ability to support Schwann cell proliferation [26]. Similarly, chitosan/PCL bio-inks have been optimized for lung tissue engineering through 3D printing, exhibiting excellent swelling, degradation, mechanical performance, and superior cell compatibility [37,38]. Blended PCL composites are also widely utilized in wound healing applications [39]. Their ability to support tissue regeneration, absorb exudate, and provide antibacterial properties renders them ideal for wound dressings. PCL/bacterial cellulose composites display high thermal stability and porosity, both advantageous for tissue repair [40]. Other blends, such as PCL/cellulose acetate/dextran with tetracycline, improve antibacterial activity while promoting blood clotting and cell proliferation [13]. Additional studies have indicated that incorporating Ag-containing nanoparticles into PCL scaffolds significantly enhances antimicrobial efficacy [41].

Due to its capacity for forming micro- and nanofibrous structures, PCL is a versatile platform for drug delivery. The large surface-area-to-volume ratio of electrospun PCL scaffolds enables high drug loading and controlled release [42,43].

2.3.2. Cosmetic, Agricultural, and Environmental Applications

PCL has also gained attention in the cosmetics industry. Owing to its biocompatibility and slow resorption profile, it is employed as a dermal filler [44]. Additionally, PCL-based encapsulating agents are utilized in skincare formulations to enhance active ingredients' targeted and sustained release, thereby improving product performance [45].

PCL's agricultural applications utilize encapsulated controlled-release fertilizers that reduce nutrient loss and enhance crop efficiency. Furthermore, PCL is employed in the development of mulch films [30,45], seed coatings [46], and eco-composites [47], all of which help minimize plastic waste and improve soil health.

In addition to biomedical and agricultural applications, PCL-based blends serve as components in biosensors and optical sensors [13]. Furthermore, modified pectin/PCL blends are being researched as flexible, compostable films for biodegradable packaging, offering an eco-friendly alternative to traditional plastics [48].

3. Biodiversity of Microorganisms Capable of PCL Biodegradation

Many microorganisms have been described in the scientific literature as capable of biodegrading PCL. Nearly all known species documented to perform such biochemical processes are either bacteria (Figure 4) or fungi (Figure 5), as archaea are rare [49].

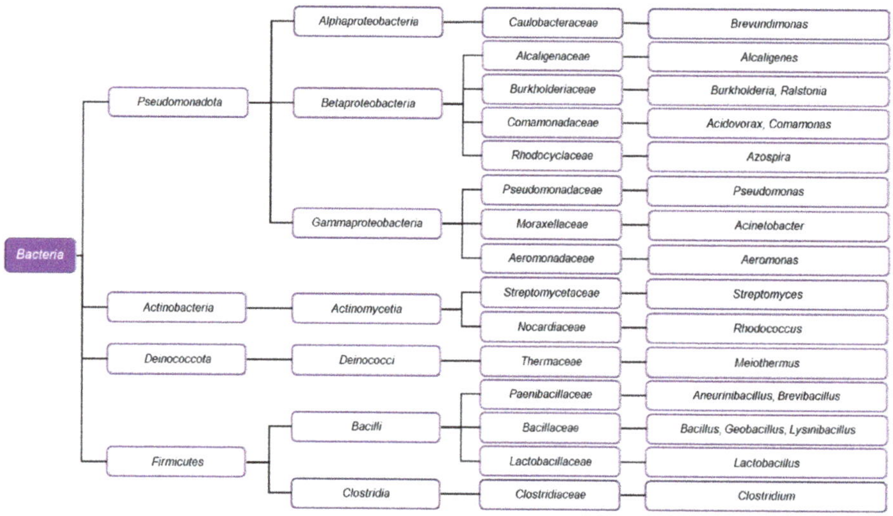

Figure 4. The phylogenetic diagram of bacterial genera capable of PCL degradation.

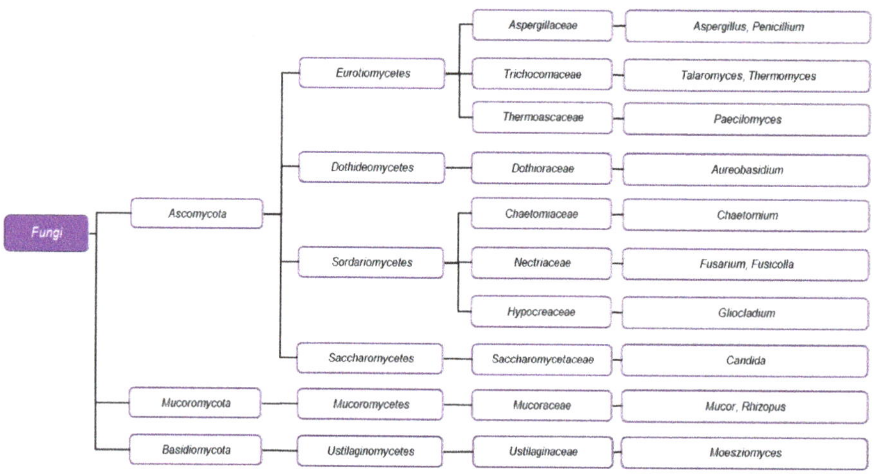

Figure 5. The phylogenetic scheme of fungal genera that are capable of PCL degradation.

3.1. PCL-Degrading Bacteria

Various classes within the *Bacteria* domain are recognized for their ability to degrade PCL, primarily represented by the phyla *Pseudomonadota*, *Actinobacteria*, *Deinococcota*, and *Firmicutes*.

3.1.1. *Pseudomonadota* (Formerly *Proteobacteria*)

At least three major lineages of the formerly known phylum *Proteobacteria* can degrade PCL: *Alphaproteobacteria*, *Betaproteobacteria*, and *Gammaproteobacteria*. The *Alphaproteobacteria* are notable for *Brevundimonas* sp., a soil-isolated bacterium recognized for producing a powerful PCL depolymerase [49,50].

Among *Betaproteobacteria*, PCL-degrading species are found in the genera *Acidovorax*, *Alcaligenes*, *Burkholderia*, *Comamonas*, *Diaphorobacter*, and *Ralstonia* (order *Burkholderiales*), and in *Azospira* (order *Rhodocyclales*) (Figure 4). *Alcaligenes faecalis* has been found to hyperproduce the enzyme PCL depolymerase when cultured in a medium containing PCL [51]. *Burkholderia cepacia* has been shown to express a lipase utilized in the bioconversion of PCL within a two-phase system [52]. The genus *Ralstonia*, initially classified as *Pseudomonas*, is a close relative of *Burkholderia* and belongs to the same family, *Burkholderiaceae* [53]. Samples from a hot spring in Pakistan contained a thermophilic strain of *Ralstonia sp.* that effectively biodegrades PCL as a biofilm [54]. The *Comamonadaceae* family is notable for its genera *Acidovorax*, *Comamonas*, and *Diaphorobacter*, which include representatives that degrade PCL. Together with *Azospira*, another Betaproteobacterium, and the unrelated *Ignavibacterium* (*Fibrobacteres-Chlorobi-Bacteroidetes* superphylum) and *Frateuria* (*Gammaproteobacteria*), these microorganisms have been employed in a specialized aquaculture system, where they carry out a heterotrophic denitrification process using PCL as the sole carbon source [55].

Among *Gammaproteobacteria* degrading PLC are the representatives of the genera *Pseudomonas* and *Acinetobacter*, which belong to the order *Pseudomonadales*. The genus *Aeromonas* has been successfully employed in nitrification and denitrification processes, where PCL serves as a carbon source for growth, suggesting its potential application in wastewater treatment systems [56]. The effects of the *Pseudomonas* lipase enzyme have also been investigated. Scanning electron microscope (SEM) images of PCL–polyglycolide nanocomposites treated with this enzyme indicate visible degradation in the material's structure [57]. In another study, a lipase from *Ps. chlororaphis* and *Streptomyces* strains grown on waste cooking oil was effective on PCL and polyhydroxyalkanoates [58]. *Ps. hydrolytica* has been found to express two different lipases, which have been successfully purified and characterized [59]. Compost isolates of *Pseudomonas pseudoalcaligenes*, currently considered a synonym of *P. oleovorans*, express cutinase when cultivated in a medium containing plant waste [60]. The genus *Acinetobacter*, a relative of *Pseudomonas*, is notable for a specific variety with biodegradative potential. Early studies with mixed cultures of *Acinetobacter calcoaceticus* var. *lwoffii* and the yeast *Cryptococcus laurentii* (syn. *Papiliotrema laurentii*) provide evidence that even polymeric chains with very high molecular weight are subjected to hydrolysis by the metabolic activity of the microorganisms [61]. Another member of the genus, *Ac. seifertii*, also demonstrates promising properties. Soil isolates, later identified as belonging to the species, exhibit a significant ability of their PCL depolymerase enzyme to degrade PCL effectively [62]. Furthermore, *Acinetobacter* has provided genes for 6-oxohexanoic acid (6-OHA) dehydrogenase, which have been utilized to develop a recombinant *Escherichia coli* strain with potential applications for eco-friendly poly-ε-caprolactone (PCL) upcycling [63].

The gammaproteobacterium *Aeromonas* attracts attention due to its ability to form biofilms on the surface of PCL and other plastic materials. A strain identified as *Aeromonas media* has been investigated in conjunction with a strain of *Rhodococcus quingshengii*. Both species showed the ability to degrade PCL with biofilm-forming properties and appeared resistant to a biocide introduced in cultured samples [64].

3.1.2. Actinobacteria

In addition to *Proteobacteria*, another major phylum—*Actinobacteria*—also includes genera with similar metabolic capabilities. A prominent example is the genus *Streptomyces*, part of the order *Streptomycetales*. A thermophilic isolate of *Str. thermoviolaceus* subsp. *thermoviolaceus*, recovered from Taiwanese soils, degrades PCL via extracellular depolymerase, as shown by agar plate assays [65]. Researchers isolated two depolymerases from this microorganism, one displaying chitinase activity. The second key actinomycete genus is *Rhodococcus*, which belongs to the *Nocardiaceae* order. Two species, *R. erythropolis* and *R. opacus*, have provided essential insights into plastic degradation research. The genomes of these organisms have been sequenced and analyzed for the presence of genetic sequences responsible for PCL biodegradation, and several genetic determinants responsible for the process have been reported [66].

3.1.3. *Firmicutes* and *Deinococcota*

Most members of the phylum *Firmicutes* that degrade PCL are limited to the class *Bacilli* and include the genera *Aneurinibacillus*, *Bacillus*, *Brevibacillus*, *Geobacillus*, *Lactobacillus*, and *Lysinibacillus*, with the genus *Clostridium* being the only exception. *Aneurinibacillus thermoaerophilus* and *Brevibacillus thermoruber*, both belonging to the family *Paenibacillaceae*, as well as the unrelated *Meiothermus cateniformans* (*Deinococcota*), are thermophilic bacterial species isolated from a plastic-contaminated hot spring in Bulgaria [67]. Through a series of morphological, microscopic, and enzymatic analyses, all three have demonstrated the ability to degrade PCL, with a strain of *Br. thermoruber* exhibiting the highest esterase activity in the collection [67]. *Bacillus* sp. strain NR4 degrades PCL alongside poly(butylene adipate-co-terephthalate), indicating that the genus may have a broader range of applications [68].

Another study shows that *B. pumilus* can also biodegrade PCL, which occurs naturally in the mangrove rhizosphere [69]. Moreover, a strain of *Bacillus* sp. has been utilized in developing bacterial fertilizers, where PCL serves as a material for creating an eco-friendly, biodegradable capsule [70]. *Geobacillus*, which belongs to the same *Bacillaceae* family, has been utilized as a source of lipase/esterase enzymes successfully engineered to degrade PCL with enhanced efficiency [71]. *Lysinibacillus* sp. also degraded PCL, as demonstrated by gel permeation chromatography [72].

PLC-degrading lactobacilli belong to the species *Levilactobacillus brevis* and *Lactiplantibacillus plantarum*. Three different lipases have been extracted from these species, as the lipase from *Lpl. plantarum* showed the highest activity [73]. Further investigation of this enzyme using an enzyme-embedded assay revealed significant physical changes to the plastic material resulting from the degradation process [74].

Clostridium is the most phylogenetically distinct example among the biodegraders within *Firmicutes*. However, at least two studies provide evidence that members of this genus exhibit this capability. Two strains of *Clostridium* sp. have been isolated from anaerobic sludge and evaluated for their biodegradative properties on PCL and two other plastic materials (polyhydroxybutyrate and the copolymer polyhydroxybutyrate-co-hydroxyvalerate). Following 16S rDNA analysis, the strains were identified as belonging to the species *C. acetobutylicum* [75]. In another study, *Clostridium* sp. isolated from aquaculture samples utilized PCL as a carbon source. The relatively high abundance of clostridia identified among the diverse array of microorganisms in the samples suggests that the genus may be essential in the biodegrading consortium [76].

3.1.4. Putative PCL Degradation by Other Bacterial Species

Research on PCL biodegradation by bacterial species from phyla other than *Proteobacteria*, *Actinobacteria*, and *Firmicutes* is limited in the scientific literature. However, the taxonomic diversity of these microorganisms may be greater than currently recognized, considering that enzymes such as lipases and esterases, which are crucial to the process, are nearly ubiquitous among bacteria [77]. Members of their enzymatic family readily act upon PCL. Additionally, polyester, a plastic material chemically related to PCL, has been found to affect soil microbial communities by increasing the presence of bacteria from the phyla *Acidobacteria*, *Verrucomicrobia*, *Bacteroidetes*, *Nitrospirae*, and *Patescibacteria*, suggesting that these microorganisms may play a role in plastic degradation. It is essential to conduct in vitro cultivation of species from these taxa, along with experiments on the impact of PCL on their growth, which is critical for enhancing our understanding of the potential of bacteria to improve the exploitation of biodegradable plastics.

3.2. PCL-Degrading Fungi

PCL-degrading fungi show less taxonomic diversity in representative phyla than bacteria (Figure 5). Except for the basidiomycete *Moesziomyces antarcticus* and two members of *Mucoromycota*, *Mucor* and *Rhizopus*, the other studied genera—*Aspergillus*, *Aureobasidium*, *Gliocladium*, *Paecilomyces*, *Penicillium*, *Chaetomium*, *Talaromyces*, *Thermomyces*, *Fusarium*, *Fusicolla*, and *Candida*—are classified within *Ascomycota*. Both *Aspergillus* and *Penicillium* belong to the *Eurotiomycetes* class, and their biodegradative potential was recognized quite early, as demonstrated by the SEM of PCL, which was subjected to the action of *Asp. flavus* and *Pen. funiculosum* [78].

Benedict et al. [79] conducted plastic degradation experiments involving *Asp. niger*, *Asp. fumigatus*, *Chaetomium globosum*, and *Fusarium* sp., demonstrating that all tested fungal strains produce low-molecular-weight material from PCL [79]. These fungi also degrade a blend of PCL and polyethylene as a consortium with two other species, *Gliocladium virens* and *Aureobasidium pullulans* [80]. Recent studies have highlighted *Chaetomium globosum*, which can fully degrade PCL in specific samples within 28 days [81].

Considering the final products of degradation, a thermotolerant strain of *Asp. fumigatus* has been discovered to produce a mixture of carboxylic acids from PCL film [82], similar to the purified lipase of *Asp. oryzae* [83]. Another *Aspergillus* strain sourced from a polluted seacoast in Korea has been studied for its ability to degrade PCL, with the authors noting that it, along with another ascomycete genus, *Talaromyces*, demonstrates higher activity than *Penicillium* [84]. Researchers have assessed the ability of *Aspergillus fumigatus* to degrade PCL in both soil and compost environments. These experiments also examined the properties of *Thermomyces lanuginosus*, another member of the *Eurotiomycetes* class, identifying it as a quick PCL microbial degrader [85]. It is noteworthy that although *Therm. lanuginosus* appears less frequently in plastic-related research, the enzymatic capabilities of this fungus warrant equal attention, given that a lipase from this species has been successfully utilized in vitro for the degradation of cross-linked PCL implants [86]. *Paecilomyces lilacinus*, also a representative of *Eurotiomycetes*, shows biodegradative capacity for both PCL and poly(3-hydroxybutyrate), as evidenced by a single study involving five different strains of filamentous fungi [87]. In contrast, the abilities of *Penicillium* species seem to be more evident in research on plastic degradation. Besides *Pen. funiculosum*, at least three other species from this genus have been studied for their effects on PCL, one being *Pen. oxalicum*. This fungus has been subjected to UV radiation mutagenesis, resulting in an effective biodegrading strain [88].

Penicillium chrysogenum, isolated from Russian soils, has been shown to degrade PCL, PLA, and polypropylene [89]. The most recently discovered fungi displaying similar activity are *Penicillium samsonianum* and *Fusicolla acetilerea*, both isolated from terrestrial samples in Korea [90]. *Fusarium*, a widely distributed soil fungus, is another genus of key importance for PCL biodegradation. In addition to being part of pioneering reports on fungal microorganisms demonstrating this ability, it has significantly contributed to understanding their enzymatic properties. *Fusarium* is a well-known plant pathogen, and as such, it relies on the activity of cutinase to lyse cutin [91], the outermost layer of many plant tissues. An analysis comparing a wild-type *F. moniliforme* and a knockout mutant of *F. solani* revealed that the cutinase functions as PCL-depolymerase [92]. Shortly after, a second depolymerase was isolated from *F. solani* and identified as a lipase distinct from the previously identified catalyst, indicating that biodegradation can be approached with different enzymatic strategies [93]. Moreover, researchers have successfully expressed a cutinase from *Fusarium solani* in the yeast *Yarrowia lipolytica* [94].

Another essential study targeting cutinase as a substrate showed that an extracellular protease can inactivate it [95]. More recent research has developed an enzymatic model for the hydrolysis of PCL through the activity of cutinase from *F. solani* and compared it to an accompanying model on the activity of a lipase from *Candida antarctica* [96], now reclassified as *Moesziomyces antarcticus*, thereby becoming the only currently known basidiomycete capable of biodegradation. The biodegradation process carried out by *M. antarcticus* has also been studied, with emphasis on the molecular weight of film-based PCL. While molecular weight does not significantly impact overall biodegradability, it does influence the size and depth of the hydrolytic pores and the time needed for the reaction to proceed [97].

Interestingly, in addition to its ability to biodegrade PCL, *M. antarcticus* has also demonstrated the ability to synthesize it, and the effects of two significant parameters on the process—temperature and organic media—have been evaluated [98].

Members of *Mucoromycotina*, the subphylum of *Mucoromycota*—including the common bread molds *Mucor* and *Rhizopus*—have also been reported to degrade PCL [99].

4. Mechanism of Microbial PCL Biodegradation

The microbial degradation of PCL proceeds through a sequence of three stages, each characterized by specific biochemical and physiological processes that convert the polymer into metabolic waste products while storing energy. These stages are as follows: (i) surface hydrolysis, where key enzymes cleave the bonds in the polymer backbone; (ii) absorption and digestion, involving the uptake and further assimilation of monomers and oligomers generated in the previous stage; and (iii) mineralization and excretion, during which the microorganism produces and releases small molecules into its environment.

4.1. Overview of Enzymes Involved in the Process of PCL Degradation

The key to microbial hydrolysis of PCL lies in the secretion of extracellular hydrolytic enzymes. Several lipases (EC 3.1.1.3) and some carboxylesterases (EC 3.1.1.1) are effective on this aliphatic polyester. Both groups of enzymes target the ester bonds in biological molecules, and depending on the specific enzyme, they participate in a broad range of biochemical processes. The precise distinction between lipases and esterases uses various criteria, including kinetic properties, structural features, substrate specificity, and hydrolysis mechanisms. Table 2 summarizes a widely accepted classification system for lipolytic enzymes, which organizes them into eight classes based on different structural and biochemical features.

4.2. Enzymes Involved in the Process of PCL Degradation by Bacteria

All esterases and lipases in bacteria share a characteristic α/β-hydrolase fold in their active sites, defining the catalytic strategy through which these enzymes carry out their functions. Most α/β-hydrolase fold proteins consist of six α-helices surrounding a central β-sheet composed of eight β-strands, nearly all of which are oriented parallel, with the only exception being strand $\beta2$, which runs antiparallel to the others. The β-sheet presents a distinctive half-barrel structure due to spatial twisting that positions the first and last strands at roughly a 90° angle to one another. Two of the α-helices flank one side of the structure, while the other four are located on the opposite side. Access to the active site for substrates is regulated by a set of secondary structures that form various lids, caps, or flaps.

These features are particularly important for the lipase class (Class I), where a lid domain is present and plays a key role in interfacial activation.

Table 2. Classification system of esterases and lipases according to Arpigny and Jaeger [100], Holmquist [101], Denessiouk et al. [102], and Cea-Rama et al. [103].

Class	Type	Substrate Preference	Notable Features
I	Lipases	Long-chain triglycerides	α/β-Hydrolase fold, Interfacial activation, Lid domain
II	Esterases	Broad	Divergent, α/β/α sandwich; SGNH conserved sequence, GDSL motif near N-terminus (most)
III	Esterases	Short-chain esters	PAF-AH-like, α/β-hydrolase fold
IV	Esterases	Short-chain esters	HSL-like, α/β-hydrolase fold
V	Esterases	Variable	Dehalogenase and haloperoxidase-like enzymes in extremophiles
VI	Cutinases	Esters	Plant cuticle degradation
VII	Esterases	Acetylated compounds	Acetylcholinesterase-specific activity
VIII	Esterases	Broad	Distinct β-lactamase-like fold, untypical catalytic triad

In contrast to most other classes, Class II esterases (SGNH, or GDSL) lack the traditional α/β-hydrolase fold and instead exhibit a highly divergent α/β/α sandwich architecture (Figure 6). A notable difference is the β-sheet, which consists of five parallel strands (flavodoxin-like) in SGNH enzymes, contrasting with the eight-stranded structure typical of α/β-hydrolases. The oxyanion hole and the catalytic triad are present in GDSL esterases, operating through a similar mechanism. However, the amino acid sequence is generally conserved as Asp-His-Ser. GDSL esterases display distinct Nuc-Oxy and Acid–Base zones, forming an extensive network of hydrogen bonds in the active site that provides additional stability during substrate transition. The so-called SHLink, a two-bond interaction between residues adjacent to the catalytic nucleophile and base, connects these zones.

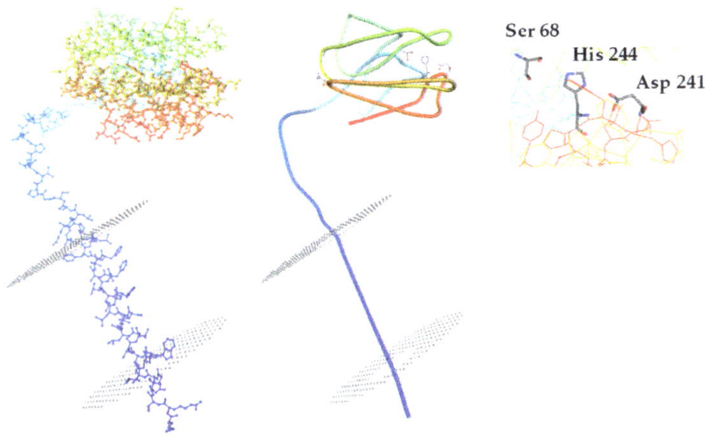

(a) (b) (c)

Figure 6. A 3D molecular model of the GDSL-type esterase/lipase family protein from *Brevibacillus thermoruber* strain 7, NCBI Reference Sequence WP_029099798.1. The sequence is deduced from genome mining of the whole genome shotgun sequence of *B. thermoruber* 423. The gene responsible, located in Scaffold 5, encodes a protein of 262 amino acids, which corresponds to the 28 kDa protein with PLA-degrading activity isolated by Atanasova et al. [67] from *Br. thermoruber* strain 7. (**a**) Ball and stick model presentation; (**b**) tube model; (**c**) catalytic triad.

Class VIII esterases represent the most distinct category of bacterial lipolytic enzymes. Unlike the α/β-hydrolase fold, these enzymes have a β-lactamase-like fold, named for its similarities to β-lactamases. These esterases feature a two-domain architecture. The large domain consists of a β-sheet surrounded by α-helices. The β-sheet significantly differs from the α/β-hydrolase fold, as all its strands are arranged antiparallelly. The small domain contains four α-helices and two β-sheets with antiparallel strands. The binding site is located between the two domains. Another notable difference is the unusual catalytic triad, Ser–Lys–Tyr, where Ser acts as the acid component while Tyr is proposed to function as the base. Table 3 presents the main properties of bacterial enzymes exhibiting activity towards PCL.

Table 3. Enzymes that bacteria use to hydrolyze PCL.

Species/Strain	Source	Enzyme	Properties	Efficiency	Reference
Brevibacillus thermoruber strain 7	Marikostinovo Hot Spring, Bulgaria	Lipase	28 kDa, T range 45–65 °C, T_{opt} 55 °C, pH range 6–9, pH_{opt} 7–8, ↑Ca^{2+}, ↓Mg^{2+}, Co^{2+}, K^+, Na^{2+}, Cu^{2+}, Mn^{2+}, Hg^{2+}, Zn^{2+}, Fe^{3+}	Clear halos on 0.1% PCL-containing agar; deep surface damages on PCL after 1 week of incubation	[67]
Brevundimonas sp. MRL-AN1	Soil sample, Pakistan	PCL-depolymerase	63.49 kDa, T range 30–37 °C, pH range 6.0–8.0, ↓Fe^{2+} and Zn^{2+}	80% of PCL film degraded in 10 days; prefers C_6-C_{10} *p*-nitrophenyl acyl esters	[51]
Alcaligenes faecalis B273	Soil and activated sludge	PCL-depolymerase	T_{opt} 40 °C, pH_{opt} 7.0, K_m PCL = 0.29 mg/mL	Prefers C_{10} *p*-nitrophenyl acyl esters and tributyrin; does not cleave PHB	[52]
Burkholderia cepacia ST8	Soil and water, Malaysia	Lipase	↑Ca^{2+} ↓Cu^{2+} and Co^{2+}	179 U/mL in medium with Tween 80	[53]
Ralstonia sp. MRL-TL	Hot spring	Esterase, serine hydrolase family	50 kDa, T_{opt} 50 °C, pH_{opt} 7.0, preferred substrate *p*-NP-caproate	50% of PCL film degraded in 10 days; it also degrades PLA, PES, PHB, and PHBV	[54]
Pseudomonas pachastrellae	Coastal seawater, Okinoshima Park, Japan	Cutinase	30 kDa, optimal at 0.51 M NaCl, operates via surface hydrolysis	Solid-state PCL-hydrolytic activity	[104]
Pseudomonas sp.	Sigma-Aldrich (Merck KGaA)	Lipase	T_{opt} 37° C in 0.05 M phosphate buffer solution (PBS) with pH 7	Visible degradation of the PCL matrix by the end of the first week	[58]
Pseudomonas hydrolytica	Lab-maintained strain, originally isolated from forest soil in China	PCLase I and PCLase II	PCLase I: 29,64 kDa; T 50 °C, pH 9; highly stable at pH 12 with 100% activity; ↑Mg^{2+}, Ca^{2+}, Fe^{3+}, and Fe^{2+}; ↓Cu^{2+} and Co^{2+} PCLase II: 33,2 kDa; T 40 °C; pH 10; ↑Mg^{2+} and Ca^{2+} at 1 Mm, ↑Co^{2+} and Cu^{2+}, ↓Mg^{2+} and Ca^{2+} at 10 mM, ↓Fe^{2+}	PCL I: degrades PCL (70% weight loss in 3 days), PBS, *p*NP ester, tributyrin, olive oil, and cutin PCL II: degrades PCL (75% weight loss in 8 days), PHB, PBS, *p*NP ester, tributyrin, and olive oil	[59]
Pseudomonas pseudoalcaligenes	Mixed-plant compost	Cutinase	32 kDa, consists of 302 amino acids, cutin-induced	Variable PCL degradation; clearing zones up to 65 mm	[61]

Table 3. *Cont.*

Species/Strain	Source	Enzyme	Properties	Efficiency	Reference
Acinetobacter seifertii	Soil samples	Esterase with PCL-depolymerase activity	30–40 °C	-	[62]
Streptomyces thermoviolaceus ssp. *thermoviolaceus*	Soil, Taiwan	Chitinase with PCL-depolymerase activity, PCL depolymerase	Chitinase with PCL-depolymerase activity: 35 kDa PCL depolymerase: 55 kDa	Both form bands in a polyacrylamide gel containing 0.1% PCL after incubation at 45 °C for 30 min	[66]
Geobacillus sp.	Lab-maintained strain, originally isolated from a Lithuanian oil well	Lipase/esterase, used for the development of engineered polyesterases: GD-95 RM and GDEst-lip in *E. coli*	GD-95 RM: LA 1400 U/mg, stable up to 85 °C GDEst-lip: 98 kDa, LA 600 U/mg; active between 5 and 90 °C, pH 6–12, both resistant to organic solvents, T_{opt} 50 °C	GD-95 RM: 264.0 mg and 280.7 mg of $PCL_{45,000}$ and $PCL_{80,000}$, 24 h at 30 °C GDEst-lip: 145.5 mg $PCL_{45,000}$ and 134.0 mg $PCL_{80,000}$, 24 h	[71]
Levilactobacillus brevis	Commercially obtained strain from IMTECH	Lipase	26 kDa, assayed at 37 °C, pH 7	10-day incubation; ~2 wt.% mass loss for 1 mg/mL lipase, increases to ~10 wt.% for 4–5 mg/mL	[73]
Lactiplantibacillus plantarum	Commercially obtained strain from IMTECH	Lipase	66 kDa, assayed at T 37 °C, pH 7; also at pH 8.1 + Tween 20 in an embedded approach	10-day incubation; ~10 wt % mass loss for 1 mg/mL lipase, increases to ~60 wt.% for 5 mg/mL; 73% mass loss in 8 days for 8% lipase-embedded PCL, crystallinity increases from 39% to 95%	[74]

A common characteristic of all lipases is that they typically target water-insoluble molecules and require a hydrophobic interface (Figure 7). Hydrophobic interactions are significant, given that lipases possess a conserved hydrophobic region of amino acids in their active site (Figure 7c,d). Well-known lipase substrates include triglycerides, cholesterol, and specific vitamins, underscoring their essential role in fat metabolism.

In contrast, esterases generally act on smaller molecules that have higher water solubility [105,106]. Consequently, they often play a significant role in nucleic acid digestion, neurotransmitter inactivation, and the metabolism of xenobiotics, including drugs and toxic organic compounds. Regardless of the classification system, enzymes from the esterase and lipase classes have shown the ability to catalyze the breakdown of the polymeric PCL molecule into smaller products. This process begins when the secreted enzyme adsorbs to the material's surface, followed by the formation of a temporary transition complex with the substrate. In the subsequent step, chain scission occurs, and the protein interacts with other molecular segments [107]. Experiments with polyethylene terephthalate, another member of the polyester family, indicate that biodegrading enzymes have a strong tendency for surface adsorption, making attempts to deliberately remove them from the plastic even more challenging [108].

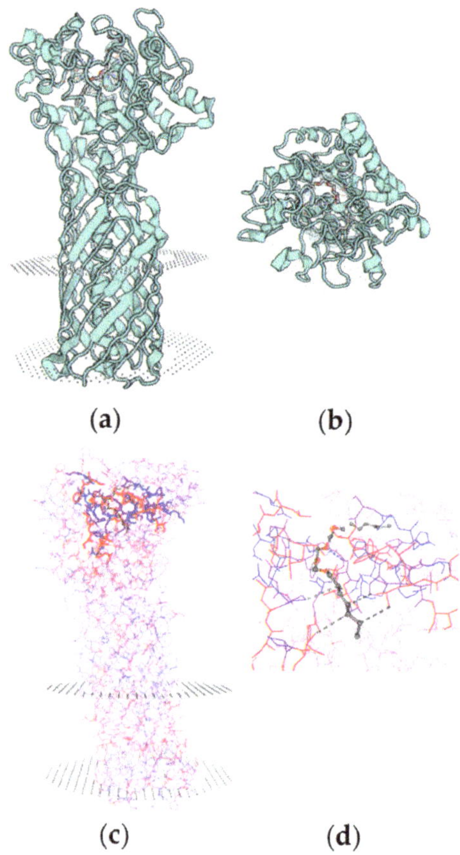

Figure 7. (**a**) Molecular model (3D) of the full-length carboxyl esterase EstA (E.C. 3.1.1.1) from *Pseudomonas aeruginosa* ATCC 15692, illustrating the protein's contact with the outer cell membrane; (**b**) 3D view of the enzyme after a 90° horizontal rotation; (**c**) amino acid residues located in the hydrophobic pocket (illustrated in blue and red); (**d**) hydrophobic interactions with a model substrate (hydroxyethyl-oxy) tri(ethyl-oxy)octane). The SWISS-MODEL tool 'Protein-Ligand Interaction Profiler' predicts that eight amino acids are involved in hydrophobic interactions: L25, P198, I242, L293, and P300; S24 and N157, which form hydrogen bonds; and G102, which forms water bridges. The SWISS-MODEL Workspace https://swissmodel.expasy.org/ (accessed on 11 May 2025) was used to create the model and generate the prediction.

4.3. Enzymes Involved in the Process of PCL Degradation by Fungi

The enzymes responsible for PCL degradation in fungi are lipases and cutinases. In addition to being subjects of scientific research, some are produced industrially (Table 4). One example is the lipase from *Candida rugosa*, although its PCL degradation level has been reported as low [89]. Similar results have been observed with the lipase produced by *Rhizopus del*emar. In contrast, the lipase from *Mucor miehei* has shown a sufficiently high degradation yield of the plastic, indicating that it may hold considerable significance for the practical application of the enzyme in this process [99].

Table 4. Enzymes involved in PCL hydrolysis by filamentous fungi.

Organism	Source	Enzyme	Properties	Efficiency	Reference
Aspergillus oryzae	Mucos Pharma [1]	Lipase	Highest activity at Topt 37 °C, pH 7; retains stability at 55 °C	10% decrease in PCL Mw for 45 days of incubation	[73]
Thermomyces lanuginosus	Sigma Aldrich [1]	Lipase	Topt: 37 °C, initial pH: 6.02, decreases during reaction	Mass loss over 5 days: ~54.8% at 0.1 mol% and ~17.3% at 3.0 mol% cross-linking. Corresponding degradation rates: 14.5%/day and 3.5%/day	[75,76]
Paecilomyces lilacinus	Soil, activated sludge	PCL depolymerase	T 30 °C; pH 3.5–4.5; expression inhibited by starch, glucose, and lactose	10% PCL degraded in 10 days	[86]
Fusarium moniliforme	University of Connecticut Culture Collection	Cutinase with PCL depolymerase activity	24 kDa, pH 9–10, induced by 16-hydroxy hexadecanoic acid	Zones of clearing on MM-PCL agar plates after 48 h	[82]
Fusarium solani f. sp. pisi 77–102	Institut fuÈr Genbiologische Forschung	Lipase/Cutinase [2]	Lipase: pH 7.8, assayed at T 30 °C Cutinase: T range 20–70 °C, T_{opt} 30–40 °C, pH range 3–11, pH_{opt} 6–9	Lipase: clear MM/PCL/agar zones with Tween 80 or tributyrin; lipase/cutinase coexpression: 0.5 g PCL degraded in 144 h	[83,84]
Candida rugosa	Sigma-Aldrich [1]	Lipase	T 40 °C, assayed at pH 7.7	PCL drops from 86,909 g/mol to 80,873 g/mol, and Mw remains at 1.48	[89]
Mucor miehei	Sigma-Aldrich [1]	Lipase	T 40 °C, retains activity up to 60 °C, assayed at pH 7.7	74% PCL hydrolyzed 24 h, Mw of PCL drops from 86,909 g/mol to 24,011 g/mol, Mn/Mw increases to 2.39	[89]
Rhizopus delemar	Fluka [1]	Lipase	T 40 °C, assayed at pH 7.7	PCL drops from 86,909 g/mol to 64,137 g/mol	[89]
Moesziomyces antarcticus	Beijing Cliscent Technology; Novozymes China Biotechnology [1]	Lipase	33 kDa, 317 amino acids, T 45 °C, pH 7.2	87% PCL weight loss in 72 h via two-phase degradation: rapid (85%, 0–12 h) and slow (86.9%, 12–20 h), independent of PCL Mw	[86,87]

[1] Commercially obtained enzyme; [2] lipase with PCL depolymerase activity; cutinase, coexpressed in *Y. lipolytica*.

Cutinases (E.C. 3.1.1.74) are serine esterases originally associated with fungal invasion of plant cells by phytopathogenic fungi. In addition to their natural substrates, cutinases show activity toward a broad range of substrates, including PCL. Their structure is highly similar among *Fusarium* and *Aspergillus* species, and Figure 8 displays the conserved regions in the enzyme molecules.

4.4. Catalytic Mechanism of Enzymatic PCL Degradation

At least two experimentally supported models of the catalytic mechanism responsible for PCL hydrolysis propose a broadly comparable concept, though with some notable differences. The first model is based on the activity of an esterase from the thermophilic archaeon *Archaeoglobus fulgidus* (AfEST). This highly thermostable enzyme operates optimally at 80 °C and can synthesize polyester.

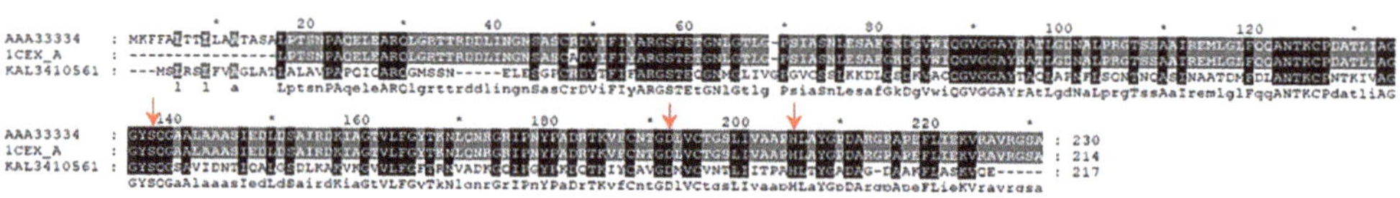

Figure 8. Comparison of amino acids in fungal cutinases (EC 3.1.1.74) using the free software GeneDoc v2.7. NCBI GenBank accession numbers and species: AAA33334, *Fusarium solani*; 1CEX_A, *F. vanettenii*; KAL3410561, *Asp. fumigatus*. Black shading indicates the conservation mode, red arrows highlight the amino acids involved in the active center, and asterisks indicate intermediate numbering of amino acids.

Three key amino acid residues, known as the catalytic triad—specifically Nucleophile–His–Acid—are responsible for the catalytic mechanism through which the α/β-hydrolase fold operates. The first and last residues may vary depending on the enzyme, while His is almost always conserved. The nucleophilic component is usually Ser, Cys, or Asp, and it is situated in the so-called nucleophilic elbow, representing a tight turn in the fold's structure between αC and β5. On the other hand, the acidic residue is usually either Asp or Glu, and its function is to stabilize His in its protonated form. The mechanism involves two steps. First, the nucleophile attacks the substrate during acylation. Then, the process passes through two transition states before completing a deacylation reaction that releases the final product. In the AfEST model, aspartate (Asp) or glutamate (Glu), histidine (His), and serine (Ser) lie between the characteristic α/β hydrolase fold and the protein's cap domain. The oxyanion hole, another crucial region, comprises three hydrophobic amino acids, two glycine (Gly) residues, and one alanine (Ala) residue. This region provides chemical stability by forming hydrogen bonds with its amide groups.

As the reaction progresses, the catalytic triad undergoes a cycle, starting with a nucleophilic attack by serine, which transfers a proton to histidine and continues with the formation of three successive intermediates: the first tetrahedral intermediate (INT-1), the acyl-enzyme intermediate (also known as enzyme-activated monomer, EAM), and the second tetrahedral intermediate (INT-2). The process concludes with the conversion to INT-2, which results in the recovery of the enzyme (RC), allowing the cycle to begin anew. The process includes four identifiable transition states: T1, T2, T3, and T4, each representing a specific step in the cycle. After the formation of INT-1 in the first step, another proton transfer occurs, cleaving a C-O bond in the substrate, leading to the release of 6-hydroxycaproic acid (6-HCA) from the PCL chain and the formation of EAM. EAM is subsequently subjected to deacylation by a nucleophilic attack from a water molecule, converting it to INT-2. In the final step, water establishes an equilibrium that drives the reaction toward releasing a second molecule, 6-HCA, and regenerating the enzyme (RC), accompanied by a decrease in Gibbs free energy, indicating that this step is exergonic. In conclusion, two 6-HCA molecules are cleaved from the PCL chain for each cycle turnover, representing the final product, which moves on to the next stage of biodegradation [109].

The second model centers on the thermophilic esterase MGS0156, identified through environmental metagenomic screening and capable of degrading various plastics, including PCL. Unlike AfEST, it operates optimally at a lower temperature (35–40 °C) and features a modified α/β hydrolase fold with a lid domain and a highly hydrophobic active site suited for binding PCL. As with AfEST, this model emphasizes the critical role of three catalytic amino acid residues in the reaction mechanism, characterized as a two-step process: triad-assisted nucleophilic attack and C-O bond cleavage. The model stresses that C-O bond cleavage is the rate-determining step, supported by a Molecular Mechanics Poisson–Boltzmann Surface Area Method (MM-PBSA). It indicates that different sets of six amino acids determine the binding strength of a PCL oligomer of a given size, noting that octamers demonstrate the highest binding energy. A valuable aspect of the study is the identification of two residues, His231 and Asp237, as targets for mutagenesis, providing implications for the bioengineering of enzymes derived from MGS0156 and their potential use as improved enzymatic tools in the biodegradation of PCL and other plastic materials [110]. Researchers have also noted other products resulting from the hydrolysis of PCL in addition to 6-HCA. An experiment comparing the abiotic and biotic degradation of PCL films, with the latter involving a mixed culture of compost microorganisms, shows that this process also leads to the release of various small molecules, namely caprolactone, a cyclic dimer, and a cyclic trimer, confirmed through a coupled analysis with gas chromatography and mass spectrometry (GC-MS). These products and 6-HCA are quickly assimilated and disappear entirely after two weeks in the biotic medium [111]. Interestingly, a GC-MS analysis of the degradation products in *Rhodococcus erythropolis* and *Rhodococcus opacus* shows a variety of other molecules, such as decanoic acid, undecanoic acid, dodecanoic acid, eicosane, pentacosane, and hexacosane, among others, but does not detect the presence of 6-HCA [112]. An earlier study examining the biodegradative potential of a thermotolerant strain of *Aspergillus* sp. identified the presence of succinic acid, butyric acid, valeric acid, and caproic acid after exposing plastic pellets to the activity of the microorganism at 50 °C. As anticipated, considering the chemical structure of its typical substrate, cutinase expressed by a strain of *Ps. pachastrellae*, isolated from coastal waters, results in 16-hydroxyhexadecanoic (Juniperic) acid, the main component of cutin, in addition to 6-HCA [113].

The post-hydrolysis stages of PCL biodegradation (Figure 9)—that is, the assimilation and further metabolization of lower-weight products followed by the excretion of waste molecules such as H_2O and CO_2 or, for some anaerobes, CH_4—remain a largely unexplored area in bioplastic research.

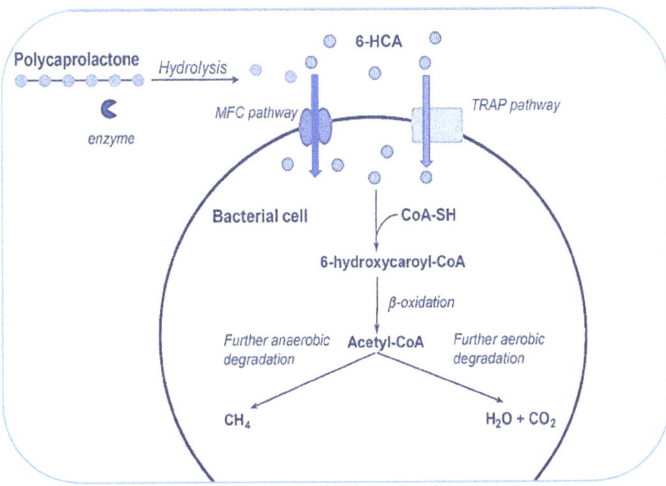

Figure 9. Proposed hypothetical model of the post-hydrolysis stage of PCL biodegradation. Designations: the thicker blue arrows indicate the possible transport pathways of 6-hydroxycaproic acid (6-HCA, blue circles) into the bacterial cell; the thin arrows indicate hypothetical biochemical conversions.

Potential candidates for transporting 6-HCA across the cell membrane can be found among proteins from the major facilitator superfamily (MFS), representing an evolutionarily ancient lineage of molecular transport systems in bacteria. The MFS routinely imports carboxylic acids, including hydroxy acids like lactate, sugar acids such as gluconate, and various short-chain fatty acids [114,115]. Another possibility is transport via a member of the tripartite ATP-independent periplasmic transporter (TRAP) family, a common prokaryote-specific protein group known to engage in the secondary transport of a diverse array of carboxylates, including typical examples such as fumarate and malate—both of which are dicarboxylates—as well as pyruvate, lactate, L-glutamate, and others [116]. Since 6-HCA is chemically related to the aforementioned compounds as a carboxylic acid and exhibits a relatively simple chemical structure, it seems reasonable to focus on MFS and TRAP when investigating its transport mechanism. The fate of 6-HCA after it enters the bacterial cytoplasm is another critical area for potential experimental research.

Degradation through a catabolic pathway, such as beta-oxidation, is plausible since this process is the primary means of breaking down carboxylic acids of various chain lengths in many organisms, including bacteria [117,118]. Enzymes from the acyl-CoA synthase family can catalyze the formation of a thioester bond with CoA, allowing them to target a wide range of carboxylic acid molecules for degradation, resulting in products like palmitoyl-CoA [119], stearoyl-CoA, 3-hydroxybutyryl-CoA [120,121], isovaleryl-CoA [122], isobutyryl-CoA [123], and caproyl-CoA. The latter represents the activated form of caproic acid, which contains the same carbon atoms as 6-HCA, differing only by a single hydroxyl group.

Most current knowledge about the genetic factors governing PCL biodegradation comes from a pivotal set of experiments conducted with strains of *R. erythropolis* and *R. opacus*, two closely related aerobic species of actinobacteria. Zampolli et al. [112] have effectively determined the biodegradation capacity of these microorganisms by evaluating established hydrolytic zones (halos) on agar plates and conducting coupled GC-MS analysis of products extracted from bacterial cultures supplied with PCL. They have also measured high esterase activity through spectrophotometric tests, performing enzymatic reactions with five different organic substrates. The substrates include p-nitrophenyl acetate, p-nitrophenyl butyrate, p-nitrophenyl octanoate, p-nitrophenyl laurate, and p-nitrophenyl palmitate. The genome of *R. erythropolis* was subsequently sequenced via Illumina MiSeq v3, annotated, and then compared with the genome of *R. opacus* using the Rapid Annotations using Subsystems Technology (RAST) server.

Furthermore, the transcription levels of genes potentially involved in PCL biodegradation have been assayed using reverse transcription quantitative real-time PCR (RT-qPCR). An automated annotation of the genome of *R. erythropolis* reveals that it contains 6165 open reading frames (ORFs) and 72 RNA genes. Sequence-based and functional comparisons indicate that the two species share a substantial degree of identity, which the authors identify as a 'common genomic core'. Interestingly, according to the RAST server, the closest neighbors of *R. erythropolis* are *R. jostii* and *Nocardia farcinica*. These findings are significant as they may indicate other microorganisms with potential biodegradation capabilities and suggest that genetic tools can be employed to identify such species. The RAST server also reveals that *R. erythropolis* and *R. opacus* share the same set of most populated functional categories, specifically amino acids and derivatives, carbohydrates, fatty acids, lipids, isoprenoids, cofactors, vitamins, prosthetic groups or pigments, and protein metabolism. The authors focus on genes involved in lipid metabolism, arguing that they are likely relevant to aliphatic polyester degradation, given the chemical similarities between the two groups of molecules. Further bioinformatic analyses have shown that this category includes a set of 10 and 9 genes encoding hypothetical proteins and lipases, which are potential objects of interest. Ultimately, RT-qPCR has revealed that three key genes from the selected group are expressed during the PCL biodegradation process: P133, which encodes a hypothetical protein; P253, which encodes a lipase; and P337, which encodes another hypothetical protein [112].

5. Conclusions and Prospects

PCL biodegradation offers a promising and sustainable approach to reducing plastic pollution while paving the way for eco-friendly material use. This review shows that various microorganisms, including bacteria, fungi, and archaea, can degrade PCL. The widespread presence of esterases and lipases in microbial communities indicates that the potential for biodegradation in nature may be significantly greater than documented. Exploring microbial diversity, particularly in extreme environments and under different ecological conditions, could reveal novel strains with enhanced biodegradative abilities.

A persistent challenge in this field is the limited understanding of the complete metabolic pathways involved in the assimilation and processing of PCL degradation products. While the enzymatic mechanisms of hydrolysis are relatively well characterized, the intracellular fate of intermediates such as 6-HCA remains largely unknown. Potential transport mechanisms, including the major facilitator superfamily (MFS) and tripartite ATP-independent periplasmic (TRAP) transport systems, are still hypothetical and have not been experimentally validated. Similarly, the specific roles of metabolic routes like β-oxidation in converting PCL-derived compounds into central metabolites require further clarification. Addressing these knowledge gaps is essential for optimizing microbial strains and enhancing biodegradation efficiency. Enzymatic models, such as those derived from *Archaeoglobus fulgidus* esterase and the metagenomic esterase MGS0156, provide mechanistic insight into substrate binding and catalysis, offering a valuable foundation for enzyme engineering. Advances in protein engineering, particularly directed evolution and rational design, could significantly enhance enzyme specificity, thermal stability, and catalytic performance, making them more suitable for large-scale bioplastic degradation and environmental remediation applications. However, translating laboratory success into large-scale biodegradation remains a considerable challenge. Environmental applications must consider microbial community dynamics, interactions with other pollutants, and abiotic factors such as temperature, pH, and nutrient availability. Utilizing complex microbial consortia instead of isolated strains may boost degradation rates while maintaining stability across varying environmental conditions.

Despite significant progress, several critical gaps remain in our understanding of microbial PCL biodegradation. Although many microorganisms and hydrolytic enzymes have been identified, most studies emphasize extracellular depolymerization under aerobic conditions, leaving anaerobic biodegradation pathways largely unexplored. The transport and intracellular metabolism of degradation products like 6-HCA are still poorly understood, particularly regarding how they are assimilated and integrated into central metabolic pathways. Furthermore, the genetic basis for the regulation and expression of PCL-degrading enzymes in various microbial taxa has yet to be fully characterized. Although genome and metagenome sequencing have revealed numerous putative enzymes, only a small percentage have undergone functional validation. Additionally, there is insufficient data on microbial consortia in complex environmental systems where PCL is present alongside other organic and inorganic contaminants.

In conclusion, a comprehensive approach that integrates microbiology, enzymology, synthetic biology, and environmental science is essential for advancing the field of PCL biodegradation. Future research should prioritize discovering novel microbial taxa, clarifying complete catabolic pathways, and developing engineered enzymes that function more effectively under environmentally relevant conditions. Addressing gaps in the existing literature will be critical for building a bridge toward scalable solutions in plastic waste management and improving the efficiency of PCL degradation in varied ecological settings.

Author Contributions: Conceptualization, N.K. and P.P.; investigation, N.A., N.K., I.B., and P.P.; resources, K.P.; writing—original draft preparation, N.K., P.P., N.A., and K.P.; writing—review and editing, I.B. and K.P.; funding acquisition, K.P. All authors have read and agreed to the published version of the manuscript.

Funding: The authors gratefully acknowledge the financial support from the National Science Fund, Republic of Bulgaria, Grant No. KP-06-PN87/12.

Conflicts of Interest: The authors declare no conflicts of interest.

References

1. Ramaraju, H.; Verga, A.S.; Steedley, B.J.; Kowblansky, A.P.; Green, G.E.; Hollister, S.J. Investigation of the biodegradation kinetics and associated mechanical properties of 3D-printed polycaprolactone during long-term preclinical testing. *Biomaterial* **2025**, *321*, 123257. [CrossRef] [PubMed]
2. Richert, A.; Kalwasińska, A.; Brzezinska, M.S.; Dąbrowska, G.B. Biodegradability of Novel Polylactide and Polycaprolactone Materials with Bacteriostatic Properties Due to Embedded Birch Tar in Different Environments. *Int. J. Mol. Sci.* **2021**, *22*, 10228. [CrossRef]
3. Injorhor, P.; Trongsatitkul, T.; Wittayakun, J.; Ruksakulpiwat, C.; Ruksakulpiwat, Y. Biodegradable Polylactic Acid-Polyhydroxyalkanoate-Based Nanocomposites with Bio-Hydroxyapatite: Preparation and Characterization. *Polymers* **2023**, *15*, 1261. [CrossRef]
4. Naser, A.Z.; Deiab, I.; Defersha, F.; Yang, S. Expanding Poly(lactic acid) (PLA) and Polyhydroxyalkanoates (PHAs) Applications: A Review on Modifications and Effects. *Polymers* **2021**, *13*, 4271. [CrossRef]
5. Pyo, S.-H.; Park, J.H.; Srebny, V.; Hatti-Kaul, R. A sustainable synthetic route for biobased 6-hydroxyhexanoic acid, adipic acid and ε-caprolactone by integrating bio- and chemical catalysis. *Green Chem.* **2020**, *22*, 4450–4455. [CrossRef]
6. Shekh, M.R.; Kumar, V. Impact of plastic pollution on ecosystems: A review of adverse effects and sustainable solutions. *Environ. Monit. Assess.* **2025**, *197*, 264. [CrossRef] [PubMed]
7. Krobot, Š.; Melčová, V.; Menčík, P.; Kontárová, S.; Rampichová, M.; Hedvičáková, V.; Mojžišová, E.; Baco, A.; Přikryl, R. Poly(3-hydroxybutyrate) (PHB) and Polycaprolactone (PCL) Based Blends for Tissue Engineering and Bone Medical Applications Processed by FDM 3D Printing. *Polymers* **2023**, *15*, 2404. [CrossRef]
8. Hottle, T.A.; Bilec, M.M.; Landis, A.E. Biopolymer production and end of life comparisons using life cycle assessment. *Resour. Conserv. Recycl.* **2017**, *122*, 295–306. [CrossRef]
9. Narancic, T.; Verstichel, S.; Chaganti, S.R.; Morales-Gamez, L.; Kenny, S.T.; De Wilde, B.; Padamati, R.B.; O'Connor, K.E. Biodegradable plastic blends create new possibilities for end-of-life management of plastics but they are not a panacea for plastic pollution. *Environ. Sci. Technol.* **2018**, *52*, 10441–10452. [CrossRef]
10. Mandic, M.; Spasic, J.; Ponjavic, M.; Nikolic, M.; Cosovic, V.R.; O'Connor, K.E.; Nikodinovic-Runic, J.; Djokic, L.; Jeremic, S. Biodegradation of poly(ε-caprolactone) (PCL) and medium chain length polyhydroxyalkanoate (mcl-PHA) using whole cells and cell free protein preparations of Pseudomonas and Streptomyces strains grown on waste cooking oil. *Polym. Degrad. Stab.* **2019**, *162*, 160–168. [CrossRef]
11. Labet, M.; Thielemans, W. Synthesis of PCL: A review. *Chem. Soc. Rev.* **2009**, *38*, 3484–3504. [CrossRef] [PubMed]
12. Martín-González, D.; de la Fuente Tagarro, C.; De Lucas, A.; Bordel, S.; Santos-Beneit, F. Genetic Modifications in Bacteria for the Degradation of Synthetic Polymers: A Review. *Int. J. Mol. Sci.* **2024**, *25*, 5536. [CrossRef] [PubMed]
13. Kayan, G.Ö.; Kayan, A. Polycaprolactone Composites/Blends and Their Applications Especially in Water Treatment. *ChemEngineering* **2023**, *7*, 104. [CrossRef]
14. Woodruff, M.A.; Hutmacher, D.W. The return of a forgotten polymer—Polycaprolactone in the 21st century. *Prog. Polym. Sci.* **2010**, *35*, 1217–1256. [CrossRef]
15. Kayan, A. Recent studies on single site metal alkoxide complexes as catalysts for ring opening polymerization of cyclic compounds. *Catal. Surv. Asia* **2020**, *24*, 87–103. [CrossRef]
16. Grobelny, Z.; Golba, S.; Jurek-Suliga, J. Mechanism of ε-caprolactone polymerization in the presence of alkali metal salts: Investigation of initiation course and determination of polymers structure by MALDI-TOF mass spectrometry. *Polym. Bull.* **2019**, *76*, 3501–3515. [CrossRef]
17. Xu, Y.; Wang, L.; Chen, C.; Huang, P.; Dai, H.; Jiang, W.; Zhou, Y. Living Cationic Polymerization of ε-Caprolactone Catalyzed by a Metal-free Lewis Acid of Trityl Tetrafluoroborate. *Macromolecules* **2023**, *56*, 501–509. [CrossRef]
18. Vidal, J.L.; Yavitt, B.M.; Wheeler, M.D.; Kolwich, J.L.; Donovan, L.N.; Sit, C.S.; Kerton, F.M. Biochar as a sustainable and renewable additive for the production of Poly (ε-caprolactone) composites. *Sustain. Chem. Pharm.* **2022**, *25*, 100586. [CrossRef]
19. Srinivasamurthy, V.S.T.; Böttcher, D.; Engel, J.; Kara, S.; Bornscheuer, U.T. A whole-cell process for the production of ε-caprolactone in aqueous media. *Proc. Biochem.* **2020**, *88*, 22–30. [CrossRef]

20. Schmidt, S.; Scherkus, C.; Muschiol, J.; Menyes, U.; Winkler, T.; Hummel, W.; Gröger, H.; Liese, A.; Herz, H.G.; Bornscheuer, U.T. An enzyme cascade synthesis of ε-caprolactone and its oligomers. *Angew. Chem. Int. Ed. Engl.* **2015**, *54*, 2784–2787. [CrossRef]
21. Dong, H.; Wang, H.; Cao, S.; Shen, J. Lipase-catalyzed polymerization of lactones and linear hydroxyesters. *Biotechnol. Lett.* **1998**, *20*, 905–908. [CrossRef]
22. Yu, W.; Song, X.; Wang, Y.; Zhang, L.; Liu, Y.; Liu, Y. Enhancing the oil/water separation efficiency of polylactic acid fiber membrane via polydimethylsiloxane-PCL copolymer. *J. Environ. Chem. Eng.* **2024**, *12*, 114738. [CrossRef]
23. Thakur, M.; Majid, I.; Hussain, S.; Nanda, V. Poly (ε-caprolactone): A potential polymer for biodegradable food packaging applications. *Packag. Technol. Sci.* **2021**, *34*, 449–461. [CrossRef]
24. Guarino, V.; Gentile, G.; Sorrentino, L.; Ambrosio, L. Polycaprolactone: Synthesis, properties, and applications. In *Encyclopedia of Polymer Science and Technology*; Wiley: Hoboken, NJ, USA, 2017; pp. 1–36. [CrossRef]
25. Goel, V.; Luthra, P.; Kapur, G.S.; Ramakumar, S.S.V. Biodegradable/bio-plastics: Myths and realities. *J. Polym. Environ.* **2021**, *29*, 3079–3104. [CrossRef]
26. Radhakrishnan, S.; Nagarajan, S.; Belaid, H.; Farha, C.; Iatsunskyi, I.; Coy, E.; Soussan, L.; Huon, V.; Bares, J.; Belkacemi, K.; et al. Fabrication of 3D printed antimicrobial PCL scaffolds for tissue engineering applications. *Mater. Sci. Eng. C* **2020**, *118*, 111525. [CrossRef]
27. Castro, J.I.; Araujo-Rodríguez, D.G.; Valencia-Llano, C.H.; Tenorio, D.L.; Saavedra, M.; Zapata, P.A.; Grande-Tovar, C.D. Biocompatibility Assessment of PCL/Polylactic Acid/Zinc Oxide Nanoparticle Composites under In Vivo Conditions for Biomedical Applications. *Pharmaceutics* **2023**, *15*, 2196. [CrossRef]
28. Zhou, L.; He, H.; Li, M.-C.; Huang, S.; Mei, C.; Wu, Q. Grafting PCL diol onto cellulose nanocrystals via click chemistry: Enhancing thermal stability and hydrophobic property. *Carbohydr. Polym.* **2018**, *189*, 331–341. [CrossRef]
29. Li, P.; Ruan, L.; Jiang, G.; Sun, Y.; Wang, R.; Gao, X.; Yunusov, K.E.; Aharodnikau, U.E.; Solomevich, S.O. Design of 3D PCL/ε-polylysine-modified chitosan fibrous scaffolds with incorporation of bioactive factors for accelerating wound healing. *Acta Biomater.* **2022**, *152*, 197–209. [CrossRef]
30. Menossi, M.; Salcedo, F.; Rivilli, N.; Nicolini, A.T.; Alvarez, V.A.; Ludueña, L.N. Biodegradable mulch films based on Starch/Poly (Lactic Acid)/Poly (E-Caprolactone) ternary blends. *J. Polym. Environ.* **2022**, *31*, 2114–2137. [CrossRef]
31. Salvekar, A.V.; Zhou, Y.; Huang, W.M.; Wong, Y.S.; Venkatraman, S.S.; Shen, Z.; Zhu, G.; Cui, H.P. Shape/temperature memory phenomena in un-crosslinked poly-ε-caprolactone (PCL). *Eur. Polym. J.* **2015**, *72*, 282–295. [CrossRef]
32. Jing, X.; Mi, H.-Y.; Huang, H.-X.; Turng, L.-S. Shape memory thermoplastic polyurethane (TPU)/poly(ε-caprolactone) (PCL) blends as self-knotting sutures. *J. Mech. Behav. Biomed. Mater.* **2016**, *64*, 94–103. [CrossRef] [PubMed]
33. Espinoza, S.M.; Patil, H.I.; San Martin Martinez, E.; Casañas Pimentel, R.; Ige, P.P. Poly-ε-caprolactone (PCL), a promising polymer for pharmaceutical and biomedical applications: Focus on nanomedicine in cancer. *Int. J. Polym. Mater. Polym. Biomater.* **2020**, *69*, 85–126. [CrossRef]
34. Homaeigohar, S.; Boccaccini, A.R. Nature-Derived and Synthetic Additives to poly (ε-Caprolactone) Nanofibrous Systems for Biomedicine; an Updated Overview. *Front. Chem.* **2022**, *9*, 809676. [CrossRef]
35. Peng, Y.; Lu, Y.; Liu, K.; Liu, S.; Fan, L.; Huang, W. Biodegradable balloon-expandable self-locking PCL stents as buckling explants for the treatment of retinal detachment: An in vitro and in vivo study. *J. Biomed. Mater. Res. A* **2012**, *101*, 167–175. [CrossRef]
36. Silva-López, M.S.; Alcántara-Quintana, L.E. The era of biomaterials: Smart implants? *ACS Appl. Bio Mater.* **2023**, *6*, 2982–2994. [CrossRef] [PubMed]
37. Stafin, K.; Śliwa, P.; Piątkowski, M. Towards Polycaprolactone-Based Scaffolds for Alveolar Bone Tissue Engineering: A Biomimetic Approach in a 3D Printing Technique. *Int. J. Mol. Sci.* **2023**, *24*, 16180. [CrossRef] [PubMed]
38. Bou-Francis, A.; Piercey, M.; Al-Qatami, O.; Mazzanti, G.; Khattab, R.; Ghanem, A. PCL blends for fracture fixation in low load-bearing applications. *J. Appl. Polym. Sci.* **2020**, *137*, 48940. [CrossRef]
39. Raina, N.; Pahwa, R.; Khosla, J.K.; Gupta, P.N.; Gupta, M. PCL-based materials in wound healing applications. *Polym. Bull.* **2021**, *79*, 7041–7063. [CrossRef]
40. Kotcharat, P.; Chuysinuan, P.; Thanyacharoen, T.; Techasakul, S.; Ummartyotin, S. Development of bacterial cellulose and polycaprolactone (PCL) based composite for medical material. *Sustain. Chem. Pharm.* **2021**, *20*, 100404. [CrossRef]
41. Ahmed, M.K.; Zayed, M.A.; El-Dek, S.I.; Hady, M.A.; El Sherbiny, D.H.; Uskoković, V. Nanofibrous ε-polycaprolactone scaffolds containing Ag-doped magnetite nanoparticles: Physicochemical characterization and biological testing for wound dressing applications in vitro and in vivo. *Bioact. Mater.* **2021**, *6*, 2070–2088. [CrossRef]
42. Baptista, C.; Azagury, A.; Shin, H.; Baker, C.M.; Ly, E.; Lee, R.; Mathiowitz, E. The effect of temperature and pressure on PCL morphology. *Polymer* **2020**, *191*, 122227. [CrossRef]
43. Musílková, J.; Beran, M.; Sedlář, A.; Slepička, P.; Bartoš, M.; Kolská, Z.; Havlíčková, Š.; Luňáčková, J.; Svobodová, L.; Froněk, M.; et al. Composite Polylactide/Polycaprolactone Foams with Hierarchical Porous Structure for Pre-Vascularized Tissue Engineering. *Int. J. Mol. Sci.* **2025**, *26*, 2974. [CrossRef]

44. Casanova, F.; Santos, L. Encapsulation of cosmetic active ingredients for topical application—A review. *J. Microencapsul.* **2015**, *33*, 1–17. [CrossRef]
45. Christen, M.-O.; Vercesi, F. PCL: How a Well-Known and Futuristic Polymer Has Become an Innovative Collagen-Stimulator in Esthetics. *Clin. Cosm. Investig. Dermatol.* **2020**, *13*, 31–48. [CrossRef]
46. Qiao, D.; Li, J.; Zhang, S.; Yang, X. Controlled release fertilizer with temperature-responsive behavior coated using polyether polyol (PPG)/PCL (PCL) blend-based polyurethane performs smart nutrient release. *Mater. Today Chem.* **2022**, *26*, 101249. [CrossRef]
47. Yang, N.; Ying, L.; Li, K.; Chen, F.; Zhao, F.; Sun, Z.; Feng, L.; Liu, J. Biodegradable mulching films based on PCL and its porous structure construction. *Polymers* **2022**, *14*, 5340. [CrossRef]
48. Chakkalakkal, N.D.; Thomas, M.; Chittillapilly, P.S.; Sujith, A.; Anjali, P.D. Electrospun polymer nanocomposite membrane as a promising seed coat for controlled release of agrichemicals and improved germination: Towards a better agricultural prospect. *J. Clean. Prod.* **2022**, *377*, 134479. [CrossRef]
49. Hamzah, M.S.A.; Austad, A.; Abd Razak, S.I.; Nayan, N.H.M. Tensile and wettability properties of electrospun polycaprolactone coated with pectin/polyaniline composite for drug delivery application. *Int. J. Struct. Integr.* **2019**, *10*, 704–713. [CrossRef]
50. Almeida, B.C.; Figueiredo, P.; Carvalho, A.T.P. PCL enzymatic hydrolysis: A Mechanistic study. *ACS Omega* **2019**, *4*, 6769–6774. [CrossRef]
51. Nawaz, A.; Hasan, F.; Shah, A.A. Degradation of poly(ε-caprolactone) (PCL) by a newly isolated *Brevundimonas* sp. strain MRL-AN1 from soil. *FEMS Microbiol. Lett.* **2015**, *362*, 1–7. [CrossRef]
52. Oda, Y.; Oida, N.; Urakami, T.; Tonomura, K. PCL depolymerase produced by the bacterium *Alcaligenes faecalis*. *FEMS Microbiol. Lett.* **2006**, *152*, 339–343. [CrossRef] [PubMed]
53. Chew, P.L.; Annuar, M.S.M.; Show, P.L.; Ling, T.C. Extractive bioconversion of poly-ϵ-caprolactone by *Burkholderia cepacia* lipase in an aqueous two-phase system. *Biochem. Eng. J.* **2015**, *101*, 9–17. [CrossRef]
54. Zhang, Y.; Qiu, S. Phylogenomic analysis of the genus *Ralstonia* based on 686 single-copy genes. *Antonie Leeuwenhoek* **2015**, *109*, 71–82. [CrossRef]
55. Shah, A.A.; Nawaz, A.; Kanwal, L.; Hasan, F.; Khan, S.; Badshah, M. Degradation of poly(ε-caprolactone) by a thermophilic bacterium *Ralstonia* sp. strain MRL-TL isolated from hot spring. *Int. Biodeterior. Biodegrad.* **2014**, *98*, 35–42. [CrossRef]
56. Luo, G.; Hou, Z.; Gao, J.; Chen, X.; Tan, H.; Fan, L. Performance of PCL-based heterotrophic denitrification for recirculating aquaculture systems with varying hydraulic retention times. *Desal. Water Treat.* **2018**, *116*, 103–111. [CrossRef]
57. Wang, X.; Wang, W.; Zhang, Y.; Sun, Z.; Zhang, J.; Chen, G.; Li, J. Simultaneous nitrification and denitrification by a novel isolated *Pseudomonas* sp. JQ-H3 using PCL as carbon source. *Biores. Technol.* **2019**, *288*, 121506. [CrossRef]
58. Spearman, S.S.; Irin, F.; Ramesh, S.; Rivero, I.V.; Green, M.J.; Harrysson, O.L.A. Effect of pseudomonas lipase enzyme on the degradation of PCL/PCL-polyglycolide fiber blended nanocomposites. *Int. J. Polym. Mater.* **2018**, *68*, 360–367. [CrossRef]
59. Li, L.; Lin, X.; Bao, J.; Xia, H.; Li, F. Two Extracellular Poly(ε-caprolactone)-Degrading Enzymes From *Pseudomonas hydrolytica* sp. DSWY01T: Purification, Characterization, and Gene Analysis. *Front. Bioeng. Biotechnol.* **2022**, *10*, 835847. [CrossRef] [PubMed]
60. Inglis, G.D.; Yanke, L.J.; Selinger, L.B. Cutinolytic esterase activity of bacteria isolated from mixed-plant compost and characterization of a cutinase gene from *Pseudomonas pseudoalcaligenes*. *Can. J. Microbiol.* **2011**, *57*, 902–913. [CrossRef]
61. Benedict, C.V.; Cameron, J.A.; Huang, S.J. PCL degradation by mixed and pure cultures of bacteria and a yeast. *J. Appl. Polym. Sci.* **1983**, *28*, 335–342. [CrossRef]
62. Budkum, J.; Thammasittirong, S.N.-R.; Thammasittirong, A. High poly ε-caprolactone biodegradation activity by a new *Acinetobacter seifertii* isolate. *Folia Microbiol.* **2022**, *67*, 659–669. [CrossRef] [PubMed]
63. Oh, Y.-R.; Jang, Y.-A.; Eom, G.T. Microbial production of adipic acid from 6-hydroxyhexanoic acid for biocatalytic upcycling of PCL. *Enz. Microb. Technol.* **2024**, *181*, 110521. [CrossRef]
64. Brzezinska, M.S.; Walczak, M.; Kalwasińska, A.; Richert, A.; Świątczak, J.; Deja-Sikora, E.; Burkowska-But, A. Biofilm formation during biodegradation of polylactide, poly (3,4 hydroxybutyrate) and poly(ε-caprolactone) in activated sludge. *Int. J. Biol. Macromol.* **2020**, *159*, 539–546. [CrossRef]
65. Chua, T.-K.; Tseng, M.; Yang, M.-K. Degradation of Poly(ε-caprolactone) by thermophilic *Streptomyces thermoviolaceus* subsp. *thermoviolaceus* 76T-2. *AMB Express* **2013**, *3*, 8. [CrossRef] [PubMed]
66. Atanasova, N.; Stoitsova, S.; Paunova-Krasteva, T.; Kambourova, M. Plastic Degradation by Extremophilic Bacteria. *Int. J. Mol. Sci.* **2021**, *22*, 5610. [CrossRef] [PubMed]
67. Atanasova, N.; Paunova-Krasteva, T.; Stoitsova, S.; Radchenkova, N.; Boyadzhieva, I.; Petrov, K.; Kambourova, M. Degradation of Poly(ε-caprolactone) by a Thermophilic Community and *Brevibacillus thermoruber* Strain 7 Isolated from Bulgarian Hot Spring. *Biomolecules* **2021**, *11*, 1488. [CrossRef]
68. Shin, N.; Kim, S.H.; Cho, J.Y.; Hwang, J.H.; Kim, H.J.; Oh, S.J.; Park, S.-H.; Park, K.; Bhatia, S.K.; Yang, Y.-H. Fast Degradation of PCL/Poly(butylene adipate-co-terephthalate) Blends by Novel *Bacillus* Strain NR4 with Broad Degrading Activity. *J. Polym. Environ.* **2023**, *32*, 898–912. [CrossRef]

69. Motiwalla, M.J.; Punyarthi, P.P.; Mehta, M.K.; D'Souza, J.S.; Kelkar-Mane, V. Studies on degradation efficiency of PCL by a naturally-occurring bacterium. *J. Environ. Biol.* **2013**, *34*, 43–49.
70. Wu, K.-J.; Wu, C.-S.; Chang, J.-S. Biodegradability and mechanical properties of PCL composites encapsulating phosphate-solubilizing bacterium *Bacillus* sp. PG01. *Process Biochem.* **2007**, *42*, 669–675. [CrossRef]
71. Malunavicius, V.; Padaiga, A.; Stankeviciute, J.; Pakalniskis, A.; Gudiukaite, R. Engineered *Geobacillus* lipolytic enzymes—Attractive polyesterases that degrade PCLs and simultaneously produce esters. *Int. J. Biol. Macromol.* **2023**, *253*, 127656. [CrossRef]
72. Xu, S.; Yamaguchi, T.; Osawa, S.; Suye, S.-I. Biodegradation of Poly (ε-Caprolactone) Film in the Presence of *Lysinibacillus* sp. 70038 and Characterization of the Degraded Film. *Biocontrol Sci.* **2007**, *12*, 119–122. [CrossRef] [PubMed]
73. Khan, I.; Dutta, J.R.; Ganesan, R. *Lactobacillus* spp. lipase mediated poly (ε-caprolactone) degradation. *Int. J. Biol. Macromol.* **2016**, *95*, 126–131. [CrossRef] [PubMed]
74. Khan, I.; Nagarjuna, R.; Dutta, J.R.; Ganesan, R. Enzyme-Embedded Degradation of Poly(ε-caprolactone) using Lipase-Derived from Probiotic *Lactobacillus plantarum*. *ACS Omega* **2019**, *4*, 2844–2852. [CrossRef] [PubMed]
75. Federle, T.W.; Barlaz, M.A.; Pettigrew, C.A.; Kerr, K.M.; Kemper, J.J.; Nuck, B.A.; Schechtman, L.A. Anaerobic Biodegradation of Aliphatic Polyesters: Poly(3-hydroxybutyrate-co-3-hydroxyoctanoate) and Poly(ε-caprolactone). *Biomacromolecules* **2002**, *3*, 813–822. [CrossRef]
76. Lopardo, C.R.; Urakawa, H. Performance and microbial diversity of bioreactors using PCL and polyhydroxyalkanoate as carbon source and biofilm carrier in a closed recirculating aquaculture system. *Aquac. Int.* **2019**, *27*, 1251–1268. [CrossRef]
77. Rüthi, J.; Rast, B.M.; Qi, W.; Perez-Mon, C.; Pardi-Comensoli, L.; Brunner, I.; Frey, B. The plastisphere microbiome in alpine soils alters the microbial genetic potential for plastic degradation and biogeochemical cycling. *J. Hazard. Mater.* **2022**, *441*, 129941. [CrossRef]
78. Cook, W.J.; Cameron, J.A.; Bell, J.P.; Huang, S.J. Scanning electron microscopic visualization of biodegradation of PCLs by fungi. *J. Polym. Sci. Polym. Lett.* **1981**, *19*, 159–165. [CrossRef]
79. Benedict, C.V.; Cook, W.J.; Jarrett, P.; Cameron, J.A.; Huang, S.J.; Bell, J.P. Fungal degradation of PCLs. *J. Appl. Polym. Sci.* **1983**, *28*, 327–334. [CrossRef]
80. Tilstra, L.; Johnsonbaugh, D. The biodegradation of blends of PCL and polyethylene exposed to a defined consortium of fungi. *J. Environ. Polym. Degrad.* **1993**, *1*, 257–267. [CrossRef]
81. Vivi, V.K.; Martins-Franchetti, S.M.; Attili-Angelis, D. Biodegradation of PCL and PVC: *Chaetomium globosum* (ATCC 16021) activity. *Folia Microbiol.* **2018**, *64*, 1–7. [CrossRef]
82. Sanchez, J.G.; Tsuchii, A.; Tokiwa, Y. Degradation of PCL at 50 °C by a thermotolerant *Aspergillus* sp. *Biotechnol. Lett.* **2000**, *22*, 849–853. [CrossRef]
83. Hermanová, S.; Omelková, J.; Voběrková, S.; Bálková, R.; Richtera, L.; Mravcová, L.; Jančář, J. The effect of processing of PCL films on degradation process initiated by *Aspergillus oryzae* Lipase. *Int. J. Polym. Anal. Charact.* **2012**, *17*, 465–475. [CrossRef]
84. Kim, S.H.; Lee, J.W.; Kim, J.S.; Lee, W.; Park, M.S.; Lim, Y.W. Plastic-inhabiting fungi in marine environments and PCL degradation activity. *Antonie Leeuwenhoek* **2022**, *115*, 1379–1392. [CrossRef]
85. Hosni, A.S.A.; Pittman, J.K.; Robson, G.D. Microbial degradation of four biodegradable polymers in soil and compost demonstrating PCL as an ideal compostable plastic. *J. Waste Manag.* **2019**, *97*, 105–114. [CrossRef] [PubMed]
86. Yang, L.; Li, J.; Jin, Y.; Li, M.; Gu, Z. In vitro enzymatic degradation of the cross-linked poly(ε-caprolactone) implants. *Polym. Degrad. Stab.* **2014**, *112*, 10–19. [CrossRef]
87. Oda, Y.; Asari, H.; Urakami, T.; Tonomura, K. Microbial degradation of poly(3-hydroxybutyrate) and PCL by filamentous fungi. *J. Ferment. Bioeng.* **1995**, *80*, 265–269. [CrossRef]
88. Li, F.; Yu, D.; Lin, X.; Liu, D.; Xia, H.; Chen, S. Biodegradation of poly(ε-caprolactone) (PCL) by a new *Penicillium oxalicum* strain DSYD05-1. *World J. Microbiol. Biotechnol.* **2012**, *28*, 2929–2935. [CrossRef]
89. Antipova, T.V.; Zhelifonova, V.P.; Zaitsev, K.V.; Nedorezova, P.M.; Aladyshev, A.M.; Klyamkina, A.N.; Kostyuk, S.V.; Danilogorskaya, A.A.; Kozlovsky, A.G. Biodegradation of poly-E-caprolactones and poly-L-lactides by fungi. *J. Polym. Environ.* **2018**, *26*, 4350–4359. [CrossRef]
90. Lee, S.-Y.; Ten, L.N.; Das, K.; You, Y.-H.; Jung, H.-Y. Biodegradative Activities of Fungal Strains Isolated from Terrestrial Environments in Korea. *Mycobiology* **2021**, *49*, 1–9. [CrossRef]
91. Urbanek, H. The role of cutinase and cell wall degrading enzymes produced by fusaria in pathogenesis. In *Elsevier eBooks Topics in Secondary Metabolism*; Elsevier: Amsterdam, The Netherlands, 1989; Volume 2, pp. 243–256. [CrossRef]
92. Murphy, C.A.; Cameron, J.A.; Huang, S.J.; Vinopal, R.T. *Fusarium* PCL depolymerase is cutinase. *Appl. Environ. Microbiol.* **1996**, *62*, 456–460. [CrossRef]
93. Murphy, C.A.; Cameron, J.A.; Huang, S.J.; Vinopal, R.T. A second PCL depolymerase from *Fusarium*, a lipase distinct from cutinase. *Appl. Microbiol. Biotechnol.* **1998**, *50*, 692–696. [CrossRef]

94. Kosiorowska, K.E.; Połomska, X.; Wang, G.; Borodina, I.; Mirończuk, A.M. Efficient Biodegradation of Aliphatic Polyester by Genetically Engineered Strains of the Yeast *Yarrowia lipolytica*. *Int. Biodeterior. Biodegrad.* **2021**, *161*, 105232. [CrossRef]
95. Murphy, C.A.; Cameron, J.A.; Huang, S.J.; Vinopal, R.T. Inactivation of PCL depolymerase (cutinase) in *Fusarium* cultures by an extracellular protease. *J. Ind. Microbiol. Biotechnol.* **1999**, *22*, 71–77. [CrossRef]
96. Shi, K.; Jing, J.; Song, L.; Su, T.; Wang, Z. Enzymatic hydrolysis of polyester: Degradation of poly(ε-caprolactone) by *Candida antarctica* lipase and *Fusarium solani* cutinase. *Int. J. Biol. Macromol.* **2019**, *144*, 183–189. [CrossRef]
97. Ma, Q.; Shi, K.; Su, T.; Wang, Z. Biodegradation of PCL (PCL) with Different Molecular Weights by *Candida antarctica* Lipase. *J. Polym. Environ.* **2020**, *28*, 2947–2955. [CrossRef]
98. Arumugasamy, S.K.; Ahmad, Z. *Candida antarctica* as catalyst for PCL synthesis: Effect of temperature and solvents. *Asia-Pac. J. Chem. Eng.* **2011**, *6*, 398–405. [CrossRef]
99. Pastorino, L.; Pioli, F.; Zilli, M.; Converti, A.; Nicolini, C. Lipase-catalyzed degradation of poly(ε-caprolactone). *Enzym. Microb. Technol.* **2004**, *35*, 321–326. [CrossRef]
100. Arpigny, J.L.; Jaeger, K.-E. Bacterial lipolytic enzymes: Classification and properties. *Biochem. J.* **1999**, *343*, 177. [CrossRef]
101. Holmquist, M. Alpha Beta-Hydrolase Fold enzymes structures, functions and mechanisms. *Curr. Prot. Pept. Sci.* **2000**, *1*, 209–235. [CrossRef]
102. Denessiouk, K.; Denesyuk, A.I.; Permyakov, S.E.; Permyakov, E.A.; Johnson, M.S.; Uversky, V.N. The active site of the SGNH hydrolase-like fold proteins: Nucleophile–oxyanion (Nuc-Oxy) and Acid–Base zones. *Curr. Res. Struct. Biol.* **2023**, *7*, 100123. [CrossRef]
103. Cea-Rama, I.; Coscolín, C.; Gonzalez-Alfonso, J.L.; Raj, J.; Vasiljević, M.; Plou, F.J.; Ferrer, M.; Sanz-Aparicio, J. Crystal structure of a family VIII β-lactamase fold hydrolase reveals the molecular mechanism for its broad substrate scope. *FEBS J.* **2022**, *289*, 6714–6730. [CrossRef] [PubMed]
104. Fojan, P.; Jonson, P.H.; Petersen, M.T.N.; Petersen, S.B. What distinguishes an esterase from a lipase: A novel structural approach. *Biochimie* **2000**, *82*, 1033–1041. [CrossRef] [PubMed]
105. Ali, Y.B.; Verger, R.; Abousalham, A. Lipases or esterases: Does it really matter? Toward a new Bio-Physico-Chemical classification. *Methods Mol. Biol.* **2012**, *861*, 31–51. [CrossRef]
106. Bartnikowski, M.; Dargaville, T.R.; Ivanovski, S.; Hutmacher, D.W. Degradation mechanisms of PCL in the context of chemistry, geometry and environment. *Prog. Polym. Sci.* **2019**, *96*, 1–20. [CrossRef]
107. Badino, S.F.; Bååth, J.A.; Borch, K.; Jensen, K.; Westh, P. Adsorption of enzymes with hydrolytic activity on polyethylene terephthalate. *Enz. Microb. Technol.* **2021**, *152*, 109937. [CrossRef] [PubMed]
108. Jaeger, K.-E.; Ransac, S.; Dijkstra, B.W.; Colson, C.; Heuvel, M.; Misset, O. Bacterial lipases. *FEMS Microbiol. Rev.* **1994**, *15*, 29–63. [CrossRef]
109. Hide, W.; Chan, L.; Li, W. Structure and evolution of the lipase superfamily. *J. Lipid Res.* **1992**, *33*, 167–178. [CrossRef]
110. Drabløs, F.; Petersen, S.B. Identification of conserved residues in family of esterase and lipase sequences. *Methods Enzymol.* **1997**, *286*, 28–61. [CrossRef]
111. Feng, S.; Yue, Y.; Chen, J.; Zhou, J.; Li, Y.; Zhang, Q. Biodegradation mechanism of PCL by a novel esterase MGS0156: A QM/MM approach. *Environ. Sci. Process. Impacts* **2020**, *22*, 2332–2344. [CrossRef]
112. Zampolli, J.; Vezzini, D.; Brocca, S.; Di Gennaro, P. Insights into the biodegradation of polycaprolactone through genomic analysis of two plastic-degrading *Rhodococcus* bacteria. *Front. Microbiol.* **2024**, *14*, 1284956. [CrossRef]
113. Suzuki, M.; Tachibana, Y.; Oba, K.; Takizawa, R.; Kasuya, K.-I. Microbial degradation of poly(ε-caprolactone) in a coastal environment. *Polym. Degr. Stab.* **2018**, *149*, 1–8. [CrossRef]
114. Drew, D.; North, R.A.; Nagarathinam, K.; Tanabe, M. Structures and general transport mechanisms by the Major Facilitator Superfamily (MFS). *Chem. Rev.* **2021**, *121*, 5289–5335. [CrossRef] [PubMed]
115. Mandrand-Berthelot, M.; Condemine, G.; Hugouvieux-Cotte-Pattat, N. Catabolism of hexuronides, hexuronates, aldonates, and aldarates. *EcoSal Plus* **2004**, *1*, 10–1128. [CrossRef]
116. Mulligan, C.; Fischer, M.; Thomas, G.H. Tripartite ATP-independent periplasmic (TRAP) transporters in bacteria and archaea. *FEMS Microbiol. Rev.* **2010**, *35*, 68–86. [CrossRef]
117. Schulz, H. Beta oxidation of fatty acids. *Biochim. Biophys. Acta—Lipids Lipid Metab.* **1991**, *1081*, 109–120. [CrossRef]
118. Wang, Z.; Hou, X.; Shang, G.; Deng, G.; Luo, K.; Peng, M. Exploring fatty acid B-Oxidation pathways in bacteria: From general mechanisms to DSF signaling and pathogenicity in *Xanthomonas*. *Curr. Microbiol.* **2024**, *81*, 336. [CrossRef]
119. Bar–Tana, J.; Rose, G.; Brandes, R.; Shapiro, B. Palmitoyl-coenzyme A synthetase. Mechanism of reaction. *Biochem. J.* **1973**, *131*, 199–209. [CrossRef]
120. Yaneva, N.; Schuster, J.; Schäfer, F.; Lede, V.; Przybylski, D.; Paproth, T.; Harms, H.; Müller, R.H.; Rohwerder, T. Bacterial Acyl-COA mutase specifically catalyzes coenzyme B12-dependent isomerization of 2-Hydroxyisobutyryl-COA and (S)-3-Hydroxybutyryl-COA. *J. Biol. Chem.* **2012**, *287*, 15502–15511. [CrossRef] [PubMed]

121. Johnson, J.; Tolar, B.B.; Tosun, B.; Yoshikuni, Y.; Francis, C.A.; Wakatsuki, S.; DeMirci, H. Crystal structure of the 4-hydroxybutyryl-CoA synthetase (ADP-forming) from *Nitrosopumilus maritimus*. *Commun. Biol.* **2024**, *7*, 1364. [CrossRef]
122. Mahmud, T.; Wenzel, S.C.; Wan, E.; Wen, K.W.; Bode, H.B.; Gaitatzis, N.; Müller, R. A biosynthetic pathway to Isovaleryl-COA in myxobacteria: The involvement of the mevalonate pathway. *Chem. BioChem.* **2004**, *6*, 322–330. [CrossRef]
123. Yang, Q.; Guo, S.; Lu, Q.; Tao, Y.; Zheng, D.; Zhou, Q.; Liu, J. Butyryl/Caproyl-CoA: Acetate CoA-transferase: Cloning, expression and characterization of the key enzyme involved in medium-chain fatty acid biosynthesis. *Biosci. Rep.* **2021**, *41*, BSR20211135. [CrossRef] [PubMed]

MDPI AG
Grosspeteranlage 5
4052 Basel
Switzerland
Tel.: +41 61 683 77 34

International Journal of Molecular Sciences Editorial Office
E-mail: ijms@mdpi.com
www.mdpi.com/journal/ijms

www.ingramcontent.com/pod-product-compliance
Lightning Source LLC
Chambersburg PA
CBHW061605210926
51850CB00009B/2218
9783725866168